Plant Resources of Arid and Semiarid Lands

A GLOBAL PERSPECTIVE

Plant Resources of Arid and Semiarid Lands

A GLOBAL PERSPECTIVE

Edited by

J. R. Goodin
Department of Biological Sciences
and International Center for
Arid and Semiarid Land Studies
Texas Tech University
Lubbock, Texas

David K. Northington
Department of Biological Sciences
Texas Tech University
Lubbock, Texas

1985

ACADEMIC PRESS, INC.
Harcourt Brace Jovanovich, Publishers
Orlando San Diego New York Austin
London Montreal Sydney Tokyo Toronto

ACADEMIC PRESS, INC.
Orlando, Florida 32887

United Kingdom Edition published by
ACADEMIC PRESS INC. (LONDON) LTD.
24–28 Oval Road, London NW1 7DX

LIBRARY OF CONGRESS CATALOGING IN PUBLICATION DATA

Main entry under title:

Plant resources of arid and semiarid lands.

Includes bibliographies and index.
1. Botany, Economic. 2. Arid regions flora.
3. Arid regions. I. Goodin, Joe R. II. Northington, David K.
SB107.P57 1985 630'.915'4 85-7385
ISBN 0-12-289745-5 (alk. paper)

PRINTED IN THE UNITED STATES OF AMERICA

85 86 87 88 9 8 7 6 5 4 3 2 1

CONTENTS

3

INDIAN SUBCONTINENT

K. M. M. Dakshini

4

THE MIDDLE EAST

Loutfy Boulos

5

NORTH AMERICA

C. M. McKell

6

PEOPLE'S REPUBLIC OF CHINA

Hsioh-Yu Hou

7

SOUTH AMERICA

Juan M. Gastó, Alfredo Olivares E., and Rolando H. Braun W.

8

UNION OF SOVIET SOCIALIST REPUBLICS

N. T. Nechayeva

9

SUMMARY

J. R. Goodin and David K. Northington

CONTRIBUTORS

Numbers in parentheses indicate the pages on which the authors' contributions begin.

EDWARD S. AYENSU (1), Office of Biological Conservation and Department of Botany, Smithsonian Institution, Washington, D.C. 20560

LOUTFY BOULOS (129), National Research Center, Cairo, Egypt

ROLANDO H. BRAUN W. (257), CONICET, Instituto Argentino de Investigaciones de las Zonas Aridas (IADIZA), Mendoza (5500), Argentina

K. M. M. DAKSHINI (69), Department of Botany, University of Delhi, Delhi 110007, India

JUAN M. GASTÓ (257), Facultad de Agronomía, Universidad Católica de Chile, Santiago, Chile

J. R. GOODIN (319), Department of Biological Sciences, and International Center for Arid and Semiarid Land Studies, Texas Tech University, Lubbock, Texas 79409

HSIOH-YU HOU (233), Laboratory of Plant Ecology, Institute of Botany, Academia Sinica, Beijing, People's Republic of China

M. LAZARIDES (35), CSIRO Division of Plant Industry, Institute of Biological Resources, Canberra City, A.C.T. 2601, Australia

C. M. McKELL (187), NPI, Salt Lake City, Utah 84108

N. T. NECHAYEVA (291), Desert Institute, Ashkhabad 744000, USSR

DAVID K. NORTHINGTON[1] (319), Department of Biological Sciences, Texas Tech University, Lubbock, Texas 79409

ALFREDO OLIVARES E. (257), Facultad de Ciencias Agrarias y Forestales, Universidad de Chile, Santiago, Chile

O. B. WILLIAMS (35), CSIRO Division of Water and Land Resources, Institute of Biological Resources, Canberra City, A.C.T. 2601, Australia

[1]Present address: National Wildflower Research Center, Austin, Texas 78725.

PREFACE

It was our intent to develop a reference work that would assess the existing native plant resources in all arid and semiarid regions of the world. This inventory was to include potential food, forage, fiber, fuel, medicinal, and industrial uses. No such inventory is currently available for the world as a whole, and since the dry areas comprise one-third of the earth's land surface, such a resource should provide a stimulating focal point for human developmental efforts in these regions.

Every effort was made to have the chapters as uniform as possible in organization and as complete as possible in content. Because of a lack of available information in given areas, however, some of the authors were unable to include parallel data on all topics. Nonetheless, the result is easily the most complete and impressive compilation of worldwide plant resource data for the arid zone.

To provide the reader with an overview of the nature, scope, and organization of information that each author attempted to provide, we include the following outline that was used in the preparation of this book:

I. A general and short paragraph introducing the chapter.

II. Brief information about the total size and the amount of area treated as arid and semiarid (km^2), and the location of these dry areas, including stability and rate and direction of change. Because of edaphic factors, seasonality, or other factors, areas not classically considered "deserts" are included because of their low productivity.

III. Annual precipitation, temperature, humidity; wind ranges and patterns; and seasons in each of the dry areas described above, including discussion of "effective" precipitation (that available to plants) and the total climatic pattern.

IV. Elevation, relief, aspect, and the effects of topography on water drainage and loss.

V. Basic soil types and their origins in these dry areas, with special reference to any unusual edaphic problems such as gypsum outcrops, high salinity or alkalinity, or serpentine soils, as well as comments on water penetration, holding capacity, runoff, and the physical features that control such factors.

VI. The availability of surface and subsurface water in the dry areas, including any past or planned modification of usage by humans, such as dam building, irrigation from diverted and underground sources, and other factors, and comments on water quality as well (including water resource projections).

VII. Current and projected growth rates for countries and for the dry areas, within each continent. Some authors discuss one or a few countries; others discuss many.

VIII. The current economic reliance on arid land productivity for each country.

IX. Balance of trade, current resources and needs, and developing resources.

X. The fate of those countries with dry regions. Can the resource and economic base be affected by the presence of these dry areas—positively and/or negatively?

XI. Emphasis on the nature of the plants to be discussed: native, with significant potential but presently *not* being developed as a "crop" plant. Introduced plants are *not* included; if such plants have potential, they are discussed with the continent of origin.

A. *Food plants.* Discussion of those plants native to the dry regions of a given area that have demonstrated potential as a source of food for human consumption. Herbs, spices, and beverage plants are included. As possible, there is commentary on the degree of usefulness, time scale of development as a crop, and potential use of the plant that is more than local (i.e., Could it become important as an exportable resource?).

B. *Forage plants.* Inventory of the herbs or shrubs that have potential as forage plants for domestic animals, including information on palatability, nutritional quality, potential problems (toxicity), and the potential degree of intensification or development as a "crop" plant, if available.

C. *Medicinal plants.* Survey of the plants having known value in the treatment of illness or ailments. This section does not include an exhaustive list of all plants, especially those having only rumored or undocumented value.

D. *Industrial plants.*

1. *Fibers.* An inventory of any plants useful as wood for building or for macerating for paper, weaving, basket making, textiles, etc.
2. *Fuels.* Sources of firewood or biomass energy from methane or alcohol production are considered, in addition to any potential sources of hydrocarbons for direct energy consumption.
3. *Oils.* A survey of those plants native to the dry areas that contain extractable oils, since industrial uses for extractable oils are very broad.
4. *Waxes, resins, and latexes.* An inventory of plants from which commercially producible waxes, resins, and latex products (such as rubber) can be extracted.

E. *Other plants.* Plants that are not potential foods, forages, or medicinals, and have no commercial or industrial value for their fibers or extractable chemical compounds but might have value as ornamentals, or be useful in dune stabilization or erosion control or land reclamation after mining or other disturbances.

XII. Discussion of the degree of habitat stability, water availability, and other factors that must be considered in developing native "crop" plants in arid and semiarid regions.

XIII. A consideration of the balance among population pressures, ecological stability, potential of the plant resources, and economic or local needs. What, philosophically, should be our overview of arid land development?

XIV. A summary of the native plants having the greatest potential as new resources from the dry regions of each continent or subcontinent. Economic worth, local need, time scale for development, etc., are considered.

The need for this book became evident at the International Conference on Arid and Semiarid Land Plant Resources, held in 1978 in Lubbock, Texas. Sponsored by the International Center for Arid and Semiarid Land Studies at Texas Tech University, this conference included 200 participants from 22 countries and resulted in the publication of the proceedings, *Arid Land Plant Resources*. Plenary talks summarizing specific geographic regions of the world were presented at this conference, but because of the length of the proceedings the presenters decided that a second publication would be appropriate.

We gratefully acknowledge the support of ICASALS and Texas Tech University, and especially thank Katina M. Clark for the preliminary editing of all the manuscripts.

1

AFRICA

Edward S. Ayensu
*Office of Biological Conservation
and Department of Botany
Smithsonian Institution
Washington, D.C.*

I. INTRODUCTION

This chapter is designed to provide a detailed perspective on the development problems facing the arid and semiarid zones of Africa. Generally, in terms of average family income, the countries that fall within these zones are among the least prosperous. They include countries in the Sahelian zones that bound the Sahara and large areas in eastern and southern Africa. The populations living in these zones also experience poor dietary standards. Their caloric intake, more than 70% of which is derived from cereals, is barely adequate; various studies by the FAO and WHO indicate that in 13 of the 17-odd countries in Africa that fall within the semiarid tropical zone, the average caloric consumption is well below acceptable levels. It is also known that because of uncertain and variable production levels and inadequate storage facilities, caloric intake just before harvest is generally 25–30% lower than it is immediately after harvest.

Regardless of cultural differences among the arid and semiarid countries of Africa, these countries share enough natural characteristics that should enable them to adopt similar policy frameworks to remedy some of the major development problems that confront them.

II. PHYSIOGRAPHY

A. Size and Location

Second in size only to Asia, Africa is joined to the Asian continent by the isthmus of Suez. Geologically, Africa was at one time also connected to Europe at what are now the straits of Gilbraltar and Tunis. The continent stretches for ~35° on either side of the equator, and is also crossed by both the tropics of Cancer and Capricorn. Thus, the bulk of the continent lies within the tropics. Additionally, its shape is such that most of the continent lies in the northern hemisphere.

B. Topography and Climate

Because its coasts are almost unbroken, Africa has been described as a compact continent. Apart from the Atlas Mountains in the northwest of the continent, Africa is mainly a plateau arising by steep escarpments from narrow coastal plains. (See Fig. 1 for the topography of the continent.)

The UNESCO (1979) *Map of the World Distribution of Arid Regions* delimits portions of Africa's 30 million km^2 area as hyperarid, arid, or semiarid. The division between arid and semiarid roughly follows the 400-mm annual rainfall contour, but the criterion actually used to separate climatic zones is the ratio of annual precipitation to evapotranspiration. The latter is estimated from insolation, atmospheric humidity, and wind data. This ratio is preferred over precipitation or the amount of available water (the difference, rather than the quotient, between precipitation and evapotranspiration) because it approximates a water balance parameter which governs vegetative dry-matter production, and does so in spite of contrasting seasons in different regions to which is it applied.

Figure 2 has been redrawn from the UNESCO map; the subhumid zone shown on the original has been omitted, and the description of temperature regimes to aridity per se has been subordinated. The hyperarid zone, ~6.5 million km^2, consists of the vast Sahara region, the north coast of Somalia, and the coast of southwest Africa. Each of its parts is bounded by broad bands of less-desiccated

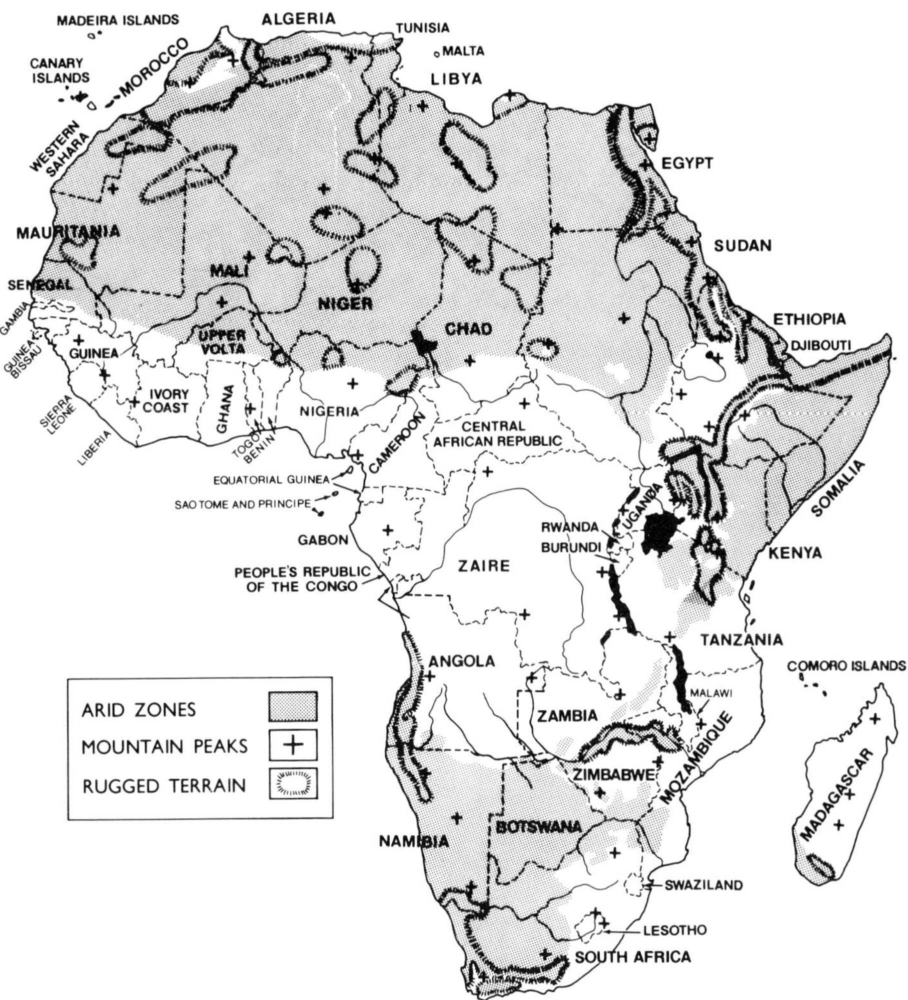

Fig. 1. Countries and topography of Africa.

land; together they make up the 6.1 million km^2 arid zone. The semiarid zone, ~4.8 million km^2 in area, surrounds drier lands and extends out from them to varying distances, depending on the topography.

Aridity, defined in terms of atmospheric parameters such as rainfall, sunlight, humidity, and the transport of air, represents a constraint on any plant–soil system. Within an arid zone, however, different soil associations and vegetative cover occur. Whereas climate is reasonably stable, vegetative change can lead to irreversible destruction of the habitat by unskilled management, and is one of the

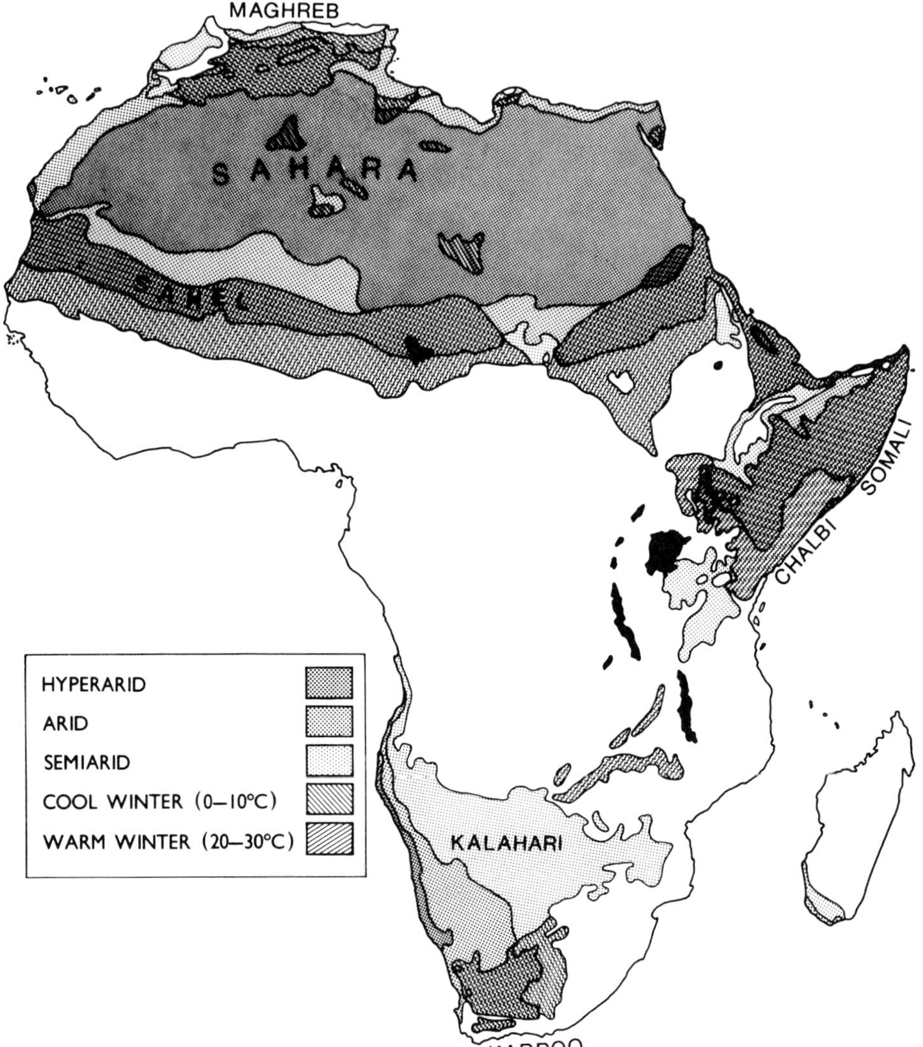

Fig. 2. Arid zones of Africa.

major and urgent problems facing the people of the continent. Soils erode because of vegetational changes or become saline through irrigation and long-term accumulation of underground water. Thus, the inhospitality of an arid zone to vegetative production increases and becomes widespread. For example, Le Houerou (1970) estimated that on an annual basis, 100 km^2 of sandy soils lose the capacity to support a *Stipa tenacissima* steppe in the four North African

Table I

Climatic Ranges of Countries That Include Arid Land

	Range of climate	Largest area
Summer drought in populous areas[a]		
Canary Islands	Semiarid	—
Western Sahara	Hyperarid to arid	Arid
Morocco	Arid to humid	Arid
Algeria	Hyperarid to humid	Hyperarid
Tunisia	Hyperarid to humid	Arid
Libya	Hyperarid to subhumid	Hyperarid
Egypt	Hyperarid to arid	Hyperarid
Winter drought: northern arid belt		
Cape Verde	Arid	—
Senegal	Arid to humid	Semiarid
Mauritania	Hyperarid to semiarid	Arid
Mali	Hyperarid to humid	Arid
Upper Volta	Arid to humid	Semiarid
Benin	Semiarid to humid	Subhumid
Niger	Hyperarid to semiarid	Arid
Nigeria	Arid to humid	Humid
Cameroon	Semiarid to humid	Humid
Chad	Hyperarid to humid	Arid
Sudan	Hyperarid to humid	Arid
Kenya	Arid to humid	Semiarid
Ethiopia	Hyperarid to humid	Arid
Djibouti	Arid	—
Somalia	Hyperarid to subhumid	Arid
Socotra Island	Arid	—
Winter drought: eastern semiarid belt		
Uganda	Semiarid to humid	Humid
Tanzania	Semiarid to humid	Subhumid
Zambia	Semiarid to humid	Humid
Madagascar	Arid to humid	Humid
Mozambique	Semiarid to humid	Subhumid
Zimbabwe	Semiarid to humid	Subhumid
Winter drought: southern arid belt		
South Africa	Hyperarid to humid	Subhumid
Botswana	Arid to subhumid	Semiarid
Namibia	Hyperarid to subhumid	Arid
Angola	Hyperarid to humid	Humid

[a]See also Sudan (coastal), Ethiopia (coastal), South Africa (western), and Djibouti (two drought seasons).

countries of Morocco, Algeria, Tunisia, and Libya. Tunisia has lost nearly half of its circum-Saharan steppe during this century.

Figure 1 also shows the political organization of the continent. Nearly all of the nations that contain parts of the arid lands have great climatic diversity. The arid zone is quite different from the other climatic zones, which are more densely populated. Table I represents the climatic ranges of African countries that contain arid lands. It should be kept in mind, however, that the table does not reflect the fact that the largest areas of a country, if arid, are almost uninhabited as compared, for example, with the exceptional narrow shore strip of the Nile, on which dwell some 38 million Egyptians. Indeed, Fig. 2 can be taken as showing the distribution of people if one remembers that the truly arid zone is sparsely populated except for Ethiopia and Somalia and in areas with surface water.

The meaning of the degrees of aridity (and therefore, that of the size of the arid areas as here defined) is implied in the land-use assessments that accompany the map from which Fig. 2 is taken. In the hyperarid zone, agriculture and grazing are generally impossible. In the arid zone, although very light pastoral use is possible, rainfed agriculture is not. In the semiarid zone, agricultural harvests are likely to be irregular, although grazing is satisfactory. Some of the dry regions of Africa, as well as some of the humid regions, experience population pressure (Hance, 1975), which of course affects larger areas in the arid zones but arises there from a base of lower population density.

Except in areas where rainfall is rare and unpredictable, one or two annual rainy seasons are discernible in the arid zones, with the months of most rain differing from place to place. Accordingly, soils have moist and dry seasons of varied durations, and it is these soil-moisture seasons that are important for plant growth, particularly in the absence of cold winters as a further constraint. Areas of long and short moist-soil periods cannot be mapped over large territories, even if known and regular, because they form a patchwork whose parts are very small. A general view of climate must therefore be related to it as an atmospheric phenomenon.

Because different degrees of aridity occur over most of Africa in contiguous, large, elongated, and well-demarcated zones, the question of how weather varies from year to year is roughly equivalent to asking how accurately the zones are located in space, as depicted in Fig. 2. Except where the boundaries are so intricately curved as to indicate that topography prescribed them, they should be perceived as mobile, with the more arid zones expanding in particularly dry years. Interannual rainfall may vary 50–100% (UNESCO, 1979) in the arid zones of the world, with averages as high as 350 mm. The normal variation is 25–50% in the semiarid zones, where averages are 300–700 mm for areas with a summer rainy season and 200–500 mm for winter, and is less than 25% outside the semiarid zones.

Excepting temperature and rainfall, climatic observations in Africa are not generally available in published form. Lebeder (1970) reported the years during which humidity, cloudiness, and wind velocity were recorded at meteorological stations on the continent. Some examples of climate with varied seasonal cycles are shown in Table II. Most of the examples are chosen for their distinct rainy seasons; less-marked seasons are common, but for only a few stations are the months of maximum precipitation different from those in the table. The geographic distribution of the several seasonal types appears in Table I; inland, two drought seasons prevail in the western countries of North Africa, giving place to erratic rainfall toward the East.

C. Edaphic Factors

Arid soils (Dregne, 1976) characteristically exhibit a pH of 7 or higher in a saturated extract of the uppermost layer, and show little evidence that clay leached from the surface has produced clay-covered aggregates in lower strata. Such soils cover 15 million km^2 (described here in relation to Fig. 3) along with a zone of more acid *alfisols* with distinct argillic (transported clay) horizons, which cover an additional 2 million km^2.

Each region sketched in the figure includes several kinds of soils, sometimes of considerable variety with one region and sometimes distributed within the region according to topography. Thus, one among the soils of an association may have developed wherever drainage is poor, or where the water table is high, and another may be found both on ridges and also on areas where alluvium from the mountains has collected, but not on terrain intermediate in character between its typical locations. Table III represents the distribution of soils within regions wherever it is known and regular.

Many of the distinctions between kinds of soils are not crucial for plant growth, and several other defining criteria for soil types merely parallel atmospheric data. Thus, *ustic* soils are those that experience drought but are moist (above the wilting point) for at least 3 months of the year or are partially moist for 6 months, unlike the drier *torric* or *arid* soils. Nevertheless, some recognized soil associations do support distinctive vegetation.

Hance (1975) provided an overview of vegetational gradients. A desert comprising the large hyperarid zone of the north, together with most of the western Sahara, is bounded by subdesert steppe; that steppe extends into the hyperarid zone in the rocky soil zone of Sudan and the shallow soil zone of Egypt and eastern Libya. Beyond the steppe, Mediterranean vegetation covers the coastal parts of Morocco, Algeria, and Tunisia as well as the two northern extensions of the Libyan coast. South of the Sahara, the steppe gradually changes, supporting trees and then giving place to dry forest and savanna in the semiarid zone, to swamp in southeastern Sudan, or to thicket-surrounded mountains that bear grasslands along the Red Sea.

Table II

Types of Seasonal Cycles: Rainfall and Mean Temperature[a]

		Jan	Feb	Mar	Apr	May	Jun	Jul	Aug	Sep	Oct	Nov	Dec	Total for the year
Two Rainy Seasons														
Summer driest														
Siguig, Morocco, el. 900 m, 32°07′N, 1°14′W	mm	8	8	18	10	5	5	0	3	8	13	18	8	104
	°C	10	12	16	20	24	30	34	33	28	21	16	11	
Marrakech, Morocco, el. 500 m, 31°40′N, 8°00′W	mm	36	37	57	24	25	19	1	7	3	22	55	40	326
	°C	12	13	16	18	21	25	29	29	26	21	16	12	
Winter driest														
Beaufort West, South Africa, el. 868 m, 32°19′S, 22°38′E	mm	12	39	30	21	10	3	10	13	12	17	29	27	223
	°C	24	23	22	18	15	12	11	12	15	18	20	22	
East London, South Africa, el. 10 m, 33°02′S, 27°55′E	mm	73	77	97	68	55	36	35	44	69	91	86	77	808
	°C	21	21	21	19	17	16	15	16	17	18	19	20	

Two droughts														
Neghelli, Ethiopia, el. 1444 m, 5°20′N, 39°44′E	mm	6	25	91	158	190	5	9	2	26	105	48	6	671
	°C	21	21	21	20	19	18	18	18	19	19	20	20	
Obbia, Somalia, el. 15 m, 5°20′N, 48°31′E	mm	13	0	8	20	33	0	<2	<2	3	38	25	25	169
	°C	26	27	28	30	28	27	25	26	26	27	28	27	
One Rainy Season														
Summer driest														
Tripoli, Libya, el. 22 m, 32°54′N, 13°11′E	mm	81	46	28	10	5	3	<2	<2	10	41	66	94	384
	°C	12	13	15	18	20	23	26	26	26	23	18	14	
El Mekili, Libya, el. 205 m, 32°10′N, 22°17′E	mm	13	8	3	3	3	<2	0	0	<2	5	5	15	55
	°C	11	12	15	19	23	27	28	27	25	22	17	13	
Winter driest														
Khartoum, Sudan, el. 385 m, 15°36′N, 32°33′E	mm	0	0	0	1	4	8	51	72	21	4	0	0	161
	°C	23	24	28	31	34	34	31	30	32	32	28	24	
Sennar, Sudan, el. 420 m, 13°33′N, 33°37′E	mm	0	0	1	2	19	67	117	160	66	17	2	0	451
	°C	21	23	24	28	29	29	26	25	26	26	25	22	

[a]Data are from Lebeder (1970).

Table III

Soil Associations

	Major types	Minor types
Vertisol zones		
Eastern	Chromusterts	—
Central	Chromusterts (lowland slopes) Pellusterts (depressions) Haplustalfs (dunes, hills)	—
Psamment zones		
Large southern	Torripsamments Paleorthids (outskirts of desert)	Torrifluvents Torriorthids Haplargids
Other	Torripsamments Haplargids (interdune)	Torrifluvents Torriorthents
Reg zone	Torriorthents (alluvium, plateaux) Calciorthids (alluvium, slopes, terraces) Rock outcrops	Torrifluvents Paleargids Haplargids
Shallow zone	Torriorthents Rock outcrops Calciorthids (alluvium, piedmont, terraces) Paleorthids	Torripsamments Torrifluvents Gypsiorthids
Rocky zone	Torriorthents Rock outcrops (upland)	Torrifluvents Haplargids Torripsamments Calciorthids
Fluvent zone	Torrifluvent Salorthids (lower valleys, depressions) Fluvaquents (deltas, swamps)	—
Mollisol zone	Ustorthents (hills, mountains) Haploxerolls (broad valleys) Rhodoxeralfs (lowland slopes)	Rendolls Argixerolls Torrifluvents Salorthids
Camborthid zone	Camborthids Torripsamments Haplargids (plateaux, piedmont, interdune)	Vertisols Salorthids Torriorthents Torrifluvents
Gypsum zone	Calciorthids Gypsiorthids (mostly coastal depressions)	Torripsamments Salorthids
Calciorthid zone	Calciorthids Torripsamments Torriorthents (ridges, plateaux, piedmont, alluvium)	Pellusterts Haplargids Torrifluvents

Table III (*Continued*)

	Major types	Minor types
Paleorthid zone	Paleorthids (terraces, plains) Calciorthids (plains, slopes) Salorthids (depressions)	Torrifluvents Torriorthents
Saline zone	Salorthids (coastal) Gypsiorthids (plains near coast) Torriorthents (plains, alluvium)	Calciorthids Torrifluvents
Alfisol zone	Haplustalfs Paleustalfs Ostuchrepts (subhumid edge of zone)	Tropaqualfs Rhodustalfs Ultisols

In the south part of Africa, the desert is the hyperarid coast. The subdesert steppe spreads far east in southern South Africa to include all of the shallow soil zone. The steppe, with trees, is the large psamment zone, and much of the shallow soil zone west of it, the Kalahari "desert," with dry forest and savanna, extends from it to the northwest, northeast, and east. Swamps exist in the fluvent areas of Botswana.

Most of eastern Africa, from the lakes to the Somalian horn, is steppe with trees; the subdesert steppe prevails, however, around the hyperarid part of Ethiopia and along the coasts, with projections inland at intervals from the gypsum zone westward to Lake Rudolf in the shallow soil zone. The arid part of Madagascar bears thickets, and the semiarid part a grassy savanna.

This broad-brush portrait of the vegetation requires the remarks on local distribution that accompany the following outline of the several soil classes.

VERTISOLS

These soils have high clay content and form deep cracks when dry. At least in the tropics, such soils develop as a consequence of multiple annual rainy seasons. The clay is of the montmorillonite series, which, together with vermiculite and allophane, exhibits high water retention and ion-exchange capacity, but (except for allophane) low permeability to water. The low-capacity, high-permeability clay is kaolinite, with illite and the septechlorite clays ranging between the extremes. Because soil particles fall into the cracks, vertisols are well churned, and, in the few uncultivated areas, hummocky.

a. Chromusterts. These are very deep, brown vertisols containing, in Africa, 40–55% clay, 0.5–1% organic matter, and many calcium carbonate nodules. Drainage is fair. At lower levels, soluble salts occur, and the ion-exchange

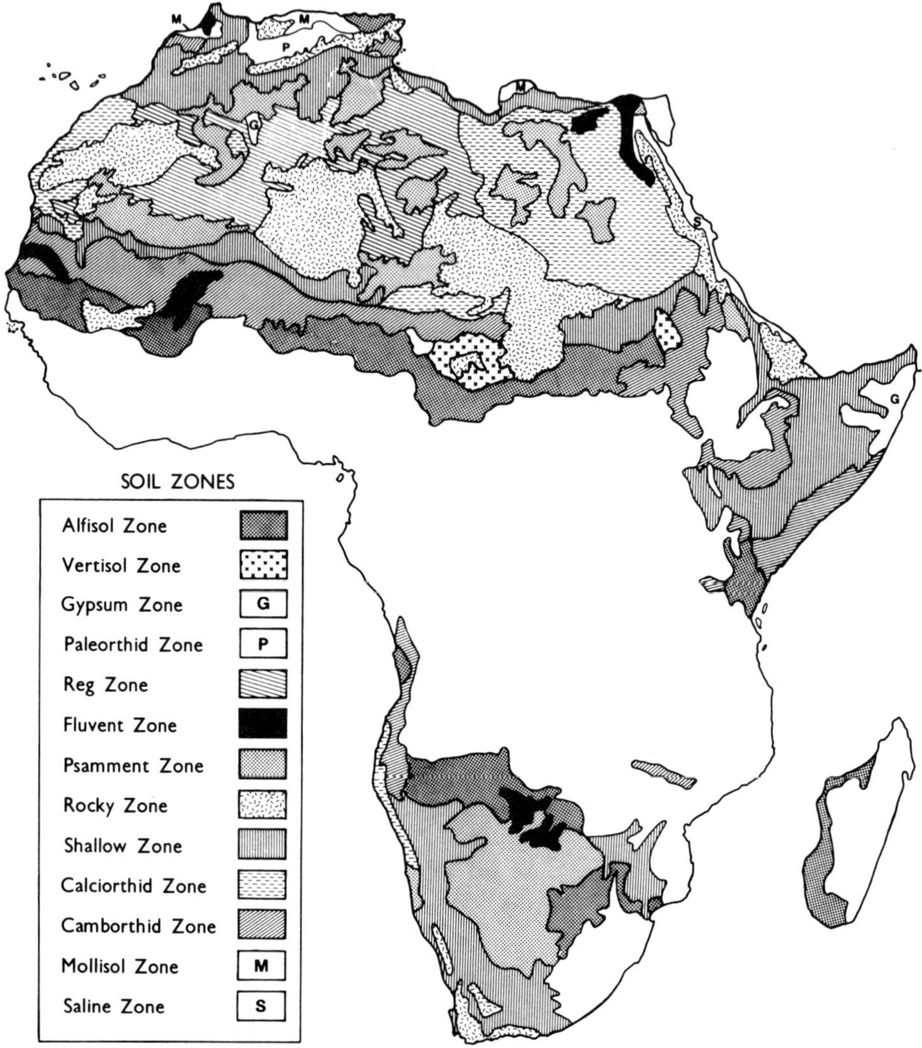

Fig. 3. Characteristic arid soils.

sites are substantially occupied by sodium. The soils can support savannas with xerophytic trees.

b. Pellusterts. These are black, with 50 to 70% clay, 0.8 to 1.4% organic matter, and little carbonate. These soils are poorly drained and generally support forest growth.

ENTISOLS AND INCEPTISOLS

These are the soils that cover a third of Africa, and that, except for flood deposits, have not developed any stratification.

a. Fluvaquents. Wet entisols of otherwise arid zones, these are gray or blue and contain rather abundant organic matter.

b. Torrifluvents. These usually are deep loam, silt loam, or clay loam, with carbonate or gypsum in occurring veins. These soils are found where flooding is frequent and drainage is not poor. (The term *loam* is used here in the American sense; it is composed of roughly equal parts of sand, silt, and clay.)

c. Torripsamments. These are very deep sandy soils. They are barren in most psamment zones, stabilized in the Kalahari, and stabilized by shrubs and xerophytic trees in the calciorthid zone where psamment areas are small and dispersed. These soils also are barren or stabilized by bunch grass and shrub in the camborthid zone, where sands (illustrating the two most climate-determined properties of soils) reach (1) the African maximum for all soils of 4% organic matter, and (2) high base saturation—that is, more than 90% of the ion-exchange sites on the clay phase occupied by Na^+, K^+, Ca^{2+}, or Mg^{2+} as opposed to H^+ or Al^{3+}, a correlate of neutral of higher pH. Soil nitrogen varies with organic matter; phosphorus, the other limiting nutrient of arid Africa, accompanies no simple indicator.

d. Torriorthents. The usual entisols, these are gravelly, though not always so in the rocky and shallow zones. Of variable depth, but deep in the saline zone, these soils are slightly to moderately calcareous, particularly in the calciorthid and rocky zones. Torriorthents are occasionally saline in the reg and calciorthid zones. There is little organic matter in the reg and shallow zones. Composition is loam to sandy loam in the reg zones, but it is more variable in the shallow zones. Vegetation is sparse in the calciorthid and saline zones, the northern shallow zone, and lowland parts of the rocky zone. The gravelly texture promotes water retention through rapid drying at the surface, as does a level terrain. Water also is retained by the presence of a pavement of pebbles on top of the soil, such as is found in orthents in the reg, calciorthid, and saline zones.

e. Ustorthents. These are primarily composed of shallow sandy loam to clay loam and support trees, shrubs, and, in open spaces, some grasses.

f. Ustochrepts. Shallow soils of open-wooded areas, these soils are sometimes calcareous below, with a low content of organic matter and only moderate base saturation.

MOLLISOLS

These are the darker soils among those whose base saturation is at least moderately high down to lower strata. The calcium-rich surface is composed of discrete particle-aggregates, which display neither continuous aggregation nor nonaggregated particles.

Haploxerolls. These are shallow loam to clay loam with up to 3% organic matter, a base saturation of at least 80%, and adequate drainage to avoid large FeMn concretions. Carbonate and higher clay content occur below. Haploxerolls make up a wet-winter soil group that mainly supports perennial grasses with scattered shrubs.

ARIDISOLS

Included in this group are any soils that do not belong to the above groups and that fit only one of the following conditions: (1) has a near-surface saline layer, (2) has an arid surface that, when dry, is not *both* hard and massive, or (3) does not fit the definition of an alfisol. The argillic layer, which otherwise would identify an alfisol, may bc prcscnt in an aridisol if the first or second condition also occurs. These soils are usually deep, have low organic content, and support sparse vegetation, except in depressions. Like some orthents, they often underlie "desert pavement," the water-retaining layer of pebbles.

a. Camborthids. These have differentiated strata, but no distinctive mineralogic character. They support xerophytic trees and shrubs and ephemeral forbs and grasses.

b. Paleorthids. Near-surface carbonate strata, these constitute a barrier to roots but not to water. They are composed of loam to silt loam, with moderate organic content, and support shrubs and grasses.

c. Salorthids. These soils do not support vegetation. At this point, saline soils must be distinguished from sodic soils. Salinity refers to soluble salts [in parts per million of saturated extract, equal to ~640 times the conductivity (mmho/cm), where a conductivity greater than 4 marks a soil unfavorable to plant growth]. The unrelated, climate-determined sodicity, on the other hand, refers to the occupation of more than 15% of the ion-exchange sites by Na^+; this soil is also unfavorable to plant growth. Sodic clay is susceptible to being dispersed in water, thus the soil is mechanically unstable. However, salinity, like Ca^{2+} on the ion-exchange sites, has the opposite effect of flocculating the clay. Salinity of the uppermost soil can increase either through the growth of halophytes, which draw salts from below, or through the presence of a high water table, which, when underground reservoirs are not pumped, can result from irrigation.

d. Gypsiorthids. In these, the gypsum is only slightly soluble at low pH (i.e., a thousandth as soluble as chlorides, $MgSO_4$, or of carbonates and sulfates of the alkali metals), but, at a neutral or basic pH, it is much more soluble than $MgCO_3$ or $CaCO_3$. A soil with a gypsum layer can support a sparse tree cover.

e. Claciorthids. In these soils, the carbonate, though forming a stratum, presents no barrier to the permeability of water. The soil, usually a gravelly sandy loam, supports xerophytic trees in the calciorthid zone, xerophytic shrubs in the gypsum zone, and shrubs and grasses in the paleorthid.

f. Haplargids. These are aridisols and have an argillic horizon and a minimum surface horizon. A thin layer of sandy loam or loamy sand overlies sandy clay, below which the soil is usually calcareous and somtimes saline. This type of soil is found in the camborthid zone and is often without plant life in the psamment.

ALFISOLS

These soils are probably always associated with a fluctuating water table.

a. Rhodoxeralfs. These represent a wet-winter soil group—that is, the African implication of "xeric." The surface organic content is ~1%. The lower argillic layer is grossly structured and has a base-saturation of 60% or more. These alfisols support open woodland with shrubs and some grasses.

b. Haplustalfs. These have a sandy loam surface with a low organic content and varied pH but are usually acidic, with a base saturation of ~50%. Near the surface is found a sandy kaolinite loam or sandy kaolinite of higher base saturation. These soils support savanna, on which, in the alfisol zone, only the trees remain during periods of drought, when the soil surface is very hard.

c. Pale Ustalfs. These are thicker, more acid soils than those previously mentioned in the group.

Because grazing is the most usual type of land use in arid zones, excessive destruction of the plant cover is the principal means by which ecologic associations become unstable. The main determinants of the rate of water erosion, in order of importance are (1) a low content of organic matter, (2) a very high silt content (as in upland loess), and (3) topography (steepness and length of a slope). Erosion by wind is inhibited only by structured organic matter; disintegrated plant remains are worse than no organic matter at all. Smooth surface, high sand content, or low silt content promote erosion by wind, as does surface calcium carbonate, except in loams and sands.

D. Hydrology and Water Resources

Cultivated or natural vegetation on arid lands must generally be exploited as a disperse resource unless irrigation is provided to concentrate spatially the harvest and to diversify the crop. Furthermore, the source of water for irrigation cannot be surface water except in a limited area.

The principal surface waters of Africa appear in Fig. 1. In Fig. 4, associated parts of some important kinds of aquifers (Department of Economic and Social Affairs, 1973) that lie within arid areas are shown. Except for karsts, these aquifers exhibit interstitial porosity; unconsolidated sands are not shown, though under them lies the Niger basin, one of the principal reservoirs of ground water in the Sahara. The Continental Terminale and the Continental Intercalaire groups consist of sandstone. The Continental Intercalaire accounts for most of the large water basin of the Sahara: those of the Great Western Erg, having a capacity of 1.5 million m^3; the Great Eastern Erg, 1.7 million m^3; the Fezzan, 0.4 million m^3; the Western Egyptian Desert, 6 million m^3; and the Chad, 3.5 million m^3 (Pritchard, 1971).

Wells driven into different aquifers of the same type may produce from 0.1 to 2000 m^3/hr of water. Thus, despite many centuries of experience with the intensive use of wells for agricultural and city needs, as well as information obtained from many recently drilled wells, it is not surprising that no pattern is yet discernible in the continent's groundwater resources. It is hardly possible, therefore, to say that plans for intensive agricultural development by irrigation are either more or less significant than the search for novel rainfed crops.

Irrigation involves two different kinds of hazards, depending on the source of water. Surface-water irrigation without deliberate drainoff can supply exogenous salt or can promote upward leaching of salts by elevating the water table. On the other hand, the use of groundwater has the opposite effect of exhausting the water supply; ill-advised pumping will likely occur, and every groundwater reservoir is thus in some danger of deterioration.

III. DEMOGRAPHY

That portion of a country's population which resides in the hyperarid, arid, and semiarid zones may constitute nearly the entire population or only a few of the people (Table IV). Benin, Cameroon, Madagascar, and Uganda are essentially humid countries. The people of Kenya and Ethiopia are overwhelmingly located outside the dry lands, but the large dry areas of those countries still contain more people than many whole nations on the continent.

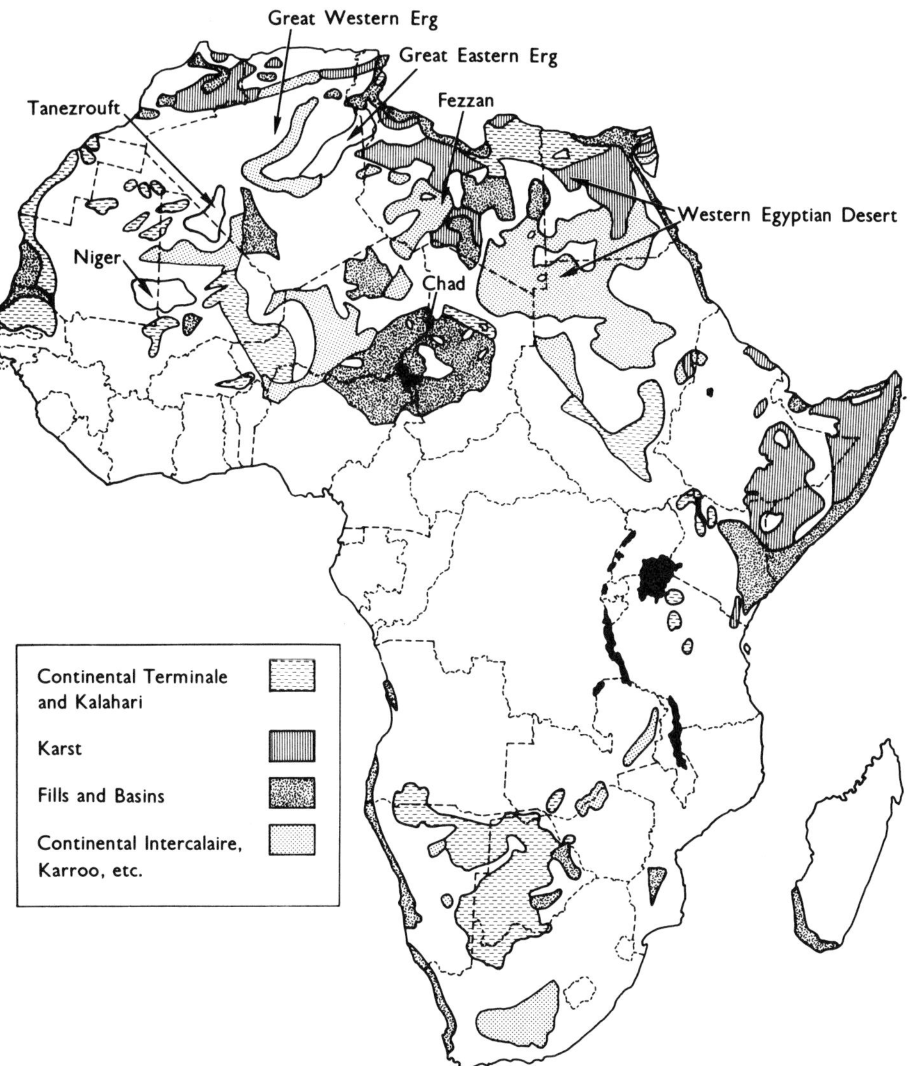

Fig. 4. Aquifers of arid areas.

Aridity, of course, is among the factors that determine the extent of arable land, although notably in the vertisol area of Sudan, part of the semiarid zone is cultivated. Plainly, the restricted local availability of water accounts for the high proportion of people to cultivated land in Egypt.

For the few African countries for which migration data are available, most show both years of net immigration and years of net departures. No doubt much

Table IV

Population Data

	Population (millions)		Annual growth[b] (%) 1970–1977	Rural residents[c] (%)		Population density[d] (km^{-2})	
	Dry lands[a]	Total[b]		1960	1980	National average	All population divided by arable area
Canary Islands	(0.9)[e]	(0.9)				(124)	
Western Sahara	(0.1)	(0.1)		55 (1974)		(1)	
Morocco	15	18.9	2.7	71	59	46	246
Algeria	6	17.7	3.2	70	39	7	262
Tunisia	4	6.1	2.0	64	48	37	144
Libya	2.65	2.7	4.1	77	48	2	117
Egypt	38.7	38.7	2.1	62	55	39	1381
Cape Verde	0.3	0.3	2.1	80 (1970)		79	862
Senegal	3.0	5.4	2.6	77	75	27	233
Mauritania	1.5	1.5	2.7	97	77	1	1188
Mali	4.0	6.3	2.5	89	80	5	292
Upper Volta	4.0	5.6	1.6	95	91	20	226
Benin	0.05	3.3	2.9	90	86	30	453
Niger	5.0	5.0	2.8	94	87	4	233

Nigeria	10.0	81.0	2.6	87	80	88	706
Cameroon	0.05	8.1	2.2	86	65	17	501
Chad	2.0	4.3	2.2	93	82	3	272
Sudan	14.0	17.4	2.6	90	75	7	469
Kenya	1.0	15.2	3.8	93	86	26	785
Ethiopia	2.0	31.0	2.5	94	85	25	193
Djibouti	0.3	0.3	3.5			14	160
Somalia	3.6	3.7	2.3	83	70	6	288
Socotra Island	(0.01)	(0.01)				(3)	
Uganda	0.05	12.4	3.0	95	88	53	450
Tanzania	0.3	16.9	3.0	95	88	18	385
Zambia	0.13	5.3	3.0	77	62	7	210
Madagascar	0.05	8.3	2.5	89	82	14	490
Mozambique	0.3	9.9	2.5	96	91	13	291
Zimbabwe	0.7	6.9	3.3	87	77	18	309
South Africa	1.0	27.7	2.7		(52)	23	238
Botswana	0.746	0.747	1.9		(88)	1	82
Namibia	0.89	0.95	2.8			1	159
Angola	0.4	6.7	2.3	90	79	5	84

[a]From map in Hance (1975).
[b]World Bank (1979).
[c]World Bank (1980).
[d]Areas from Instituto Geográfico de Agostisni (1976).
[e]Parenthetic data mostly from United Nations (1978).

of this movement is in response to recent military conflicts in conjunction with famines. For some countries, long-term migrants numbered in the thousands or low tens of thousands annually.

Urbanization is occurring in nearly all of the countries for which data are available. In recent years, the percentage of the population that is rural has dropped rapidly. On a shorter time scale than appears in Table IV, the average annual growth of the urban population from 1970 to 1975 among the arid-land countries, as tabulated by the World Bank (1980), exceeded 4% in all but Upper Volta, Egypt, Senegal, Tunisia, and South Africa.

IV. SOCIOECONOMIC FACTORS

Among the 33 political entities considered in Table V are 27 of the 45 mainland nations, including all the economic giants: Nigeria, South Africa, Algeria, Libya, Egypt, and Morocco on a per capita basis; however, only Libya, Gabon, and the island of Réunion are prosperous by northern standards, and among these three only Libya is part of our sample. The large number of very poor countries also is shown in Table V.

Except for migrant labor in Upper Volta and Botswana, electricity in Uganda, and canned fruit in South Africa, principal exports are unprocessed material goods, few of which are ready for consumption. Note that many of the products mentioned in Tables V and VI are characteristic of humid-zone agriculture. Thus, even the table of agricultural data shows the countries' economic status and reveals little about the present use of arid lands.

For areas where agriculture is now practiced regularly, Pereira (1977) has presented a brief, optimistic review. Acknowledging that past low densities of population have fostered tribal traditions inimical to the conservation of soil and water, he cited the successful adoption of American planning for tillage and diversion of runoff in Kenya. On the other hand, he also cited experiences that show that only uniform and mandatory controls for a whole watershed have effected much improvement. More modest irrigation schemes have been useful for capital-intensive irrigation; Pereira suggested that they hold more promise in Africa today than in the United States of 30 years ago, but must be preceded by a fully commercial economy in the area of their use.

To link irrigation with tillage is to forget that the land under consideration is formidably arid. Date trees are irrigated. Pasture is irrigated. The need for water prompted attempts to impound large reservoirs in Egypt thousands of years before adequate technology existed for the effort to succeed. Water newly brought from rivers or from underground will enter an economy already dependent on cisterns (some of them ancient) or artificial watershed divides (Kassas, 1970).

Table V

Economic Data

	Per capita GNP[a] (U.S. dollars)		Balance of payments[b] (million U.S. dollars) 1978		
	1977	Annual growth (%) 1970–1977	Interest, public	Other	Main exports[c]
Canary Islands	—	—	—	—	Bananas
Morocco	610	4.2	−252	−1040	Phosphate, eggs, fiber
Algeria	1140	2.1	−561	−2977	Oil, wine, citrus, iron
Tunisia	840	6.5	−95	−411	Oil, phosphate, olive oil
Libya	6520	−4.5		+1024	Oil
Egypt	340	5.2	−386	−540	Cotton, rice, oil, citrus
Cape Verde	150	−2.1			Fish, salt, bananas, oilseed
Senegal	380	0.4	−31	−114	Peanuts
Mauritania	270	−0.2	−10	−65	Iron
Mali	120	1.9	−3	−72	Cotton, peanuts, livestock
Upper Volta	140	1.6	−3	−79	Cattle, labor, cotton, shea nuts
Benin	210	0.5	−3	−70	Peanuts, cotton
Niger	190	−1.8	−4	−81	Uranium
Nigeria	510	4.4	−75	−3696	Oil, peanuts, cocoa, palm oil
Cameroon	420	1.0	−46	−112	Coffee, cocoa
Chad	130	−1.0	−3	−188	Cotton
Sudan	330	2.5	−36	−54	Cotton, peanuts, gum, cattle
Kenya	290	0.9	−45	−474	Coffee, tea
Ethiopia	110	0.2	−13	−98	Coffee
Djibouti	450	−0.3			Coffee
Somalia	120	−1.1	−2	−63	Cattle, bananas, sugar
Uganda		(−2.0)	−4	−129	Cotton, coffee, oilseed, electricity
Tanzania	210	2.1	−18	−442	Coffee, cotton, cloves
Zambia	460	−0.2	−46	−191	Copper
Madagascar	230	−2.7	−6	−51	Cloves, vanilla, coffee
Mozambique	140	−4.3			Cashew nuts, cotton, sugar
Zimbabwe	460	−0.1	(−2)		Asbestos, chrome, copper, tobacco

(continued)

Table V (*Continued*)

	Per capita GNP[a] (U.S. dollars)		Balance of payments[b] (million U.S. dollars) 1978		
	1977	Annual growth (%) 1970–1977	Interest, public	Other	Main exports[c]
South Africa	1400	1.1	−366	+2010	Maize, fruit, wool, gold, coal
Botswana	540	16.1			Labor, cattle, nickel, diamonds
Namibia	1030	0.8			Diamonds, lead, copper
Angola	280	−3.4			Oil, diamonds, coffee

[a]World Bank (1979).
[b]World Bank (1980).
[c]Mostly from International Economics Division (1979).

V. PLANT RESOURCES

Arid and semiarid zones have few natural plant resources that can be relied on on a continuous basis. Resources of this harsh environment are always at the mercy of severe climatic changes, and some species survive only precariously. Further, some species have been overexploited to the point of becoming endangered, or are at least vulnerable to the pressures of man.

For example, *Cupressus dupreziana,* the Saharan cypress, is a living legacy from the more humid past of the Sahara; some individuals are 2000 years old (Stewart, 1969). One of the most drought-resistant species known, this shade tree is seriously threatened by browsing livestock that consume new growth and by the indiscriminate cutting of branches for firewood.

Medemia argun and *Cyperus papyrus hadidii* both have been greatly depleted for their fibrous qualities and are further threatened by habitat destruction. *Dracaena ombet,* the Nubian dragon tree, another good source of fiber, is succumbing to the demands for firewood and is also overgrazed. *Cordeauxia edulis* (the Yeheb nut), *Olea laperrinei* (closely related to the cultivated olive), and *Euphorbia cameronii,* all potential food crops, and *Wissmannia carinensis,* a potential ornamental plant, are all similarly threatened by the pressures of livestock.

Proper husbandry of these endangered species and all natural resources should be a cardinal consideration in the management of arid and semiarid regions. Where possible, species should be protected, *in situ,* and also cultivated in botanic gardens in order to preserve genetic material. These species are magnifi-

Table VI

Agriculture

	Percentage of GDP[a] 1978	Grain production[b] (million kg) 1978–1979	Annual growth (%) 1970–1979	Net import 1978–1979	Available per capita (kg) 1978–1979	Food production growth per capita (%) 1965–1978	Other major crops (million kg)[c] Root	Other
Morocco	18	4,662	0.4	1,503	322	−4	375	Citrus 1,010, sugar 375
Algeria	8	1,800 (wheat)				−22	575	Citrus 487
Tunisia	18	901	1.9	815	290	26	105	Citrus 128
Libya	2	95 (wheat)				29	85	
Egypt	29	7,413	1.2	5,761	327	−5	1,040	Citrus 827, sugar 695, cotton 406
Cape Verde		15		53	209			
Senegal	24	925	4.3	370	222	−20	140	Peanuts 1,100, cotton 14
Mauritania	(26)[d]	70		110		(−32)		
Mali	37	871	0.6	10	423	−18	228	Peanuts 125, cotton 41
Upper Volta	38	1,051	1.1	35	168	−35	78	Peanuts 65, cotton 16
Benin	31	265	−1.5	40	90	−16	2,226	Cotton 14
Niger	43	,452	4.7	50	302	−37	210	Peanuts 100
Nigeria	(34)	9,038	1.8	1,760	155	−15	33,910	Bananas 1,440, peanuts 600, cocoa 140, cotton 42
Cameroon	32	750	2.4	80	123	−5	2,205	Bananas 800, coffee 100, cocoa 115
Chad	(52)	607		0	142	(−24)		

(*continued*)

Table VI *(Continued)*

	Percentage of GDP[a] 1978	Grain production[b] (million kg) 1978–1979	Annual growth (%) 1970–1979	Net import 1978–1979	Available per capita (kg) 1978–1979	Food production growth per capita (%) 1965–1978	Other major crops (million kg)[c] Root	Other
Sudan	(43)	3,070	4.4	250	199	14	175	Sugar 225, peanuts 900, cotton 152
Kenya	(41)	2,435	2.3	−93	169	−5	370	Sugar 784, coffee 74
Ethiopia	54	2,578	−4.9	101	89	−39	135	Sugar 145, coffee 190, cotton 22
Somalia	60	234	10.4	155	115	(−9)		
Uganda	57	1,380	−0.4		108	−27	2,000	Bananas 355, peanuts 217, coffee 132
Tanzania	51	2,672	2.1	112	162	4	3,445	Sugar 120, coffee 50, cotton 54
Zambia	17	620	0.1	325	203	41	228	Sugar 95
Madagascar	(38)	1,303	−0.7	163		−6	1,750	Bananas 400, sugar 104, coffee 72
Mozambique	45	315	−8.4	377	71	(−5)		
Zimbabwe	(20)	1,710	0.1	−90	232	−19	74	Sugar 275, peanuts 120, cotton 32
South Africa	8	10,054 (maize)				6	744	Citrus 679, sugar 2,050, peanuts 298, cotton 50
Botswana	(43)							
Angola	50	300 (maize)				−49	955	Coffee 90

[a] World Bank (1980).
[b] International Economics Division (1980).
[c] International Economics Division (1979).
[d] Parenthetic data from miscellaneous sources or for an earlier year.

cently adapted to their environment, and with judicious management can be utilized on a sustained-yield basis.

A. Food Plants

Anacardiaceae

Pistacia atlantica—fruit; Sahara; tree; firewood, construction material

Capparidaceae

Capparis spinosa—fruit, seeds; Sahara; half-shrub or shrub

Chenopodiaceae

Arthrocnenum glaucum—Sahara; shrub
Cornulaca monacantha—foliage; Sahara; shrub; animal food

Cucurbitaceae

Acanthosicyos horrida—foliage; Kalahari–Namib; shrub
Citrullus vulgaris—fruit; Sahara; vine

Elaeagnaceae

Elaeagnus angustifolia—fruit; Sahara; tree; animal food, sand binding, construction material

Geraniaceae

Erodium hirtum—Sahara; shrub; animal food

Gramineae

Panicum turgidum—seeds; Somali–Chalbi, Sahara; perennial; animal food

Leguminosae

Alhagi maurorum—Sahara; half-shrub; animal food
Cordeauxia edulis—nuts; Ethiopia, Somalia; shrub; source of good dyes
Glycyrrhiza glabra—Sahara; perennial

Oleaceae

Olea europaea—fruit; Sahara; tree or shrub; animal food, construction material
Olea laperrinei—fruit; Sahara, Sahel; small tree

Palmae

Phoenix dactylifera—fruit; Somali–Chalbi, Sahara; fiber

Polygonaceae

Rumex roseus—foliage; Sahara; annual; animal food

Rhamnaceae

Zizyphus lotus—fruit; Sahara; shrub; firewood
Zizyphus spina-christi—fruit, Sahara; shrub or tree

Rosaceae

Prunus amygdalus—fruit; Sahara; tree

Salvadoraceae

Salvadora persica—foliage; Somali–Chalbi, Sahara; shrub or tree; medicine, construction material

Solanaceae

Lycium arabicum—Sahara; shrub

Zygophyllaceae

Balanites aegyptiaca—foliage; Sahara; shrub; construction material

B. Forage Plants

Chenopodiaceae

Anabasis articulata—foliage; Sahara; succulent shrub; medicine
Atriplex halimus—foliage; Sahara; shrub
Cornulaca monacantha—foliage; Sahara; shrub; human food
Eurotia ceratoides—Sahara; half-shrub; sand binding
Haloxylon articulatum—Sahara; shrub
Salsola aphylla—Kalahari–Namib; shrub
Salsola foetida—Somali–Chalbi, Sahara; shrub
Salsola zeyheri—Kalahari–Namib; shrub
Suaeda fruticosa—foliage; Somali–Chalbi, Sahara, Kalahari–Namib; shrub
Suaeda vermiculata—foliage; Sahara; shrub

Compositae

Artemisia herba-alba—foliage; Sahara; shrub; medicine
Pentzia virgata—Kalahari–Namib; shrub

Convolvulaceae

Convolvulus lanatus—foliage; Sahara; shrub

Cruciferae

Eremobium lineare—foliage; Sahara; annual

Elaeagnaceae

Elaeagnus angustifolia—Sahara; tree; human food, sand binding, construction material

Euphorbiaceae

Euphorbia cameronii—Somalia; shrub; can provide moisture in dry periods

Geraniaceae

Erodium hirtum—foliage; Sahara; shrub; human food

Gramineae

Aristida adscensionis—Sahara
Aristida plumosa—foliage; Sahara
Aristida pungens—seeds; Sahara; perennial
Danthonia forskalii—foliage; Sahara
Distichlis scoparia—Kalahari–Namib
Lasiurus hirsutus—foliage; Somali–Chalbi, Sahara; perennial
Panicum turgidum—foliage; Somali–Chalbi, Sahara; perennial; human food
Phragmites communis—foliage; Kalahari–Namib; fiber
Schismus barbatus—foliage; Sahara; annual
Stipa tenacissima—foliage; Sahara; fiber
Stipa tortilis—foliage; Sahara; annual

Leguminosae

Acacia albida—foliage; Sahara, Kalahari–Namib; tree
Acacia giraffae—seeds; Kalahari–Namib; tree or shrub; tannin
Alhagi maurorum—foliage; Sahara; half-shrub; human food
Genista saharae—Sahara; shrub

Oleaceae

Olea europaea—Sahara; tree or shrub; human food, construction material

Plantaginaceae

Plantago ovata—Sahara; annual; medicine

Polygonaceae

Calligonum comosum—foliage; Sahara; shrub; sand binding
Rumex roseus—foliage; Sahara; annual; human food

Sapotaceae

Argania spinosa—foliage, fruit; Sahara; tree

Umbelliferae

Deverra scoparia—foliage; Sahara; shrub
Pituranthos tortuosus—Sahara; annual; medicine

Zygophyllaceae

Zygophyllum simplex—foliage; Somali–Chalbi, Kalahari–Namib; annual

C. Medicinal Plants

Amaranthaceae

Aerva tomentosa (A. javanica)—Somali–Chalbi, Sahara; toxic shrub

Apocynaceae

Nerium oleander—Sahara; toxic shrub

Asclepiadaceae

Calotropic procera—Somali–Chalbi, Sahara; shrub; fiber

Chenopodiaceae

Anabasis articulata—Sahara; succulent shrub; animal food

Compositae

Achillea santolina—Sahara; perennial
Artemisia herba-alba—Sahara; shrub; animal food

Cucurbitaceae

Citrullus colocynthis—Sahara; toxic annual

Gramineae

Cymbopogon schoenanthus—Somali–Chalbi, Sahara

Leguminosae

Cassia obovata—Sahara; shrub

Liliaceae

Aloe dichotoma—Namib; succulent tree

Melianthaceae

Melianthus comosus—Kalahari–Namib; shrub

Plantaginaceae

Plantago ovata—Sahara; annual; animal food

Salvadoraceae

Salvadora persica—Somali–Chalbi, Sahara; shrub or tree; human food, construction material

Solanaceae

Hyoscyamus muticus—Sahara; toxic

Tamaricaceae

Tamarix articulata—Somali–Chalbi, Sahara, Kalahari–Namib; shrub or tree; construction material, tannin

Umbelliferae

Pithuranthos tortuosus—Sahara; annual; animal food

Zygophyllaceae

Peganum harmala—Sahara; shrub; oil, tannin
Zygophyllum album—Sahara; shrub

D. Industrial Plants

1. FIBER PLANTS

Asclepiadaceae

Calotropis procera—Somali–Chalbi, Sahara; shrub; medicine

Cyperaceae

Cyperus papyrus subsp. hadidii—Sahara; giant sedge; presently used for paper; used for food, medicine, mats, boats, and sandals in ancient Egypt

Gramineae

Eragrostis binnata—Sahara
Phragmites communis—Kalahari–Namib; animal food
Stipa tenacissima—Sahara; animal food

Liliaceae

Dracaena ombet—Djibouti, Ethiopia, Sudan; tree; trunk cut for firewood

Palmae

Medemia argun—Egypt, Sudan; leaves used for mats
Phoenix dactylifera—Somali–Chalbi, Sahara; tree; human food

2. FUEL WOOD

Anacardiaceae

Pistacia atlantica—Sahara; tree; human food, construction material

Cupressaceae

Cupressus dupreziana—Sahara; tree; shade

Rhamnaceae

Zizyphus lotus—Sahara; shrub; human food

Salicaceae

Populus euphratica—Sahara; tree; construction material

3. OIL PLANTS

Sapindaceae

Pappea capensis—Kalahari–Namib; tree

Zygophyllaceae

Peganum harmala—Sahara; shrub; medicine, tannin

4. WAX PLANTS

Euphorbiaceae

Euphorbia antisiphylitica—Kalahari–Namib; toxic succulent

5. GUM PLANTS

Leguminosae

Acacia gummifera—Sahara; tree; tannin
Acacia seyal—Sahara; tree; tannin

6. TANNIN PLANTS

Leguminosae

7. CONSTRUCTION-MATERIAL PLANTS

Anacardiaceae

Pistacia atlantica—Sahara; tree; human food, firewood

Ebenaceae

Euclea pseudebenus—Kalahari–Namib; tree

Elaeagnaceae

Elaeagnus angustifolia—Sahara; tree; animal and human food, sand binding

Oleaceae

Olea europaea—Sahara; tree or shrub; animal and human food

Salicaceae

Populus alba—Sahara; tree
Populus euphratica—Sahara; tree; firewood

Salvadoraceae

Salvadora persica—Somali–Chalbi, Sahara; shrub or tree; human food, medicine

Tamaricaceae

Tamarix articulata—Somali–Chalbi, Sahara, Kalahari–Namib; shrub or tree; medicine, construction material, tannin

Zygophyllaceae

Balanites aegyptiaca—Sahara; shrub; human food

E. Other Plants

1. SAND-BINDING PLANTS

Chenopodiaceae

Eurotia ceratoides—Sahara; half-shrub; animal food

Compositae

Artemisia monosperma—Sahara; shrub

Elaeagmaceae

Elaeagnus angustifolia—Sahara; tree; animal and human food, construction material

Gramineae

Aristida obtusa—Sahara

Leguminosae

Retama raetam—Sahara; shrub

Polygonaceae

Calligonum comosum—Sahara; shrub; animal food

2. SHADE PLANTS

Cupressaceae

Cupressus dupreziana—Sahara; tree; firewood

VI. SUMMARY

The information presented in this chapter dictates that steps need to be taken immediately by the responsible African governments to alleviate the vulnerable position in which their people find themselves. It is mandatory that very imaginative and innovative programs be designed to help arrest further displacements of the people. There are certain basic research projects in crop sciences that must be given serious consideration. The first consideration is to conduct research on the adaptation of major crop plants to drought stress. Currently, very little information exists on the morphological, physiological, genetic, and biochemical pathways inherent in drought resistance in some of the important food crops in the arid and semiarid regions. For example, climatic behavioral models covering some of the major farming systems will have to be simulated in such experiments in order to evolve alternative cropping strategies to suit different growing conditions.

Another activity that requires investigation, to raise the yield of potential of crops in arid and semiarid regions, is the exploitation of heterosis or hybrid vigor, namely, the modification of plant architecture and growth rhythm so that the response of the genotype to irrigation and organic or chemical fertilizer application may be enhanced. Similarly, intensive experiments on raising the yield ceiling without affecting stability of performance are yet to be tried in rainfed crops grown in arid and semiarid regions. Such experiments could achieve early seedling vigor and rapid root growth through heterosis. Suitable F_1 hybrids may help to elevate and stabilize yields in rainfed crops through early seedling establishment as well as early maturity.

In this regard, it is imperative that the collection and preservation of both crop and animal genotypes which are adapted to drought stress conditions be undertaken before they become extinct in irrigated areas. It will also be most useful for the relatively few well-equipped facilities in Africa to use radioisotope techniques in estimating heterosis for root growth. Because of the importance of legumes in the diet of people living in the arid and semiarid regions of Africa, it is necessary that studies of the bioenergetics of plant mass production, as well as microbiological and physiological aspects of nitrogen fixation, be encouraged in the laboratories that are equipped to undertake such studies.

Because of the excessive pressures on trees for firewood and other wood uses in the arid and semiarid regions of Africa, programs need to be initiated to assist in the establishment of new fast-growing plant communities that can serve as multipurpose trees for fodder and firewood, as well as for soil conservation. In this regard it is highly recommended that sound land-use policies be undertaken together with soil and water management techniques. Recent technologies such as remote sensing from both aerial and satellite photography are now available to arid and semiarid regions of Africa and must be exploited for the full benefit of the people.

The production of high-yield crops should not be the only major concern in these areas. Problems relating to postharvest crop production should be given equal attention. It seems most unfortunate that because of the lack of expertise, large quantities of crops go to waste after harvest in these regions. This is largely due to the lack of attention paid to the chemical composition of the harvested crops, and the seed anatomy and physiology as they affect the nutritional qualities of the produce. It has been repeatedly demonstrated that few data, comparable to that known about the composition and structure of wheat, exist to guide plant breeders in the selection of high-yield millet, sorghum, and dryland legumes that combine nutritional and functional properties of such crops. In addition, there is little understanding of the biochemical and biophysical characteristics that influence the useful properties of grains and how they are affected during storage. Surely, in order to change the current unsatisfactory situation, research in these and other areas of postharvest food conservation should be given the highest priority.

ACKNOWLEDGMENTS

I am most grateful to William Ricc, who assisted me in the preparation of this chapter and particularly in generating the data for the tables and maps. My appreciation is extended to Alice Tangerini for drawing the maps, and to Robert DeFilipps and Mary Mangone for their most helpful suggestions.

REFERENCES

Department of Economic and Social Affairs (1973). "Ground Water in Africa." United Nations, New York.

Dregne, H. E. (1976). "Soils of Arid Regions." Elsevier Sci. Publ., Amsterdam.

Hance, W. A. (1975). "The Geography of Modern Africa," 2nd ed. Columbia Univ. Press, New York.

International Economic Division (1979). "Agricultural Situation: Review of 1978 and Outlook for 1979." U.S. Dep. Agric., Washington, D.C. [Africa and West Asia]

Instituto Geográfico de Agostisni (1976). "World Atlas of Agriculture." Inst. Geograf. Agostisni, Novara.

IUCN (1978). "The IUCN Plant Red Data Book." IUCN, Morges, Switzerland.

Kassas, M. (1970). Desertification versus potential recovery in circum-Saharan territories. *In* "Arid Lands in Transition" (H. E. Dregne, ed.). Am. Assoc. Adv. Sci., Washington, D.C.

Lebeder, A. N. (1970). "The Climate of Africa," Part I. Israel Program for Scientific Translations, Jerusalem.

Le Houerou, H. N. (1970). "North Africa: Past, Present, Future." *In* "Arid Lands in Transition" (H. E. Dregne, ed.). Am. Assoc. Adv. Sci., Washington, D.C.

Pereira, H. C. (1977). Land use in semi-arid southern Africa. *In* "Resource Development in Semi-arid Lands" (J. B. Hutchinson *et al.*, eds.). Royal Soc., London.

Pritchard, J. D. (1971). "Africa: The Geography of a Changing Continent." Africana Publ., New York.

Stewart, J. M. (1969). *Cupressus dupreziana,* threatened conifer of the Sahara. *Biol. Conserv.* **2,** 10–12.

UNESCO (1979). "Map of the World Distribution of Arid Regions." United Nations Educational, Scientific and Cultural Org., Paris.

United Nations (1978). "Demographic Yearbook 1977." United Nations, New York.

World Bank (1979). "1979 World Bank Atlas." World Bank, Washington, D.C.

World Bank (1980). "World Development Report, 1980." World Bank, Washington, D.C.

2

AUSTRALIA

O. B. Williams
CSIRO Division of Water and Land Resources
Institute of Biological Resources
Canberra City, A.C.T.
Australia

M. Lazarides
CSIRO Division of Plant Industry
Institute of Biological Resources
Australian National Herbarium
Canberra City, A.C.T.
Australia

I. INTRODUCTION

The aim of this chapter is to describe briefly the environment and the plant resources of Australia's arid interior, to outline the present status of development of these resources in both commercial and research aspects, and to supply an inventory of useful plants. The potential use of plant resources is discussed in terms of the "real world": national and international economic, demographic, and political factors that influence plant resource development in the Australian arid lands.

The reader who wishes to proceed further into the Australian arid scene can do so with the assistance of various contributors to the literature of Australian geography and landforms (Jeans, 1977; Mabbutt, 1977), and more than 450 references chosen by Williams (1979) to illustrate Australian arid zone biology.

In the 1980s there are few contrasts as great as the one described here, which shows the limited opportunity for plant resource use and the much-publicized discovery and development of substantial mineral resources. The development of existing mineral resources, the search for more, and the acquisition and expansion of Australian companies in the manufacturing, retailing, and service sectors have promoted massive capital inflows from overseas akin to those that flowed into the Australian pastoral industry and the Australian economy at the beginning of the 1880s.

II. PHYSIOGRAPHY

A. Size and Location

Although the United States and Australia were invaded and colonized by Europeans at about the same time and have similar land areas and similarities in their short histories of development toward the modern state, there is a marked difference in their continental resource base. The magnitude of this difference is frequently misunderstood outside Australia, and is just as frequently not appreciated within Australia.

This resource inequality can be seen if we place Australia upside down over the position occupied by the United States, as in Fig. 1, but keeping in mind that Australia's Cape York, at 11°S, is being fitted over Florida at 25°N, and that the southern part of the Australian mainland, at 39°S, is 10° above the U.S.–Canadian border. The boundaries of the major Australian mainland states and territories are shown.

B. Topography and Climate

The natural resources of the desert area have been reviewed (Williams, 1979), but a brief summary is presented here.

The desert area covers 4.2 million km^2 or 37% of the continental land mass. The plant communities (Fig. 2) and their approximate areas are *Acacia* shrublands, 1.6 million km^2; hummock grasslands (*Triodia* spp.), 1.6 million

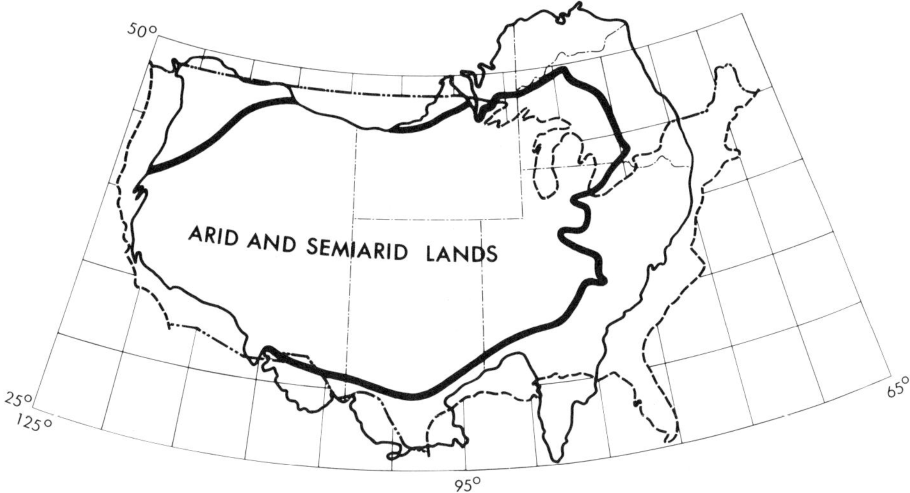

Fig. 1. A comparison of the resource base for Australia and North America.

km^2; tussock grasslands (*Astrebla* spp.), 0.5 million km^2; and shrub steppe (Chenopodiaceae), 0.4 million km^2.

The semiarid area adjoining the desert covers another 1.1 million km^2; the semiarid plant communities are low woodlands of 0.3 million km^2, shrub woodlands occupying 0.4 million km^2, and eucalypt shrub woodlands (*mallee*) covering 0.4 million km^2.

The topography is subdued, with mountain ranges and associated plateau (Precambrian, some Paleozoic) covering 580,000 km^2. Elevations range up to 1800 m, with relief ranging from 300 to 800 m. The mountains do not have snow, although the Macdonnell ranges enhance rainfall and lower temperatures, which might also improve the water balance. Runoff from the mountains is immediate and short-lived; water seldom flows far into the plains. The so-called shield deserts of Precambrian origin cover 750,000 km^2, and the stony deserts of the younger sedimentary basins cover 400,000 km^2. The riverine and clay plains cover 630,000 km^2 in a great arc from north to south, and the sand deserts cover the remaining 1 million km^2.

Climate in the desert and its surroundings is dominated during summer by a high pressure (anticyclonic) belt; the high-pressure cells move slowly eastward, well to the south of the interior itself, but their effect is paramount. At the same time, the tropical low-pressure (cyclonic) belt may influence weather in the northern desert, and swaths of heavy rainfall may cut across the desert margins and into the desert.

Around the autumn equinox, both of these systems commence their northward latitudinal shift so that during winter the subtropical high pressure belt operates

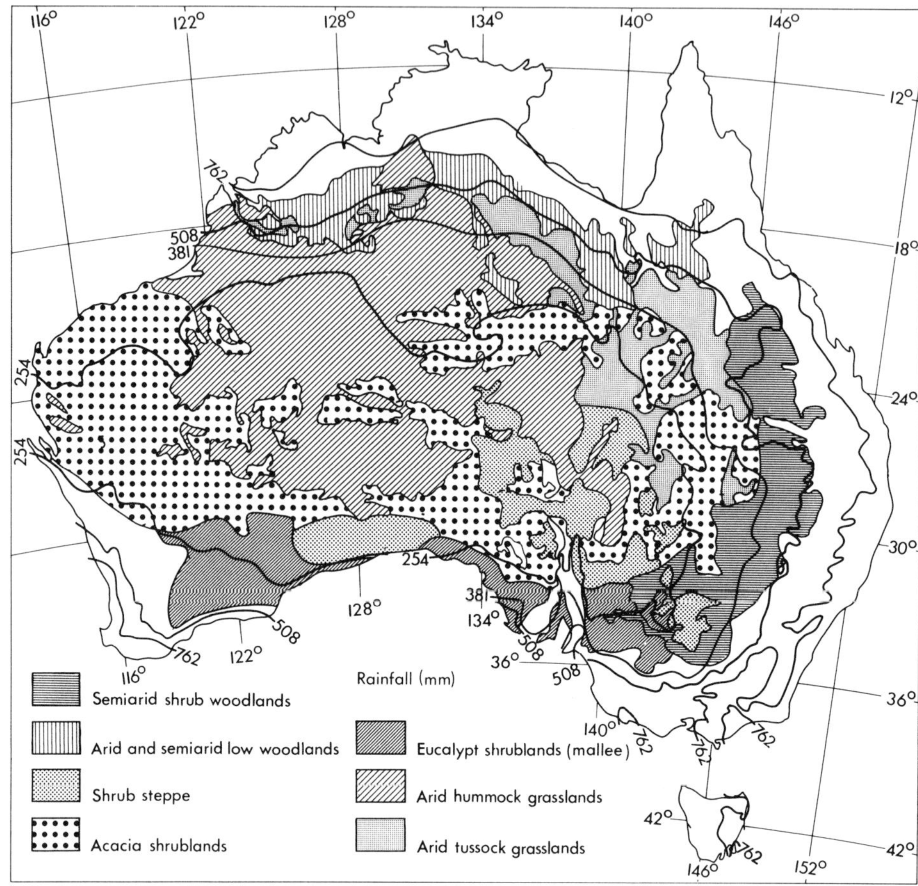

Fig. 2. The major plant communities of the Australian desert.

within a strip bounded by latitudes 27 and 30°S across the center of the desert. At the time of the spring equinox these belts move southward.

From year to year there is considerable variation in the date of the onset and speed of movement of these latitudinal shifts as well as in the size, pressure, direction, and speed of the cells within these belts. For example, in recent years, the blocking effect of massive high-pressure cells in the Tasman Sea between eastern Australia and New Zealand has forced many rainbearing low-pressure cells in a southeasterly direction below the continent, while at the same time inducing cyclonic-type rainbearing low-pressure cells to cross the desert from northwest to southeast. The rainfall pattern is characteristically of the midsummer type, but it occurs in winter.

In addition to the possibilities of rain from the major systems, there is the chance of rain from local convectional disturbances. Indeed, the variability in precipitation and the observed "clumping" of both rainfall and dry weather can be understood best by an analysis of rainfall contributions on a synoptic basis.

In arid lands, the occurrence of consecutive years of rainfall in the upper quartile levels, at frequencies of one in 50 years or more, initiates the growth of substantial populations of long-lived plants. Although the dry periods receive most attention, these wet years are probably of greater ecological significance than long dry periods.

Temperatures in the desert and its surroundings are high in summer, with a mean of 30°C, but there is little variability between and within regions. In contrast, winter temperatures are not uniform over the inland, with the mean monthly midwinter temperature ranging from 11°C in the south to 19°C in the north. The frost frequency is from 12 to zero per annum. Weather systems that produce substantial summer rain also tend to depress summer temperatures for 3 or 4 days afterward, and so enhance the germination and establishment of plants.

Dew point is seldom reached because atmospheric humidity is low; for example, at Alice Springs, a relative humidity of 30% would require a night temperature of 0°C before the dew point would be reached.

Radiation levels are not typical of the desert and its semiarid regions. Winter values are latitudinal on a continental basis and summer levels tend to be highest in the mesic southeast of the continent. This is due largely to the cloudiness in the north of the continent, which is attributable to the tropical low-pressure belt.

Evaporation does not vary greatly across the desert and surroundings. Values from a class A pan would approximate 360 cm annually, and monthly midsummer and midwinter values would approximate 50 and 12 cm, respectively.

Winds are light and have a small erosion potential over most of the desert. In the southwest Simpson Desert, however, strong winds blow in the opposite direction of the prevailing winds; thus, some dune fields are mobile.

Figure 1, with Australia reversed and superimposed over the United States to allow a better comparison of the areas, shows an immense area marked as arid and semiarid grazing lands. Although a few mountain ranges are in the west and small zones of woodland are in the wetter eastern region, most of arid Australia is a complex of sparse bunch grass, desert shrubland without trees (similar in appearance to the intermountain rangeland of southern Utah), and desert landscapes with scattered small trees similar in appearance to the rangelands northeast of El Paso, Texas.

C. Edaphic Factors

The dominant soils are sands, loams, and cracking clays with uniform texture profiles, soils with gradational texture profiles such as calcareous and noncalcareous earths, soils with texture-contrast profiles, and red siliceous sands.

The permeable soils exhibit more leaching than might be expected in a desert, possibly because weather conditions in the past either affected the soils of old landscapes directly or affected young soils indirectly through their leached parent materials. The rare runs of above-average rainfall years can accentuate the leaching process, and temporary freshwater lakes have been observed in the swales between the sand dunes in the Simpson Desert. Impermeable soils show a small amount of leaching, some salt accumulation, but little profile development. The potential erodibility of these various soils under pastoral use is high in light-textured and texture contrast or duplex soils, and low in heavy-textured soils. The severe use characteristic on texture-contrast soils leads to their frequent representation in erosion categories.

The chemical characteristics of the soils in arid Australia that are of particular relevance to plant growth are the availability of the nutrients nitrogen and phosphorus. These are generally low; total nitrogen (percent oven-dry soil) ranges from 0.01 in siliceous sands to 0.06 in red earths. Levels of phosphorus are low, with values of 0.01 to 0.05%; the range in available phosphorous is from 3 to 22 ppm, with levels in some clays ranging up to 40 ppm. Trace-element deficiencies also occur, but their visible expression is usually noted in the growth of exotic species and in plants under irrigation. Soluble salts are lower in the north than in the south, where in certain clays the levels exceed 1.6% and exchangeable sodium can be in excess of 15%. This northern salt is of cyclic origin.

There is little erosion in the desert, even though there exists an erosion potential should the landscape be disturbed. There is no general expansion of Australian deserts and no desert front advancing across the landscape. The semidesert areas close to the mountain ranges of the interior (Condon *et al.*, 1969) and semiarid grazing lands (Williams *et al.*, 1980) show evidence of widespread and substantial soil disturbance associated with grazing use by the pastoral industry. Rate measures for "desertification" are not available. Methods of assessing range condition are being tested, however, and at some future date these should enable one to estimate an improving or deteriorating condition. The occurrence of runs of good-rainfall years and of droughts makes a trend difficult to see, but because we can successfully discern small rangeland areas in the improving category and note their limited size and rarity, we consider that the underlying trend is either downward from a degraded condition—attributed largely to heavy utilization in the past—or stable at that degraded condition. If demonstrable desertification is to be seen anywhere, it is in the nonrecruitment of desirable shrubs and small trees, particularly in the southern part of semiarid and arid Australia.

D. Hydrology and Water Resources

Groundwater is the major, and often the only, source of water in arid Australia; large reserves of groundwater occur in sedimentary basins such as the

Great Artesian, Murray, Canning, Amadeus, and Georgiana basins. The Great Artesian Basin supplies water for livestock and limited domestic use by means of some 20,000 bores over an area of 1.74 million km^2. The quality of water varies considerably over these basins; sodium, calcium, carbonate, high sulfate, and high fluoride restrict the variety of uses. Salts accumulating in gardens are not always removed by precipitation.

Flow is intermittent for most small rivers along the eastern coastline, and the larger rivers exhibit marked fluctuations in flow. There are no permanent rivers in the interior, apart from the Darling–Murray system, which draws its water from the eastern divide and crosses the eastern semiarid region to discharge into the Southern Ocean. Lakes in the desert are better classified as saline and non-saline playas. Few atlases, apart from the *Reader's Digest Atlas of Australia* (Anon., 1977), showing an abundance of stony and sandy deserts, dunes, playas, and intermittent streams, depict the interior as it really is; to the innocents abroad there are myriad streams and hundreds of substantial lakes awaiting exploitation and development.

III. DEMOGRAPHY

The human population of Australia is approaching 15 million or ~6% of the present U.S. population, with one small city of 800,000 to 900,000 inhabitants in each of Western Australia, South Australia, and Queensland, and two large cities of 2.5 to 3 million inhabitants in each of Victoria and New South Wales. Urbanization has been a continuing process for at least 100 years, and even between 1954 and 1971 the percentage of rural residents fell from 21.3% of the population to 14.4%. It is significant that, for example, 73.1% of the overseas-born residents lived in capital cities in 1966 compared with 54.8% of the Australian-born residents. Much of the area in Fig. 1, marked as arid and semiarid grazing lands, has not been occupied by Europeans, and some has been occupied, then abandoned. Favored parts of the arid interior, such as Alice Springs near the Macdonnell ranges, together with the adjoining semiarid grazing lands, are permanently occupied. The European invasion of Australia, which followed the post-1788 establishment of the British penal colonies, was made possible by the first industrial revolution; continued occupation of the continent by a population with northern European standards of living depends largely on the present second industrial revolution. The arid and semiarid grazing lands exploited in the first phase generally had an adequate resource base for the profitable production of livestock and livestock products. Continued severe use and subsequent degradation made an intensification phase impossible, however, even with further capital and technological investment.

Population density on the wetter margins of the semiarid region is one person to 8 km^2 at best, and on the drier margins is one person to more than 64 km^2.

Omitting mining settlements, one can anticipate that population densities in these lands will decrease further (Rowland, 1979; Young, 1980). This decrease is likely to accelerate because population densities, and income densities, in the range of 64 to 8 km^2 per person are barely able to sustain the present simple infrastructure. However, the numbers of people involved will be few.

From the viewpoint of recreation, Australia is notable for thousands of kilometers of public beaches of excellent quality. Approximately 80% of the population, including most of Australia's recent immigrants, live within 200 km of the coastline within each reach of its beaches, bays, and estuaries. The population in this coastal zone increased by 2,070,000 between 1961 and 1971, as compared with an increase in the inland population of only 176,000, of which more than half were in the national capital, Canberra.

IV. SOCIOECONOMIC FACTORS

The states making up the Australian Commonwealth are sovereign states, and no politically aware Australian would ever refer to them as the "United States of Australia"; it is not farfetched to consider Australia as an "antipodean European Economic Community" in which the inhabitants speak English. Canberra is the antipodean Brussels. This political structure, and somewhat irreverent interpretation, is of more than passing interest because the control of land, including the arid lands and most matters pertaining to them, is within the jurisdiction of the individual Australian states and not that of the Commonwealth or federal administration. Laws governing the use and conservation of arid lands can and do differ substantially among states (Young, 1979a,b). However, this supposedly clear separation of state and federal jurisdiction is made exceedingly complex in practice by the overriding power of federal laws governing mineral exports, land rights of the aboriginal people, and the seldom-used but existing rights of the military to exercise its power, as well as federal financial power over the states. The federal government has first rights to income tax and levies it at rates sufficiently high to preclude entry of the states into this lucrative field.

V. PLANT RESOURCES

Seven vegetation formations cover an area equivalent to that of Europe (Fig. 2). Briefly, these are

1. Semiarid woodlands (385,000 km^2) consisting of a *Eucalyptus* spp. overstory with a ground layer dominated by grasses of the genera *Aristida, Bothriochloa,* and *Enneapogon*

2. Arid and semiarid low woodlands (272,000 km^2) consisting of *Acacia aneura* and of *Casuarina, Heterodendrum, Callitris,* and dwarf *Eucalyptus* spp.

3. Shrub steppe or low shrubland (434,000 km^2) dominated by Chenopodiaceae of genera such as *Atriplex, Sclerolaena, Maireana, Chenopodium,* and *Rhagodia*

4. Shrublands dominated by *Acacia aneura* (1.6 million km^2), but with a large number of associated *Acacia* spp. and a ground layer of grasses and some forbs, mainly of the family Malvaceae (The complexity of these shrublands can be gauged from a vegetation survey in the Alice Springs district in which one upper-story community had up to 18 distinct ground-story communities; the most common of these ground-story communities recurred with 12 other upper-story communities.)

5. Arid *mallee* (410,000 km^2) of *Eucalyptus*-dominant woodland with an understory of the hummock grass *Triodia* on sandy soils, and *Maireana* and *Striplex* spp. where the soils are of medium texture

6. Hummock grasslands (1.6 million km^2) dominated by various species of *Triodia* and by *Plechtrachne*

7. Arid tussock grasslands (5 million km^2) dominated by various species of *Astrebla*

Families with a signifcant number of desert species include Gramineae, Chenopodiaceae, and Compositae. The major genera of grasses include *Triodia, Zygochloa, Aristida, Enneapogon, Eragrostis, Astrebla, Plechtrachne, Dichanthium, Panicum,* and *Chrysopogon.* The major genera of Leguminosae are *Acacia* (only one with spines) and *Swainsona, Indigofera, Tephrosia, Glycine, Sesbania,* and *Astragalus* (introduced). *Clianthus* is the only monospecific genus in the Leguminosae, but a number of monospecific genera occur in the grasses. *Zygochloa, Plagiosetum, Paractaenum, Paraneurachne, Spathia,* and *Uranthoecium* are monotypic and endemic to Australian arid regions; *Triraphis, Tripogon, Tragus,* and *Perotis* are monotypic in Australia, but the genera and some of the species occur elsewhere. The Leguminosae harbor some outstanding toxic plants. The Chenopodiaceae have large numbers of species in genera such as *Sclerolaena, Maireana,* and *Atriplex.* The Asteraceae are represented by many annual species in the genera *Helipterum* and *Helichrysum,* although the perennials should not be overlooked.

The arid communities do not appear to be diverse. Published figures from two surveys put the number of species at one per 51 km^2 in central Australia and one per 240 km^2 in southwest Queensland, where 615 plant species were in 74 families and 241 genera. Forty-two percent of these species were in the Gramineae and Leguminosae. However, a figure of one per 0.4 km^2 was obtained in a central Australian study area over a period of 5 years.

A. Ecology

The major use of plant resources in arid Australia is made by the pastoral industry (Wilson and Graetz, 1979): Merino wool-growing sheep are raised across the south, southeast, and southwest, and cattle are raised primarily in the central regions and in the north. An estimated 7000 sheep holdings with 18.7 million sheep occupy 1.7 million km^2; cattle holdings are estimated at 9000, with 2.1 million cattle occupying a further 3 million km^2. Flocks and herds in arid Australia make up ~10% of the Australian total. Other minor uses of arid lands are occupancy as homeland by aboriginals, mining, tourism and recreation, national parks, and military and telecommunication facilities. Plant and seed collection are insignificant activities.

There is considerable literature covering pastoral use, its settlement, exploitation, animal production systems, and rangeland. The present chapter will concentrate on describing the factors involved in the evolution and management of the plant resource.

The great plate of which the Australian continent is a part has moved a considerable distance since it parted from Africa and Antarctica. Further, the range of shifting global and latitudinal climates imposed on this moving land mass has resulted in speciation and elimination among plants and animals. Some superlative examples of discontinuous distribution are known in plants, and this particular field is being actively researched.

In broad terms, a very large number of species within a few genera such as *Eucalyptus* and *Acacia* have representatives in both the desert and far-distant alpine communities. Remnants of former rain forest blocks in central Australia are still found in a few favored sites in the Macdonnell ranges. The extremes in aridity would appear to have been so severe as to confine relict areas to these favored sites, mainly in the Hammersley and Flinders ranges and what is now the humid southeast of the continent.

A second important factor in shaping the vegetation was the absence of the ruminant, beasts of burden, and flocks or herds of the dominant marsupials. Ruminant-like marsupials were present but were restricted to the few reliable water sources. One should keep in mind that permanent surface water was almost nonexistent on the continent. Vertebrate herbivores did not have access to water in sand soaks and the small rock cisterns (so-called native wells) that supplied the small bands of aboriginal people. Dry-matter accumulation was controlled by fire resulting from lightning and the skilled use of fire by the aboriginal people (Gill *et al.*, 1981), who have been residents of the continent since 40,000 BP. Recently, fire has come to be considered an ecological tool, though as yet there is little research and even less deliberate application in arid Australia. Research has commenced, however, and European man has set about acquiring the manipulative skills necessary to handle fire for management of the biota.

A third factor in shaping the vegetation was the nature of the weather systems;

precipitation is extremely variable, both seasonally and yearly. The perennial plants of the desert and its surroundings are long-lived, with irregular flowering, fruiting, and germination. Recruitment once in 50 or 100 years has maintained optimum populations of trees and some shrubs, and depending on the species, the environment may have offered only 2–10 such opportunities in each 100 years. Under the influence of aridity, animal and plant species were disappearing from the desert and its environs well before the invasion by European man and his animals and plants.

The European invasion commenced in the late 1700s, was maximal by 1850, and was complete by 1890. The European capital, derived from overseas investments and released by the bursting of the railway boom in Britain, fueled this pastoral program and massive migration in the 1880s and early 1890s. The resulting outputs of wool and meat entered the European industrial system.

The European rabbit invasion and massive droughts in the 1890s precipitated a disaster that would have happened anyway. The great development period came to a close. In hindsight, this era occurred within a high-rainfall period in the southern hemisphere, at the end of the European little ice age, and at the time of a European population explosion that ultimately led to the migration of 75 million people to the Americas and Australia.

The present plant resources are composed of the original indigenous species, reduced in density or eliminated from extensive tracts of country, and around the southern semiarid margin, an ingress of Mediterranean-type annual plants. There is a small resident population of the rabbit in the desert southwards from the tropic of Capricorn, and this population, together with sheep and cattle, is influential in removing recruits of some desirable trees and shrubs so that these species exist as sparse geriatrics. These species are the long-lived perennial plants that maintain leaf production during long rainless periods and so enable maintenance of breeding stock. Succulent annual plants also permit the flocks and herds to produce as well as reproduce.

Plants in the long-lived perennial group with excellent fodder characteristics or soil stabilizing qualities have a potential for use in semiarid and arid environments once their limitations and virtues are recognized. For example, it is difficult to obtain seed, let alone genotypes of defined species, and information is rudimentary so far as cultural and management practices are concerned. Most important, however, is the recognition that these species need up to 10 years before commercial grazing can be started. Once started, however, the longevity of these plants exceeds 300 to 500 years. An important step toward educating Australians about the plants of their semiarid grazing lands has been made by Cunningham *et al.* (1981) in their *Plants of Western New South Wales.* The recent establishment of national (i.e., state) parks in these lands will lead inevitably to detailed descriptions of their flora, publications about the species growing there, and commercial possibilities.

The use of the arid plant resource by other than pastoral enterprises is slight.

The small resident aboriginal population makes an impact similar to a small white community; mining activity takes place in remote areas, and in generally nonpicturesque scenery. Tourism is not heavy but is location-specific and has caused marked degeneration at several sites. National parks are not heavily used, and the military and telecommunication usage has not, so far, resulted in widespread damage.

Applied research has been centered on animal and plant production, with accompanying ecological surveys and small but significant programs on soil erosion and erosion prevention, revegetation, management of various plant communities that are under pressure from sheep and cattle grazing, introduction of plants, and factors involved in plant distribution. Pioneer studies in what has become known as plant demography have been expanded, and there is a better understanding of the changes that occur in the composition of plant communities than previously. Investigations have begun on the use of native plants for direct consumption by the aboriginal people.

B. Food Plants

With the exception of honey, most of the plants listed in this section are of primary importance to the aboriginal people.

1. ROOTS AND TUBERS

The roots and tubers group includes the tubers of *Ipomoea* spp., bulbils of *Cyperus bulbosus* Vahl (*nalgoo*), swollen roots of *Vigna lanceolata* Benth. (*maloga* bean), *Haemodorum corymbosum* Vahl (bloodroot), *Boerhavia diffusa* L. (tarvine), *Tribulus, Erythrina,* and *Clerodendurm.*

2. FRUITS

Species of *Solanum,* including *S. ellipticum* R. Br. (potato bush) and *S. esuriale* Lindl. (*quena*) are very important food plants, and the berries are eaten whole. Many other fruits are eaten; the most important include *Owenia acidula* F. Muell. (emu apple), *Ficus, Santalum, Cucumis, Leichardtia, Carissa lanecolata* R. Br. (Conkerberry), *Enchylaena tomentosa* R. Br. (ruby saltbush), and the pulp and seed of *Capparis mitchellii* (F. Muell.) Benth. (bumble or wild orange).

3. SEEDS

Seeds require considerable effort to collect and prepare but nevertheless are an important food item. Grasses are an important source and include species of *Panicum, Eragrostis,* and *Brachiaria* as well as *Dactyloctenium radulans* (R.

Br.) Beauv. (button grass). Seeds from many species of *Acacia* are crushed and eaten; most important are *A. aneura* F. Muell. *ex* Benth. (*mulga*) and *A. oswaldii* F. Muell. (*nelia*). The fleshy bulbs containing small seeds of *Tecticornia verrucosa* P. G. Wils. and *T. arborea* P. G. Wils. are eaten directly from the plant. The sporocarp of *Marsilea drummondii* A. Br. is eaten after preparation. Seeds are utilized from a wide range of other plants, including *Calandrinia, Portulaca, Amaranthus, Brachychiton, Chenopodium, Sida,* and *Heliotropium*. Usually, the minute but abundant seeds of plants such as *Portulaca* are ground into a cake and cooked.

4. GREENS

After steaming, green plants such as *Lapidium* spp., *Calandrinia, Convolvulus, Cynanchum,* and *Marsdenia* are utilized as plant foods in times of scarcity.

5. HOST PLANTS

The roots and trunks of *Acacia kempeana* F. Muell. (witchetty bush), *Salsola kali* L. (prickly saltwort), *Atalaya hemiglauca* (F. Muell.) Benth., and *Codonocarpus continifolius* (Desf.) F. Muell. (desert poplar) are a source of grubs. The leaves of *Eucalyptus* spp. and branchlets of *Acacia* spp. often provide lerp insects. Edible gum or sap is obtained from the trunks of *Acacia kempeana* F. Muell., *A. lingulata* Benth. (umbrella bush), *Atalaya hemiglauca* (F. Muell.) Benth., *Ventilage viminalis* Hook. (supplejack), and a variety of other plants. *Eucalyptus* spp. of the "bloodwood" group provide insect galls.

6. HONEY

Honey was a much-sought food supplement by the aboriginals and was derived from the flowers of certain plants, especially *Grevillea juncifolia* Hook. and other species, *Eucalyptus* spp., and *Eremophila latrobei* F. Muell. (native fuchsia).

Honey and beeswax are commercially produced on a large scale, mainly from *Eucalyptus* spp. Most species, especially the "gums," yield nectar and pollen suitable for gathering by bees. However, the "box" and "ironbark" groups, though foremost in yields of excellent quality nectars, produce pollens that are unattractive to bees. Since pollen is essential for brood-rearing and for maintaining hive strength, associated pollen-bearing plants are necessary. Among the most important of these are trees and shrubs belonging to *Acacia, Melaleuca, Hakea, Leucopogon,* and naturalized pasture weeds such as *Arctotheca calendula* (L.) Levyns (cape weed) and *Echium lycoposis* L. (Paterson's curse).

Important honey- and pollen-producing plants include the following species:

Eucalyptus camaldulensis Dehnh. (river red gum), a major source of nectar and pollen; *E. dumosa* A. Cunn. *ex* Schau. (white *mallee*), a good to medium producer of nectar and pollen; *E. eremophila* (Diels) Maid. (sand *mallee*), which yields abundant nectar and good quantities of pollen; *E. gardneri* Maid. (blue *mallee*), from which is derived excellent nectar and pollen yields; *E. largiflorens* F. Muell. (black box), an infrequent producer of good-quality honey but good pollen yields; *E. loxophleba* Benth. (York gum), a very important source from which an excellent light amber honey is derived, but giving poor pollen yields; *E. platypus* Hook. (*moort*), an excellent source of high-quality honey and pollen; *E. salubris* F. Muell. (gimlet), which produces extra-light amber honey and abundant pollen; *Acacia pulchella* R. Br. (western prickly moses), from which no honey is produced but has very good pollen yields; *A. doratoxylon* A. Cunn., from which no honey is produced but is the most prolific pollen producer; *Angophora floribunda* (Sm.) Sweet (roughbark apple), a medium producer of dark amber honey and a major pollen source; *Brachychiton populneum* (Schott and Endl.) R. Br. (*kurrajong*), the source of a dark strong-flavored honey and abundant pollen; *Bursaria spinosa* Cav. (Australian blackthorn), an average producer of dark amber honey with medium to heavy quantities of good-quality pollen; and *Melaleuca lanceolata* Otto (rottnest teatree), a medium source of honey and a major pollen source.

C. Forage Plants

1. TOP-FEED

Edible trees and shrubs (top-feed) become important as fodder for domestic grazing animals, particularly in areas with 150 to 300 mm average annual rainfall. The chief value of top-feed is as a food reserve for times of drought and as a nutrient supplement when the herbaceous fodder is dry and low in nutritive value. In the diet of cattle, top-feed usually comprises less than 10%, but can be increased to 20% or more in dry or drought periods.

Australian top-feed plants of importance include the following species: *Acacia aneura* F. Muell. (*mulga*), a valuable drought reserve plant, leaves and stems, up to 5 mm in diameter, eaten by cattle, branches cut for foliage for sheep and also used as mineral supplement for maintenance of livestock, and resprouts after correct cutting; *A. kempeana* F. Muell. (witchetty bush), low nutritive value but an important drought reserve plant; *A. estrophiolata* F. Muell. (desert ironwood), highly palatable and eaten at all times; *A. georginae* F. M. Bail. (georgina *gidyea*), leaves and pods freely eaten but highly toxic (fluoracetate) in certain circumstances; *A. victoriae* Benth. (prickly wattle), prolifically produced pods are protein source for cattle; *Atalaya hemiglauca* F. Muell. *ex* Benth. (whitewood), leaves highly acceptable to cattle and horses, but fruit and sucker growth suspected of being toxic; *Canthium latifolium* F. Muell. *ex* Benth. (native current), moderately acceptable when young, low in nutritive value but high in fiber content; *Capparis mitchellii* (Lindl. *ex* F. Muell.) Benth. (wild orange), moderate in acceptability and nutritive value; *C. umbonata* Lindl. (northern wild orange), foliage and immature fruit moderately nutritious and acceptable; *Ehretia saligna* R. Br. (*coonta*), highly acceptable and moderately nutritious; *Eremophila longifolia* (R. Br.) F. Muell. (berrigan), fruit and foliage acceptable to sheep and cattle and high in protein and phosphorus; *Heterodendrum oleifolium* Desf. (*boonery*), acceptable to sheep and cattle, but containing toxic amounts of hydrocyanic acid during flowering; *Pittosporum phylliraeoides* DC. (butterbush), highly acceptable and nutritious to cattle and subject to preferential grazing; *Santalum acuminatum* (R. Br.) A. DC. (sweet *quandong*), highly acceptable and nutritous; *S. lanceolatum* R. Br. (plumbush), highly acceptable and nutritious to sheep and cattle; and *Ventilago viminalis* Hook. (supplejack), highly acceptable and moderately nutritious to sheep and cattle.

All of the above species are slow-growing native plants, and regeneration depends on seasons and livestock use. *Acacia aneura* regenerates well in the northern arid and semiarid areas when rainfall permits; conservation measures have been devised for it in Queensland. No plantations have been established for *A. aneura* or any of the other species listed. Research has been concentrated on toxic properties, supplemental measures designed to reach at least maintenance level for domestic ruminants, and dietary selection.

2. GRASSES

Grasses are a major item in the diet of cattle in arid Australia and contribute 80–100% of the volume of fodder. Sheep also strongly select grasses, particularly the short succulent species. As fodder plants, grasses can be classified into three groups as follows:

a. Perennial Drought-Resistant Species. The chief constituents of this group are strongly lignified sclerophyllous plants, low in acceptability and nutritive value, and are widely known as "spinifex" grasses although they are not in the genus *Spinifex*. The more acceptable species provide subsistence fodder for stock, and some spinifex-dominated areas are utilized as drought reserves. Livestock reproduction can be increased by the use of suitable burning and grazing regimes in the management of this group of plants. Associated soils have a very low nutrient status. Noteworthy species include *Triodia basedowii* E. Pritz. (loped spinifex), a "hard" spinifex but grazed to a limited extent because of its predominance over very extensive areas; *T. pungens* R. Br. (gummy spinifex), a "soft" spinifex more acceptable as fodder than most of its relatives, the sheaths, and sometimes the blades, exude a viscid resin that appears to be unattractive to stock; *T. irritans* R. Br. (porcupine grass); *T. longiceps* J. M. Black (giant gray spinifex), an extremely robust plant that forms large, impenetrable clumps; *Plectrachne schinzii* Henr. (feathertop spinifex), foliage and large feathery panicles palatable to stock; and *P. pungens* (R. Br.) C. E. Hubb. (curly spinifex), a "soft" spinifex regarded as more useful than most of its relatives.

These grasses can reach ages in excess of 20 years, but as they are highly flammable and "spinifex" country is frequently burned, most plants do not reach this age. Regeneration is prompt, but excessive burning and grazing can lead to deteroiration. Research has shown that if the low nutrient status of the soils is corrected to encourage the growth of exotic grasses, the low nutrient status is quickly regained and these grasses die.

Before the advent of air conditioning, *Triodia* plants were used as insulation between wire netting for roofs and walls of shade houses in northwest Western Australia; water was trickled through the walls for additional cooling.

b. Perennial Drought-Evading Species. This group comprises tussock-forming, long-lived plants that are deciduous in that new growth takes place from vegetative buds after a period of dormancy during the dry season. Large, old plants reach ages of 15 to 30 years. As fodder, they provide acceptable forage in the summer wet season and standing dry feed of little more than maintenance value in dry to drought periods. Important species in the group, some of whose members extend well into the southern cool-season rainfall region, include *Astrebla pectinata* (Lindl.) F. Muell. *ex* Benth. (barley Mitchell), *A. elymoides* F. Muell. *ex* F. M. Bail. (hoop Mitchell), *A. lappacea* (Lindl.) Domin (curly Mitchell), *A. squarrosa* C. E. Hubb. (bull Mitchell), *Bothriochloa ewartiana* (Domin) C. E. Hubb. (desert bluegrass), *B. bladhii* (Retz.) S. T. Blake (forest bluegrass), *Dichanthium fecundum* S. T. Blake (curly bluegrass), *D. tenuiculum* (Steud.) S. T. Blake (tassel bluegrass), *D. coenicola* (F. Muell.) Hughes (finger panic), *Eragrostic eriopoda* Benth. (naked wollybutt), *E. setifolia* Nees (narrow-leaf neverfail), *E. xerophila* Domin (knottybutt neverfail), *Enteropogon acicularis* (Lindl.) Lazar. (curly windmill), *Eulalia fulva* (R. Br.) O. Kuntze (silky browntop), *Chrysopogon fallax* S. T. Blake (golden beard grass), *Eriachne helmsii* (Domin) W. Harl. (woollybutt wanderrie), *E. benthamii* W. Hartl. (swamp wanderrie), *Monchather paradoxa* Steud. (wallaby grass), and *Thyridolepis mitchelliana* (Nees) S. T. Blake (*mulga* Mitchell).

The higher nutritive value of these grasses, compared with those in the previous section, reflects the soils of moderate fertility on which they grow. However, the fiber content is still substantial in mature green leaves so that voluntary food intake and resultant animal production are restricted. Species of *Astrebla* have been sown; regeneration is a function of early summer rainfall. Levels of dry matter production and nutritional quality are not sufficient to warrant further investigations.

A small group of perennial drought-evading grasses grow in the southern arid region where precipitation is more usual during the cool season. These include *Danthonia caespitosa* (Gaudich (white top), *Stipa aristiglumis* F. Muell. (plains grass), *S. falcata* Hughes (slender spear grass), and *S. variabilis* Hughes (variable spear grass).

The forage value of *Danthonia caespitosa* is high, it is acceptable to sheep, and can withstand severe defoliation; *Stipa elegantissima* is heavily selected, but the remaining species of *Stipa* are fibrous, possess sharp awns, and are less acceptable to herbivores.

c. Ephemeral Drought-Evading Species. The plants of this group are ephemerals, annuals, or short-lived perennials, all of which regenerate readily from seed. For short periods after rainfall, they often produce abundant fodder which is highly acceptable, nutritious, and, particularly in association with acceptable forbs, capable of producing cattle in prime condition and enhancing wool growth rates in sheep. Important representatives in the warm-season group include the following species: *Arsitida contorta* F. Muell. (bunched kerosene grass), *Brachiaria miliiformis* (Presl) Chase (green summer grass), *B. piligera* (Benth.) Hughes (hairy armgrass), *Chloris truncata* R. Br. (umbrella grass), *Dactyloctenium radulans* (R. Br.) Beauv. (button grass), *Digitaria brownii* (R. and S.) Hughes (cotton panic), *Enneapogon avenaceus* (Lindl.) C. E. Hubb. (bottle washers), *E. polyphyllus* (Domin) N. T. Burb. (leafy nineawn), *E. cylindricus* N. T. Burb. (jointed nineawn), *Eriachne aristidea* F. Muell. (threeawn wanderrie), *Iseilema vaginiflorum* Domin (red Flinders), *I. macratherum* Domin (bull Flinders), *I. membranaceum* (Lindl.) Domin (small Flinders), *Panicum decompositum* R. Br. (native millet), *P. whitei* J. M. Black (pepper grass), *Perotis rara* R. Br. (comet grass), *Sporobolus caroli* Mez. (fairy grass), *Tragus australianus* S. T. Blake (small burr grass), and *Tripogon loliiformis* (F. Muell.) C. E. Hubb. (five-minute grass).

Apart from the occurrence of these species in run-on areas where water from light rainfalls is concentrated, they cannot be relied on for forage even though their nutritional quality is satisfactory. Landscapes in which these species are the only component lack stability and are open to temporary occupation by weedy species that are of little value to livestock. *Panicum decompositum* can be a long-lived perennial; its seed has been collected for cultivation with unknown results.

Many species that are important in the southern cool-season region are naturalized immigrants from Mediterranean-type environments; there are no native grass species of importance.

3. FORBS

The forb group of forage plants comprises annual and perennial non-gramineous herbs belonging to a wide range of plant families. In contrast to most native grasses in arid Australia, their growth is promoted chiefly by winter rainfall. A particular forb species does not appear after every effective rain, nor does it always inhabit the same site in the landscape. By volume, their proportion of the diet of cattle in arid regions rarely amounts to more than 15%, and more usually it is less than 10%. However, sheep select forbs when they are available, including strongly aromatic species of Compositae such as *Pterigeron* sp.

Species lists of native forbs eaten by herbivores, in dietary studies conducted in the pasture, do not offer encouragement for plant collection; it is the diversity of this group in terms of habitat, growth habit and structure, seasonal response, and nutritional performance that is important rather than its individual species (excepting representatives of the Fabaceae). The following species are among the more important: *Abutilon malvifolium* (Benth.) J. M. Black, *A. otocarpum* F. Muell. (desert Chinese lantern), *Alternanthera angustifolia* R. Br., *Atriplex* spp. (saltbushes), *Blennodia* spp. (mustards), *Boerhavia diffusa* L. (tarvine), *Brachycome curvicarpa* G. L. Davis, *B. marginata* Benth, *B. melanocarpa* (Sond.) F. Muell, *Calandrinia balonensis* Lindl. (broad-leaf *parakeelya*), *Calotis erinacea* Steez, *Chenopodium desertorum* (J. M. Black) J. M. Black, *Convolvulus erubescens* Sims, *Craspedia chrysantha* (Schlechdt.) Benth. (golden billy buttons), *Crotalaria eremaea* F. Muell. (bluebush pea), *Daucus glochidiatus* (Labill.) Fisch. *et al.*, Desmodium spp., *Erodium cygnorum* spp. *glandulosum* Carolin, *Goodenia havilandii* var. *pauperata* J. M. Black, *Helichrysum apeculatum* (Labill.) DC., *Helipterum charsleyae* F. Muell., *H. floribundum* DC., *H. molle* (A. Cunn. *ex* DC.) P. G. Wils., *Ixiolaena leptolepis* (DC.) Benth., *Lepidium phlebopetalum* (F. Muell.) F. Muell., *Lotus cruentus* Court., *Maireana* spp., *Malvastrum americanum* (L.) Torr., *Marsilea* spp., *Portulaca oleracea* L. (pigweed), *Ptilotus exaltatus* Nees var. *exaltatus, P. obovatus* (Gaudich.) F. Muell., *Salsola kali* L. (roly poly), *Sclerolaena eriacantha* (F. Muell.) R. H. Anders., *Sida corrugata* Lindl. (corrugated *Sida*), *S. rhombifolia* L. (common *Sida*), *S. trichopoda* F. Muell. (high *Sida*), *Sonchus oleraceus* L. (milk thistle), *Stenopetalum nutans* F. Muell., *Swainsona phacoides* Benth., *S. campylantha* F. Muell., *Tribulus terrestris* L. (caltrop), and *Trigonella suavissima* Lindl. (Cooper clover).

Few of these species. or other forbs in the genera in this list, have a forage potential to exploit in Australia. In summary, these forbs are usually abundant under conditions where the general level of forage availability is already high.

4. BROWSE SHRUBS

This group of forage plants comprises shrubs and undershrubs that are characteristic components of the arid-zone flora in the shrub steppe or low shrubland formation and that have a role in the diet of sheep similar to that of top-feed for cattle. In some areas of low winter rainfall and highly saline conditions, browse shrubs constitute the principal or sole source of fodder; elsewhere there may be perennial tussock grasses, annual grasses, and forbs between and under the spaced browse plants. Representative plants fall readily into two groups of species:

a. Chenopods. This major group is composed of numerous predominantly Australian endemics in the family Chenopodiaceae, which range from herbaceous or woody perennials to small or large shrubs, frequently with fleshy or succulent leaves and stems. Several are tolerant to drought, and salt or gypsum. When compared with nonchenopods, these plants are high in nitrogen, sodium, potassium, and chloride salts. For more than 100 years, the seed of various perennial chenopods listed in this inventory has been sent overseas to many arid countries in response to their requests, and plantations have been established. There are no commercial plantations in Australia, and these species, although valued as forage by the pastoral industry, are not the subject of active conservation or management procedures. The most important representatives are *Atriplex,* a genus of ~42 species, the most widespread being *A. vesicaria* Benth. (bladder saltbush); *A. nummularia* Lindl. (old-man saltbush), a valuable browse shrub but now infrequent, upright form and susceptibility to uncontrolled grazing militate against its use in the Australian pastoral context; *A. paludosa* R. Br. (marsh saltbush) and *A. rhagodioides* (river saltbush), recommended for sowing in salinized wheatlands in Western Australia, grazing experiments have been conducted; *Chenopodium auricomum* Lindl. (Queensland bluebush); *C. nitrariaceum* (F. Muell.) Benth. (nitre goosefoot); *Enchylaena tomentosa* R. Br. (ruby saltbush); *Maireana,* a genus of ~58 spp. (with several heavily grazed by stock), among which the most common and widespread in arid regions are *M. aphylla* (R. Br.) P. G. Wils. (cotton bush), *M. astrotricha* (L. Johns.) P. G. Wils. (southern bluebush), *M. campanulata* P. G. Wils., *M. coronata* (J. M. Blacks) P. G. Wils., *M. georgei* (Diels) P. G. Wils, *M. planifolia* (F. Muell.) P. G. Wils. (low bluebush), *M. pyramidata* (Benth.) P. G. Wils. (shrubby bluebush), *M. sedifolia* (F. Muell.) P. G. Wils. (pearl bluebush), *M. tomentosa* Moq., *M. triptera* (Benth.) P. G. Wils. (threewing saltbush), *M. brevifolia* (R. Br.) P. G. Wils. (*yanga* saltbush), recommended for sowing in salinized wheatlands in Western Australia, grazing experiments have been conducted; *Rhagodia nutans* R. Br. (climbing saltbush); *R. spinescens* R. Br. (spiny saltbush); and *Salsola kali* L. (prickly saltwort); *Sclerolaena,* comprising ~80 endemic species, many of which are characterized by stout, spiny fruit; the more acceptable species tend to be perennial herbs or undershrubs, such as *S. convexula* (R. H. Anders.) A. J. Scott, *S. costata* (R. H. Anders.) A. J. Scott, *S. diacantha* (Nees) Benth. (gray copper burr), and *S. lanicuspis* (F. Muell.) Benth. (spinach burr)

Salicornieae, a tribe of six genera and 36 species, represent the "samphire" group of chenopodiaceous plants, which are characterized by succulent, articulate branches and much-reduced leaves. Though generally less valuable than other chenopodiaceous browse plants, samphires thrive in extremely saline sites where few other plants survive. Widespread representatives include *Halosarcia calyptrata* P. G. Wils., *H. halocnemoides* (Nees) P. G. Wils. (blackseed samphire), *H. indica* (Willd.) P. G. Wils., *H. pergranulata* (J. M. Black) P. G. Wils., *Pachycornia triandra* (F. Muell.) J. M. Black, *Sclerostegia medullosa* P. G. Wils., *S. tenuis* (Benth.) P. G. Wils., and *Tecticornia verrucosa* P. G. Wils.

b. Nonchenopods. Important browse plants occur in a variety of non-chenopodiaceous plant families; these are indicated in the following list: *Carissa lanceolata* R. Br. (conkerberry), a leafy, highly acceptable and moderately nutritious shrub that also tends to be drought resistant and fire tolerant; *Cassia artemisioides* Gaudich. *ex* DC. (silver *Cassia*), a widespread, rapid-growing, relatively short-lived shrub, high in protein and phosphorus and low in fiber, but only moderately acceptable to cattle; *C. nemophila* A. Cunn. *ex* Vogel (desert *Cassia*), leaves and pods eaten, and responds well to grazing or lopping; *Dodonaea attenuata* A. Cunn. (narrowleaf hopbush), moderately valuable in nutrient-deficient "spinifex" areas; *Eremophila latrobei* F. Muell. (native fuchsia), a widespread adaptable shrub, acceptable to sheep and cattle, and susceptible to preferential grazing; *Muehlenbeckia cumminghamii* (Meisn.) F. Muell. (lignum), a sprawling shrub forming compact, tangled clumps; small leaves (present only in young growth) and branchlets moderately acceptable to sheep and cattle; and *Spartothamnella teucriiflora* (F. Muell.) Mold., a strongly branched, semiscandent, small-leaved shrub of low nutritive value, but often heavily grazed to ground level.

Acceptability of these browse species to domestic herbivores is determined mainly by the availability of other more acceptable forages. If these alternative forages are absent, as is usually the case in normal rainless periods or in drought, the browse species listed in this inventory are heavily utilized; they form a major part of the important "bridge" group of species that, at appropriate stocking rates, permit breeding stock to be maintained in numbers and at a levels of condition sufficient for successful animal reproduction during extended dry periods.

D. Medicinal Plants

A wide variety of plant substances are used for medicinal purposes by the aborigines. The majority are applied externally rather than ingested. Several species of *Eremophila* and *Acacia* are important in this respect. The latex of several plants, including *Sarcostemma australe* R. Br. (caustic vine) and *Euphorbia* spp., contains substances that assist the healing of wounds and sores. Extracts of the bark and roots of *Cassia,* which contains the laxative senna, have been used for the treatment of skin diseases and stomach disorders. A considerable proportion of the species used, including *Stemodia viscosa* Roxb. and *Pterocaulon,* are viscid and/or aromatic.

Duboisia hopwoodii (F. Muell.) F. Muell. (*pituri*) and *Nicotiana* spp. are highly regarded as chewing tobaccos. When not available, species of *Goodenia* provide poor substitutes. *Duboisia* is used also as a poison in the watering holes of emus; species of *Tephrosia* that contain rotenone are used as a fish poison.

E. Timber Plants

Though not a principal source of timber for industry, the forest resources of arid Australia are of considerable local importance and also supply products for use throughout Australia. The most important uses are as fuelwood and materials for fencing and railway sleepers. Timbers of the arid region are derived chiefly from species of *Eucalyptus* (especially those belonging to the ironbark, box, and gum groups) and species of *Callitris* (cypress pines). With the exception of the latter, most are hard, dense, heavy, and durable, but are difficult to split, saw, or work by hand because of their interlocking grain.

The most important timbers, and their uses, include the following species:

Eucalyptus camaldulensis Dehnh. (river red gum), firewood, sleepers, and flooring, especially parquetry; *E. dichromophloia* F. Muell. (gum-topped bloodwood), firewood, and fencing; *E. dumosa* A. Dunn. *ex* Schau. (white *mallee*), garden and horticultural stakes, and firewood obtained from roots; *E. gardneri* Maid. (blue *mallee*), mining timbers, round construction, firewood; *E. largiflorens* F. Muell. (black box), fencing, fuelwood; *E. longicornis* (F. Muell.) F. Muell. *ex* Maid. (red morrel), mining timbers, round construction, fuel; *E. microtheca* F. Muell. (*coolibah*), termite resistant, used for fuel, fencing; *E. polycarpa* F. Muell. (longfruit bloodwood), fencing, fuel; *E. wandoo* Blakely (*wandoo*), extremely durable timber suitable for use as poles, for heavy construction, high-quality flooring, and heavy joinery; and *E. woollsiana* R. T. Baker (inland gray box), round construction, sleepers, fencing, and excellent fuel.

Acacia aneura F. Muell. (*mulga*), fencing, small ornaments; *A. cambagei* R. T. Baker (*gidgee*), fencing, firewood, ornaments; *A. excelsa* Benth. (ironwood), mining timbers, fencing, ornaments; *A. harpophylla* F. Muell. *ex* Benth. (brigalow), fencing, good fuel, source of charcoal, suitable for fishing rods, cabinet making, and other ornamental work requiring a fine finish and a high polish; *A. omalophylla* A. Cunn. *ex* Benth. (*yarran*), timber faintly aromatic, well suited for cabinet and ornamental work; *A. pendula* A. Cunn. *ex* G. Don (*myall*), timber strongly aromatic, an excellent fuel, and useful for making ornaments; *A. peuce* F. Muell. (waddywood), very heavy, hard, and termite-resistant timber used for fencing material and stockyard construction; *A. salicina* Lindl. (*cooba*), furniture and ornaments; *A. shirleyi* Maid. (lancewood), stockyard rails; *A. stenophylla* A. Cunn. *ex* Benth. (*belalie*), fencing and similar purposes, excellent firewood, dark, close-grained wood suitable for furniture and ornamental work.

Other timber plants are *Casuarina cristata* Miq. (*belah*), suitable for split posts, rails, and similar farm purposes, used for tool handles, shingles, cabinet work, and excellent fuel wood; *Eremophila mitchellii* Benth. (*buddah,* false sandalwood), termite-resistant timber used for fence posts in the round, an excellent fuel with a sweet, pleasant aroma when burning, used as a substitute for true sandalwood; *Grevillea striata* R. Br. (beefwood), posts, shingles, house blocks, and similar farm purposes; *Hakea leucoptera* R. Br. (needlewood), smoking pipes and similar small ornamental objects; and *Callitris columellaris* F. Muell. (white cypress pine), valuable because of resistance to termites and decay, and small shrinkage on seasoning, suitable for building scantlings, flooring, linings, and weatherboards, also used in the round for fencing material and small poles.

F. Industrial Plants

1. FIBER PLANTS

Several grasses produce fibers suitable for the manufacture of paper pulp, of varying quality, and associated by-products. Australian arid species used for this purpose elsewhere, or with pulp-making potential, include *Phragmites karka* (Retz.) Steud (tropical reed), *Heteropogon contortus* (L.) Beauv. *ex* R. and S. (bunch speargrass), and *Themeda avenacea* (F. Muell.) Maid. and Betche (native oat).

The plant families Malvaceae and Tiliaceae, which contain the major fiber-producing plants, such as cotton (*Gossypium*), jute (*Corchorus*), and kenaf (*Hibiscus*), are well represented in the Australian arid flora by a number of species belonging to these and allied genera. *Hibiscus cannabinus* L. (kenaf) is considered to be a viable crop for growing in monsoonal Australia as a source of softwood and hardwood pulps for papermaking; it also has a protein-rich juice, is a nutritive stock feed, and a high-grade vegetable oil. It is possible that related arid species have a similar potential.

2. GUMS

Xanthorrhoea spp. (grasstree, blackboy, *yacca*) are known to be a source of yacca gum, used principally in the manufacture of linoleum and low-grade varnishes.

3. TANNINS

Australian sources of natural tannins include the bark of many *Acacia,* and both bark and wood of some *Eucalyptus* species. Notable species with potential value in arid regions include *A. pycnantha* Benth. (golden wattle), bark of the trunk of mature trees yields usually 30–40% and sometimes higher amounts of tannin, and this species is considered to be one of the richest sources; *E. gardneri* Maid. (blue *mallee*), bark has tannin content of 22 to 30%; and *E. wandoo* Blakely (*wandoo*), a source of tannin extract, obtained by reducing the wood and bark to chips, and leaching with water.

4. ESSENTIAL OILS

The leaves of *Eucalyptus* spp. contain essential oils that have been used chiefly for medicinal purposes but also in nontoxic disinfectants, as solvents for substances such as tar, grease, and paint, in perfumery, and for mineral recovery and other industrial purposes. Only some species, particularly those of the "peppermint" (not represented in arid regions) and *mallee* groups, produce useful quantities from distillation. *Eucalyptus viridis* R. T. Baker (green *mallee*) is one of the more important species. The yield at the time of collection from young leaves and terminal branchlets is 1.5% by weight, and the valuable cineol content of the crude oil may be as high as 80%.

The grass genus *Cymbopogon* is cultivated in some parts of the world as a commercial source of commercial aromatic oils such as lemon oil, citronella oil, palmarosa, and ginger-grass oils. Also, some species have medicinal value, while others yield moderate amounts of quality material for the manufacture of paper pulp. Australian arid relatives with potential value include *C. ambiguus* A. Camus (scent grass), *C. obtectus* S. T. Blake (silkyheads), and *C. bombycinus* (R. Br.) Domin (silky oilgrass). All are tussock-forming, long-lived perennials with citronella-scented roots, which are most aromatic in the fresh state.

Species of *Bothriochloa* are also oil producing, and the oil of *B. bladhii* (Retz.) S. T. Blake (forest bluegrass) contains acetic and butyric acids.

G. Other Plants

1. PLANTS FOR SHADE, SHELTERBELTS, ORNAMENTAL AND AMENITY PLANTING, AND SOIL STABILITY

The provision of shade for the benefit of animals is necessary throughout the pastoral areas of Australia, particularly in arid parts where summer temperatures are high. This applies also to shelter, which affords protection to both animals and crops, including pasture. The Australian arid flora includes several trees and shrubs that are suitable for these purposes, or as ornamental and amenity plants that have an important role in the amelioration of local microclimates and landscapes. Other native species, including grasses, are useful in providing a plant cover in unstable and problem areas (such as loose sands and saline soils). The list includes the following:

Acacia beckleri Tindale (Barrier Range wattle), ornament, low hedges, shelterbelts; *A. coriacea* DC. (wirewood), soil conservation, light shade and shelter; *A. cyanophylla* Lindl. (orange wattle), ornament, shade, soil and sand stability, and moderately salt tolerant; *A. estrophiolata* F. Muell. (desert ironwood), shade and shelter; *A. excelsa* Benth. (ironwood), ornament, shade, shelter; *A. omalophylla* A. Cunn. *ex* Benth. (*yarran*), ornament, shade, shelter; *A. iteaphylla* F. Muell. *ex* Benth. (Flinders Range wattle), ornament, hedges, two-tier shelterbelts; *A. ligulata* A. Cunn. *ex* Benth. (umbrella bush), soil conservation, low windbreaks, and stabilizes drifting sands; *A. notabilis* F. Muell. (*notabilis* wattle), low shelter, soil protection; *A. pendula* A. Cunn. *ex* G. Don (*myall*), ornament, shade and shelterbelts; *A. salicina* Lindl. (*cooba*), ornament, shade, shelter, soil conservation, park planting; *A. sowdenii* Maid. (western *myall*), hedges, windbreaks; *A. tetragonophylla* F. Muell. (dead finish), shade, soil conservation; *A. victoriae* Benth. (*gundabluie*), low shelterbelts, soil conservation, regenerates freely and is a potential pest around waterholes.

Apophyllum anomalum F. Muell. (warrior bush), shelter, shade, ornament; *Atriplex nummularia* Lindl. (oldman saltbush), shelterbelts, a fire-retarding species tolerant to high salinity and growing on a wide range of soils, responsive to phosphatic and nitrogenous fertilizers; *Lysiphyllum carronii* (F. Muell.) Pedley (northern beantree), ornament, park planting, shade, shelter; *L. cunninghamii* (Benth.) de Wit (''bauhinia''), ornament, shade; *Brachychiton gregorii* F. Muell. (desert *kurrajong*), shade, ornament; *Callitris columellaris* F. Muell. (white cypress pine), ornament, shelterbelts, avenue planting; and *Capparis mitchellii* (Lindl. *ex* F. Muell.) Benth. (bumble, wild orange), shade, shelter, and soil protection.

Cassia is represented in Australia by more than 50 mainly native and endemic species, most of which, in arid regions, are bushy shrubs that are highly suitable for ornaments and low shelterbelts. Features include copious yellow flowers, and often, pruinose or bluish foliage. Common arid species include *C. artemisioides* Gaudich. *ex* DC. (silver *Cassia*), *C. desolata* F. Muell. (gray *Cassia*), *C. glutinosa* DC., *C. helmsii* Symon (crinkled *Cassia*), *C. notabilis* F. Muell. (cockroach bush), *C. nemophila* A. Cunn. *ex* Vogel (desert *Cassia*), *C. phyllodinea* R. Br. (silver *Cassia*), *C. pleurocarpa* R. Muell. (ribfruit senna), *C. pruinosa* F. Muell. (white *Cassia*), *C. oligophylla* F. Muell., *C. sturtii* R. Br. (gray *Cassia*), and *C. venusta* F. Muell.

Casuarina (she oak) is represented in arid regions by bushy shrubs or, more often, by small to large trees with a characteristic pyramidal habit; they are useful as ornaments, shelter, and for conservation. Common species include *C. campestris* Diels, used for low shelter and two-story shelterbelts; *C. cristata* Miq. (*belah*), moderately alkaline tolerant, used for light shade, shelter, and avenue planting; *C. decaisneana* F. Muell. (desert oak), a graceful pendulous tree attaining a height of 16 m in the most arid regions, useful for shade and ornament; *C.*

dielsiana C. A. Gardn. (northern she oak), a large shrub or small tree providing good shade and shelter when young, useful in multirow shelterbelts; *C. huegeliana* Miq. (rock she oak), used as ornamental, for shade, and two-story shelterbelts in poor, shallow, sandy soils and inhospitable outcropping sites.

Codonocarpus cotinifolius (Desf.) F. Muell. (desert poplar), a small, short-lived tree with conspicuous form, which is suitable for temporary shelterbelts and ornamental planting.

Dodonaea consists of species of compact shrubs suitable as ornamentals, and for low, dense shelterbelts. They have attractive foliage and papery, often colored, wings on the fruit. Common arid species include *Dodonaea attenuata* A. Cunn. (narrowleaf hopbush), *D. lobulata* F. Muell., *D. microzyga* F. Muell., and *D. viscosa* Jacq. (sticky hopbush).

Eremocitrus glauca (Lindl.) Swingle (desert lime), a large shrub with a thorny, intricately branched habit during juvenile and sucker stages. It is suitable for screen belts when young, and as an ornamental at maturity.

Eremophila is an Australian endemic genus of more than 100 species, characteristically distributed in arid and semiarid regions. The great majority are compact, strongly branched shrubs that are useful for planting as ornamentals, low shelterbelts, in parks, and in reserves. They feature showy, many-colored flowers. Common widespread species include *E. frelingii* F. Muell. (limestone fuchsia), *E. gilesii* F. Muell. (Charleville turkey bush), *E. glabra* (R. Br.) Ostenf. (black fuchsia), *E. latrobei* F. Muell. (native fuchsia), *E. longifolia* (R. Br.) F. Muell. (Berrigan), *E. maculata* (Ker) F. Muell. (fuchsia bush), and *E. sturtii* R. Br. (turpentine bush).

Eucalyptus angulosa Schau., used in coastal and sand dune conservation, and two-tier shelterbelts; *E. brockwayi* C. A. Gardn. (Dundas mahogany), ornamental, street and park plantings, shade, multirow shelterbelts; *E. camaldulensis* Dehnh. (river red gum), ornamental, excellent for shade, and used in mixed-species shelterbelts; *E. campaspe* A. Moore (silvertop gimlet), street and park plantings, mixed shelterbelts, moderately alkaline tolerant; *E. cladocalyx* R. Muell. (sugar gum), extensively planted for shade, shelter, and ornament; *E. dumosa* A. Cunn. *ex* Schau. (white *mallee*), used for soil conservation, shelter; *E. dundasii* Maid. (Dundas blackbutt), used for street and ornamental shade planting, tolerant of salt in soil; *E. eremophila* Maid. (sand *mallee*), ornamental, used in small parks, hedges, low shelterbelts; *E. flocktoniae* (Maid.) Maid. (*merrit*), ornamental, used for street and park planting; *E. forrestiana* Diels (Forrest's *mallee*), ornamental, hedge planting; *Eucalyptus gamophylla* F. Muell. (blue *mallee*), highly suitable for dune stability, soil protection, ground shelter, and as an ornamental; *E. gillii* Maid. (curly *mallee*), ornamental, used as a low shelter, also for soil conservation; *E. gracilis* F. Muell. (*yorrell*), used for shelterbelts, shade, and street planting, and moderately salt tolerant; *E. intertexta* R. T. Baker (inland red box), park planting, shade, multirow shelterbelts, moderately salt tolerant; *E. kondininensis* Maid. and Blakely (Kondinin blackbutt), shade, ornamental, foremost of the salt-tolerant species; *E. lesouefii* Maid. (Goldfield's blackbutt), ornamental, park planting, and moderately alkaline tolerant; *E. papuana* F. Muell. (ghost gum), popular as an ornamental; *E. salmonophloia* F. Muell. (salmon gum), popular ornamental, shade tree, used in large parks, avenue planting, shelterbelts; *E. salubris* F. Muell. (*gimlet*), ornamental, multirow shelterbelts, tolerant to salt in soil; *E. sargentii* Maid. (salt river gum), ornamental, street and park planting, shelterbelts, highly salt tolerant; *E. spathulata* Hook. (swamp *mallee*), low shelterbelts, street planting; *E. torquata* Lehm. (coral gum), popular ornamental for shade and park planting, highly alkaline tolerant; *E. viridis* R. T. Baker (green *mallee*), used for soil conservation, low shelterbelts.

Grevillea juncifolia Hook., ornamental; *Heterodendrum oleifolium* Desf. (*boonery*), shade, shelter, park planting.

Melaleuca, a genus of more than 100 predominantly Australian endemic species, is well represented in arid regions by various-sized shrubs and small trees with a dense, bushy habit, and is highly recommended as a low shelter or for use in multirow shelterbelts. The more useful species include *M. bracteata* F. Muell. (river teatree); *M. glomerata* F. Muell.; *M. pauperiflora* F. Muell. (*boree*); *M. lanceolata* Otto (Rottnest teatree), which is salt tolerant; and *M. halmaturorum* Miq. (South Australian swamp paperbark), which is suitable for saline areas with a shallow water table.

Myoporum montanum R. Br. (*boobialla*), ornamental, used in small parks, and as low shelter; *M. platycarpum* R. Br. (sugarwood), shade tree used in soil conservation, and as ornamental; *Owenia acidula* F. Muell. (emu apple), excellent ornamental and shade plant, used for park planting; *Parkinsonia aculeata* L. (Jerusalem thorn), useful as a shade tree on cracking clay plains, where trees are difficult to establish, prevalent around bores, creek lines, and swamps or Barkly Tableland, and becoming a troublesome weed in some areas; *Pittosporum phylliraeoides* DC. (butterbush), ornamental, small parks.

Gramineae includes several native perennial grasses that because of their rhizomatous or stoloniferous, spreading, or mat-forming habit are considered suitable as soil binders or as ground cover plants for sites susceptible to soil erosion. Noteworthy species include *Cynodon dactylon* (L.) Pers. (couch), rhizomatous and stoloniferous, a vigorous perennial tolerant to an extremely wide climatic and edaphic range, also a turf and lawn plant; *Brachyachne convergens* (F. Muell.) Stapf (native couch); *Chloris truncata* R. Br. (windmill grass); *Eragrostis australasica* (Steud.) C. E. Hubb. (canegrass), robust, shrubby, probably salt-tolerant perennial, especially suited for claypans and clayey floodouts; *Pseudoraphis spinescens* (R. Br.) J. Vickery (spiny mudgrass), semiaquatic stoloniferous perennial, mat forming in shallow water or on floor of dry waterholes and similar depressions; *Sporobolus mitchellii* (Trin.) C. E. Hubb. *ex* S. T.

Blake (rat's tail couch), decumbent mat-forming perennial, especially suited for alluvial or seasonally flooded, heavy-textured soils; *S. virginicus* (L.) Kunth (sand or marine couch), salt-tolerant, rhizomatous perennial of coastal and inland sites such as tidal flats and seasonal lakes; *Zygochloa paradoxa* (R. Br.) S. T. Blake (sandhill canegrass), a characteristic species of the Simpson Desert with a dioecious, shrubby, rhizomatous, clump-forming habit, well adapted for growing in loose, drifting sands of dunes and sandhills.

2. MISCELLANEOUS USEFUL PLANTS

Wood, fibers, and other materials provided by a variety of plants are, and were, used by the aborigines to make implements, weapons, and similar articles. Spear shafts were made from *Pandorea doratoxylon* (J. M. Black) J. M. Black (spearwood bush), *Acacia kempeana* F. Muell. (witchetty bush), *Ventilago viminalis* Hook. (supplejack), and to a lesser extent, *Acacia aneura* F. Muell. *ex* Benth. (*mulga*), which has a heavy wood reputed to contain a virulent poisonous principle. Boomerangs and spears were made from *Acacia omalophylla* Benth. (*yarran*) and *A. salicina* Lindl. (*cooba*). The resinous exudate of *Triodia* spp. provided the cementing substance necessary for fixing stones to weapons and implements. The light wood of *Erythrina vesperitilio* Benth. (bat's wing coral tree) is suitable for making food and water receptacles, and the stronger wood of *Eucalyptus camaldulensis* Dehnh. (river red gum) was used for scooping and digging implements. Fibers were derived from *Crotalaria* spp. and from *Brachychiton populneum* (Schott and Endl.) R. Br. (*kurrajong*), which has bark containing a strong fiber that is suitable for making fishing nets. Also, the roots of *Brachychiton* spp. provided a storage source for water. For decorative purposes, the red seeds of *Erythrina,* the capsules of *Eucalyptus dichromophloia* F. Muell. (gum-topped bloodwood), and the flowers of *Cassia, Eremophila,* and *Helichrysum* were used.

H. Harmful Plants

The flora of arid Australia contains several plants that are harmful, either by chemical or mechanical means, to domestic animals.

1. TOXIC PLANTS

The poisonous chemicals in toxic plants are commonly alkaloids, glucosides, including saponins and cyanide-producing glucosides, or metal salts such as oxalates and nitrates. The effect of toxicity varies considerably in relation to a number of plant and animal factors, such as genetic or environmental variation within a species, parts of the plant and its growth stage (e.g., juvenile or mature foliage, fruit and seed or foliage, flowering or nonflowering period, suckers), relative quantities consumed, associated plants in the diet, health of the animal and its sensitivity to different plants, pressures due to traveling or stress, and drought or favorable conditions, including timing of access to water.

Though legumes are among the most dangerous, toxic plants occur in a wide range of plant families, as indicated by the following list of species:

Crotalaria dissitiflora Benth. (gray rattlepod), alkaloid, most dangerous to sheep; *C. retusa* L. (wedgeleaf rattlepod), alkaloid causing Kimberley horse disease, which is fatal to horses; *C. novae-hollandiae* DC. (New Holland rattlepod), similar to the above species in toxicity.

Gastrolobium grandiflorum F. Muell. (desert poison), leaves and flowers highly toxic and fatal to cattle; *Indigofera linnaei* Ali (Birdsville indigo), cause of Birdsville disease, which is fatal to horses, young growth is the most toxic; *Isotropis atropurpurea* F. Muell. (poison sage), extremely toxic and fatal to cattle; *Lotus australis* Andr. (austral trefoil), contains prussic acid, especially in young plants, but also in pods and seeds, traveling sheep most susceptible to poisoning.

Acacia georginae F. M. Bail. (Georgina *gidgee*), cause of Georgina River poison, which is fatal to cattle and sheep, pods contain greater amounts of the toxic polysaccharide than leaves, death often occurs near watering points, traveling or worked cattle most susceptible to poisoning; *Cassia pleurocarpa* F. Muell. (ribfruit senna), purgative in effect, losses occur in traveling cattle.

Oxalis corniculata L. (yellow wood sorrel), fatal to traveling sheep, dangerous in summer and autumn, toxic principle probably an oxalate or oxalic acid; *Tribulus terrestris* L. (caltrop), fatal to sheep due to inorganic nitrite; *Atalaya hemiglauca* (F. Muell.) Benth. (whitewood), leaves, which contain saponins, and fruit cause death in horses; *Heterodendrum oleifolium* Desf. (*boonery*), foliage, especially young growth, contains toxic amounts of prussic acid during flowering in summer, fatal to sheep and cattle; *Sarcostemma australe* R. Br. (caustic vine), fatal to horses, cattle, and sheep; contains saponins, which probably contribute to toxic principle; *Duboisia hopwoodii* (F. Muell.) F. Muell. (*pituri*), leaves contain alkaloids nicotine and/or nornicotine; fatal to horses, cattle, sheep, goats, and camels.

Solanum esueiale Lindl. (*quena*), toxic to sheep, cattle, and horses; *S. nigrum* L. (blackberry nightshade), toxic to sheep and cattle, green immature fruits contain the alkaloid solanine; *S. quadriloculatum* F. Muell. (tomato bush), toxic to sheep (green and mature berries, especially seeds, are strongly alkaloid); *S. sturtianum* F. Muell. (*thargomindah* nightshade), fatal to cattle and sheep (traveling sheep are highly susceptible, and drinking soon after eating appears to be a contributing factor), berries most toxic part of plant.

Eremophila maculata (Ker) F. Muell. (fuchsia bush), fatal to traveling sheep, plant contains large amounts of the glucoside prunasin, which produces prussic acid when acted on by an enzyme present only in the fruits; *Cheilanthes ten-*

uifolia (Burn. f.) Sw. (rock fern), mature dry plants toxic to sheep and, to a lesser extent, cattle; *Brachyachne convergens* (F. Muell) Stapf (common native couch), plants contain prussic acid in latter months of summer, which has been fatal to sheep; *Dactyloctenium radulans* (R. Br.) Beauv. (button grass), contains prussic acid at all times except winter, has been fatal to sheep and cattle at certain times.

2. PLANTS CAUSING MECHANICAL INJURY

The chief causal organs of mechanical injury are modified fruits bearing spines, bristles, awns, or pungent basal calli. Usually such fruits become dangerous only on hardening with maturity; often in the immature state, they are grazed by stock along with the foliage of the plant without harmful effects.

Plants with burrs ranging from two to many stout spines, which can lame grazing stock—especially horses and sheep—include the following species: *Tribulus terestris* L. (caltrop), *Cenchrus echinatus* L. (Mossmann River grass), *Sclerolaena bicornis* Lindl. (goathead burr), *S. birchii* (F. Muell.) Domin (galvanized burr), *S. divaricata* (R. Br.) Domin (tangled copper burr), *S. longicuspis* (F. Muell.) A. J. Scott, *S. tricuspis* (F. Muell.) Ulbrich (giant redburr), and *S. muricata* (Moq.) Domin.

Several grasses are characterized by awned, spearlike fruits that cause injury to grazing animals by penetrating the tender parts of mouth, skin, and hooves. Furthermore, the quality of fleece and hides is reduced by infestation of seeds and awns adapted to penetrate or adhere to wool, hair, and the coats of animals. Such adaptations include retrorse hairs and bristles, viscid exudates, and spirally twisted, articulate, one- or three-fid awns. Although they serve as useful aids in seed dispersal, in soil penetration for protective cover during unfavorable periods, and for germination in favorable periods, these fruits can cause mechanical injury to livestock.

Grasses and other plants belonging to this category include the following species: *Perotis rara* R. Br. (comet grass), *Heteropogon contortus* (L.) Beauv. *ex* R. and S. (bunch spear grass), *Aristida* spp. (threeawn or wire grasses), *Stipa* spp. (spear grasses), *Themeda* spp., *Chrysopogon* spp., *Sorghum* spp., *Calotis hispidula* (F. Muell.) F. Muell. (bogan flea), *Achyranthes aspera* L. (chaff flower), and *Alternanthera pungens* Kunth (khaki weed).

VI. SUMMARY

Compared with many of the world's arid regions, the Australian arid lands exhibit stable habitats, and even the more pessimistic of global scenarios leaves the largely roadless, trackless, waterless, unoccupied deserts untouched. Deteriorating climates in the northern hemisphere could only slightly ameliorate the climate of arid Australia. The semiarid regions of Australia do exhibit damage attributable to European man and his introduced herbivores; although the worst influences are in the past, the attrition of rangelands due to rabbits and sheep still continues. Again, pessimistic scenarios do not doom the Australian semiarid regions to destruction so much as to continuing depopulation.

It is not difficult to envision the aboriginal people in arid Australia developing a stable "outstation" movement that has the important spiritual and cultural elements of their original lifestyle, and with crop plants (not to be confused with crops) growing unattended in small favored locations. By way of contrast, the semiarid pastoral zone will continue to be a major part of the inland Australian culture of pastoralism versus cropping, and will be increasingly differentiated from the urban and coastal internationalism.

Projections

A realistic assessment of use requires the definition of the likely economic framework. Capital has been withdrawn progressively from the semiarid grazing lands for the past 15 years, and this trend is expected to continue. These regions are high risk in terms of climate and return of capital, with outputs that are predominantly for export in an uncertain global economy. Transport costs are high; personal transportation is expensive and is likely to be increasingly so. People do not wish to live in these regions.

The Australian population of 15 million has almost attained zero population growth; population increase is generated mainly from an excess of births over deaths, and migrant arrivals have exceeded departures by only 20,000 to 56,000 per annum. With most agricultural and pastoral products in excess of domestic demand, and export prices depressed by a subdued world commodity demand, the recent foray of cereal-cropping enterprises into the semiarid areas is likely to be short-lived. A contraction of arable farming to the higher rainfall region is likely; already there is a move from cereal growing to Merino sheep production because of a change in relative profitability.

Present research groups are small and have an industry objective, modified in recent years by a landscape stability component, and there is considerable self-examination in terms of function and effectiveness. There are a number of reasons for maintaining effective and talented research groups: maintenance, in the sense that nations maintain military forces, maintenance for export or loan of expert personnel, maintenance for assistance in training Australians and non-Australians, or the "museum–art gallery" approach as a "good" thing.

In more general terms, the arid environments and their biota have proven to be productive laboratories from which ideas have flowed into mainstream biological research. Currently there appears to be no diminution in either the amount or range of individual research interest exhibited. Recent research gives us a salutary reminder that, in the past, arid Australia has been more arid than it is now. A significant increase in the severity of aridity, or in the area affected, apparently requires little in the way of change in current temperature or precipitation levels.

The role of commerce in the utilization of plant products in arid Australia depends on profitability in regions that possess high internal transport costs, among other disabilities. With global economic policies such as high protectionism in its various forms, balance-of-payments problems, population growth in some countries and zero growth in others, the vagaries of climate, increasing energy costs, and migration of the technologically skilled, one can argue that prediction of plant use and development within Australia's arid lands is hazardous, at least. Our assessment is for continued rangeland use of favored areas by livestock, and for wood and energy products to be harvested in limited amounts and collected at strategically placed sites in favored semiarid areas close to transport systems and end-users. Transport subsidies will be part of a trade-off between rural and urban population.

One prediction of high probability is that Australia's semiarid and arid plants will have a greater value internationally than nationally. This is recognized internationally to an embarrassing degree, but in the present Australian commercial environment, which is dominated by pastoralism and mining, such a contribution by Australia has yet to be consummated.

Recommendations pertaining to Australia are few and general. Some of the plants and simple cultural techniques used by the aboriginal people are being investigated; these are unlikely to be of commercial value in Australia or overseas. The browse and shelterbelt categories have species of potential—research has been done on some of them—but in the extensive export-oriented grazing industries that are characteristic of Australia, this potential is unlikely to be realized. It is in the areas of browse, shelterbelt, ornamental, and timber species for export, and in the expertise in the management of arid and semiarid resources, that the Australian contribution has been most successful and should be continued.

ACKNOWLEDGMENTS

W. F. Goodwin, CSIRO Division of Land Use Research, Canberra, contributed Figs. 1 and 2. Facilities and the environment for writing this chapter were provided by the Department of Plant Husbandry. I am indebted to the Swedish Agricultural University, Uppsala. Financial support to attend the International Arid Lands Conference on Plant Resources, Texas Tech University, in Lubbock, was provided by the university's International Center for Arid and Semi-Arid Land Studies (ICASALS).

REFERENCES

Anon. (1977). "The Reader's Digest Atlas of Australia." Reader's Digest Services, Ltd., Sydney, in conjunction with Div. Natl. Mapping, Dep. Natl. Resources, Canberra.

*Askew, K., and Mitchell, A. S. (1978). "The Fodder Trees and Shrubs of the Northern Territory." Div. Primary Industry, Extension Bull. No. 16.
*Chippendale, G. M. (1968). "A Study of the Diet of Cattle in Central Australia as Determined by Rumen Samples." Primary Industries Branch, North. Territory Administration. Tech. Bull. No. 1.
*Chippendale, G. M., and Murray, L. R. (1963). "Poisonous Plants of the Northern Territory." Animal Industry Branch, North. Territory Administration. Extension Article No. 2.
*Cleland, J. B., and Johnston, T. H. (1933). The ecology of the aborigines of central Australia: botanical notes. *Trans. Proc. Roy. Soc. South Aust. J.* **57,** 113–124.
Condon, R. W., Newman, J. C., and Cunningham, G. M. (1969). Soil erosion and pasture degeneration in central Australia. Part IV. *J. Soil Conserv. Serv. N.S.W.* **25,** 295–321.
Cunningham, G. M., Mulham, W. E., Milthorpe, P. L., and Leigh, J. H. (1981). "Plants of Western New South Wales," New South Wales Government Printing Office in association with the Soil Conservation Service of N.S.W., Sydney.
*Everist, S. L. (1974). "Poisonous Plants of Australia." Angus and Robertson, Sydney.
*Gardner, C. A., and Bennetts, H. W. (1956). "The Toxic Plants of Western Australia." W. Aust. Newspapers Ltd., Perth.
Gill, A. M., Groves, R. H., and Noble, I. R. (1981). "Fire and the Australian Biota." Aust. Acad. Sci., Canberra.
*Hall, N., *et al.* (1972). "The Use of Trees and Shrubs in the Dry Country of Australia." Aust. Gov. Publ. Serv., Canberra.
*Hartley, W. (1979). "A Checklist of Economic Plants in Australia." CSIRO, Melbourne, Australia.
Jeans, D. W. (1977). "Australia: A Geography." Sydney Univ. Press, Sydney.
*Latz, P. K., and Griffen, G. F. (1978). Changes in aboriginal land management in relation to fire and to food plants in central Australia. *In* "The Nutrition of Aborigines in Relation to the Ecosystem of Central Australia" (B. S. Hetzel and H. J. Frith, eds.). CSIRO, Melbourne, Australia.
*Lazarides, M. (1970). "The Grasses of Central Australia." Aust. Nat. Univ. Press, Canberra.
*Leigh, J. H. (1974). Diet selection and the effects of grazing on the composition and structure of arid and semi-arid vegetation. *In* "Studies of the Australian Arid Zone. II. Animal Production" (A. D. Wilson, ed.). CSIRO, Melbourne, Australia.
Mabbutt, J. A. (1977). "Desert Landforms." Aust. Natl. Univ. Press, Canberra.
*Moore, R. M., and Perry, R. A. (1970). Vegetation. *In* "Australian Grasslands" (R. M. Moore, ed.), pp. 59–73. Aust. Natl. Univ. Press, Canberra.
*Perry, R. A. (1960). "Pasture Lands of the Northern Territory, Australia." CSIRO, Australia, Land Res. Series No. 5. Canberra.
Rowland, D. T. (1979). "Internal Migration in Australia." Aust. Bureau of Statistics (Census Monograph Series), Aust. Gov. Publ. Serv., Canberra.
Rye, B. L., Hopper, S. D., and Watson, L. E. (1980). "Community Exploited Vascular Plants Native in Western Australia: Census, Atlas and Preliminary Assessment of Conservation Status." Dep. Fish. Wildl., Western Australia, Rep. No. 40.
Williams, O. B. (1979). Ecosystems of Australia. *In* "Arid-Land Ecosystems: Structure, Functioning, and Management," Vol. 1 (R. A. Perry and D. W. Goodall, eds.), pp. 145–212. Cambridge Univ. Press, Cambridge.
Williams, O. B., Suijdendorp, H., and Wilcox, D. G. (1980). The Gascoyne Basin. *In* "Desertification," Vol. 12. Environmental Sciences and Applications (M. R. and A. K. Biswas, eds.), pp. 3–106. Pergamon, Oxford.

*General references for further information.

Wilson, A. D., and Graetz, R. D. (1979). Management of the semi-arid and arid rangelands of Australia. *In* "Management of Semi-Arid Ecosystems" (B. H. Walker, ed.), pp. 83–111. Elsevier, Amsterdam.
Young, M. D. (1979a). "Differences Between States in Arid Land Administration." Div. Land Resour. Manage., Land Resour. Manage. Series No. 4. CSIRO, Melbourne, Australia.
Young, M. D. (1979b). Influencing land use in pastoral Australia. *J. Arid Environ.* **2,** 279–288.
Young, M. D. (1980). Population decline and socio-economic adjustment in the central poplar box (*Eucalyptus populnea*) lands: responses to changes in the grazing industry.

3

INDIAN SUBCONTINENT

K. M. M. Dakshini
Department of Botany
University of Delhi
Delhi, India

I. INTRODUCTION

The arid and semiarid regions as defined by Thornthwaite (1948), Pramanik *et al.* (1952), Bharucha (1955), and Krishnan (1968a) cover the whole of Pakistan (Fig. 1) and approximately one-third of the surface area of the Indian Republic (Fig. 2). In this chapter all the arid and semiarid regions of India and the "region south-east of Indus Valley including the dreary steppes of Sind Sagar Doab and Cholistan in Sind in Pakistan" (Pakistan, 1977–1978) have been considered for discussion.

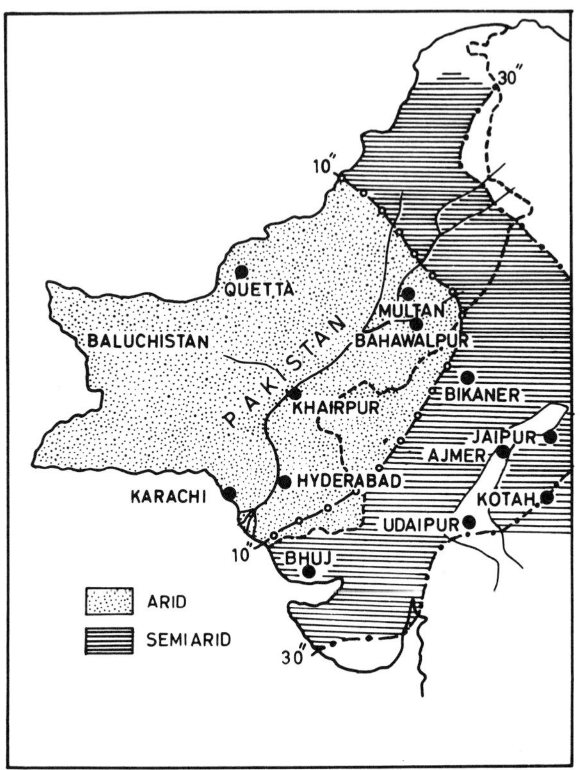

Fig. 1. Distribution of arid and semiarid zones in Pakistan and northwest India. After Bharucha (1955).

Evolution of arid and semiarid regions is an oft-recurring geological phenomenon found within the stratigraphic records of all parts of the world. In general, these regions are formed when the surface profiles get reduced and worn due to erosion by denudation. Such an ecological phenomenon brings about an interruption of water and air circulation systems, which in turn leads to the total extinction of drainage courses (Wadia, 1960). It is likely that similar ecological events have occurred during the evolution of these arid and semiarid tracts in India. It has been suggested (Wadia, 1960) that the Thar Desert was a well-watered and cultivated country before the Christian era. The discovery of fossils of *Mesua, Garcinia,* and *Calophyllum* (Lakhanpal and Bose, 1951) from the fuller's earth in the Barmer District, *Cocos* from Kapurdi in the Jodhpur District of Rajasthan (Kaul, 1951) during the Eocene age, the existence of cities like Mohenjodaro and Harappa, and the occurrence in the past of animals such as rhinoceros, water buffalo, and elephant in these regions suggests the existence of a very different type of climate and physical relief (Wadia, 1960).

Thus, desert conditions seem to have gradually set in well after humans

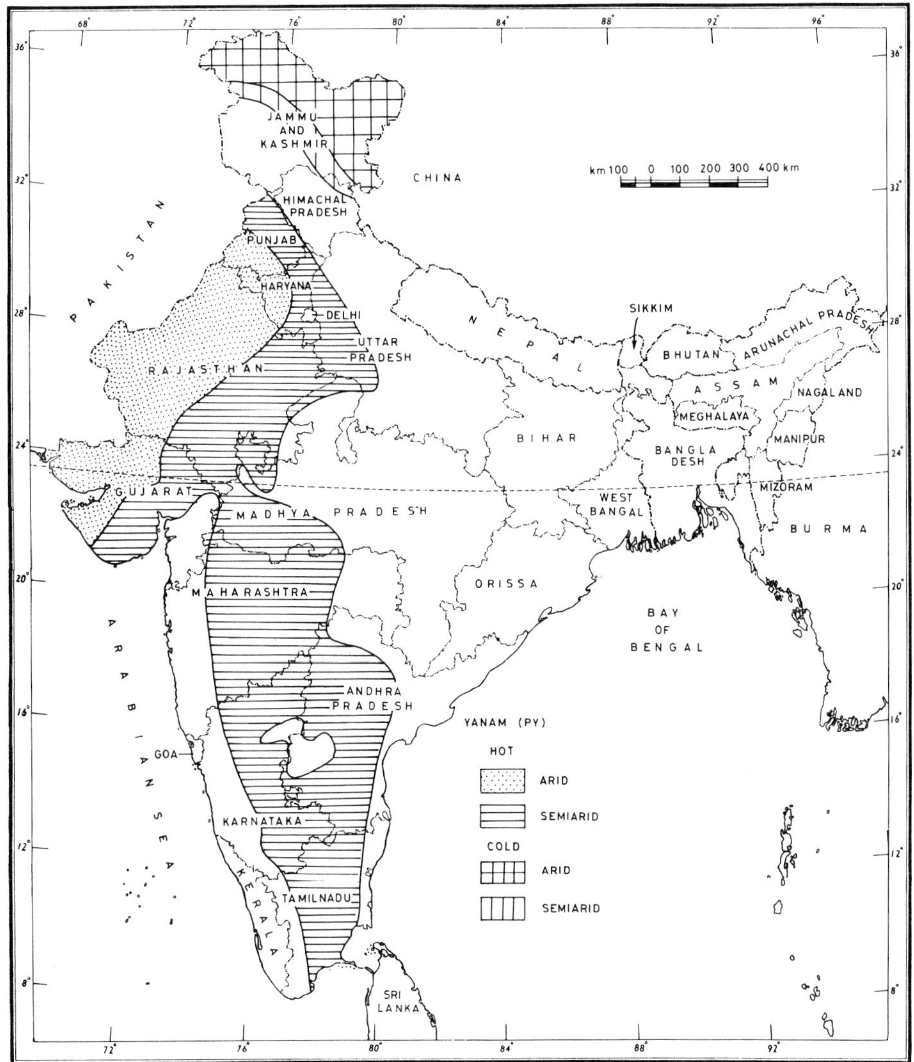

Fig. 2. Distribution of arid and semiarid zones in the Indian Republic. Survey of India outline map. Copyright 1980 Government of India.

appeared, possibly during subrecent times (M. Krishnan, 1952), and perhaps during the Harappan culture period, from 2300 to 1750 BC (Ghosh, 1952; Wadia, 1960; Wheeler, 1960). The paleobotanical evidence indicates an increasing trend in aridity since the Miocene in western India (Vishnu-Mittre, 1977). It was most probably during the early to mid-Quaternary that repeated fluctuations in temperature and precipitation caused the formation of desert and the overall aridity in this portion of the subcontinent (Vishnu-Mittre, 1977). Further, it was perhaps during this period that the Saharo–Sudanian and African floristic elements began to invade this region at the expense—the decline and eventual extinction—of the dense evergreen and semievergreen forests of western Indian and Rajasthan (Vishnu-Mittre, 1977). G. Singh (1977) suggested that the whole process of desertification has taken place in two steps: (1) the activation of sand dunes due to desiccation and dry winds, and (2) further drying of the substratum, leading to the accumulation of salt in the sand and the consequent disappearance of the mesophytic vegetation. The evolutionary history of arid and semiarid regions in the peninsular region of India, however, is not so well documented. In southern parts of India, location has played a major role in the origin and maintenance of these regions. The whole zone not only lies in the rain-shadow area, but also has a poorly developed drainage system. The nature of the igneous rocks has controlled the groundwater table, the process of weathering, and the development of homogeneity in the substratum.

An assessment of certain physiographic and topographic features and meteorological records of the previous 70 to 100 years has shown no significant changes in rainfall patterns, humidity values, and wind velocity in these regions, especially of the northwestern parts (Pramanik *et al.*, 1952; Rao, 1958, 1963; G. Singh, 1977; Meher-Homji, 1977). It may, however, be noted that during this period minor climatic shifts and fluctuations in the aridity index line have occurred within these regions, especially in the northwestern part of the Rajasthan Desert area (Jagannathan *et al.*, 1978). Thus, the limits of arid and semiarid regions in the Indian subcontinent have not changed during the past 100 years or so, and even now there are few signs of desertification of new areas; there has been no expansion of these regions throughout the subcontinent (Sen and Mann, 1977; Ghosh *et al.*, 1977a,b). The main influencing factor, which has maintained this stability, is the meteorological influence of the Himalayas (Krishnan, 1977a,b). Vishnu-Mittre (1977) has suggested that by controlling precipitation and feeding the rivers, the Himalayas have protected central India from becoming more arid than the northwestern part of the country. Increases in human and livestock populations, destruction of vegetational cover, and degradation of soil conditions have been occuring rapidly, however, and this has caused continual fresh deposition of sand within these zones, leading to further deterioration and desert conditions (Sen and Mann, 1977; Mann, 1977a). This is more alarming than the actual expansion of the arid and semiarid regions.

II. PHYSIOGRAPHY

A. Size and Location

The arid and semiarid regions of the Indian subcontinent total ~1,525,670 km^2. Such regions in India are found in all the 11 states situated in the western

Table I

Arid and Semiarid Regions in the Indian Subcontinent: Size and Location[a]

Country and state/province	Arid (km^2)		Percentage of total area of the state/province	
	Arid	Semiarid	Arid	Semiarid
India				
Andhra Pradesh	21,550	138,670	7	15
Gujarat	62,180	90,520	20	9
Haryana	12,840	26,880	4	3
Jammu and Kashmir cold desert	70,300	13,780	32	6
Madhya Pradesh	—[b]	59,470	—	6
Maharashtra	1,290	189,580	0.4	19
Mysore	8,570	139,360	—	15
Punjab	14,510	31,770	5	3
Rajasthan	196,150	121,020	61	13
Tamil Nadu	—	95,250	—	10
Uttar Pradesh	—	64,230	—	7
	387,390	970,530		
Pakistan				
Punjab	85,000	—	41	—
Sindh	82,750	—	59	—

[a]From Krishnan (1968a); Gupta and Prakash (1975a); Pakistan (1977–1978).
[b]Not represented.

half of the country (Fig. 2). The total spread of arid and semiarid regions in India is 387,390 and 970,530 km^2, respectively (Table I). In Pakistan, the entire area under discussion, 167,750 km^2, is arid (Table I; Gupta and Prakash, 1975a). This arid region of Pakistan and the contiguous regions in the states of Punjab, Haryana, Rajasthan, and Gujarat, and parts of the Indian Republic all along its border with Pakistan, form the Thar Desert, the main arid region of the subcontinent. The Indian part of the Thar Desert represents ~88% of the total arid area of the country. The area covered by the Thar Desert in Pakistan and India is 167,750 and 278,330 km^2, respectively.

B. Climate

The Indian subcontinent is primarily a land of tropical monsoon climate with four seasons: the northeast and southwest monsoon seasons (December–March and June–September, respectively), the transitional hot weather (April–May), and the transitional period of the retracting southwest monsson (October–November) (Spate and Learmonth, 1967). In arid and semiarid regions of the subcontinent, the climatic elements also follow this general seasonal rhythm, but

Table II

Climatic Data for Arid and Semiarid Regions of the Indian Republic[a]

	Mean maximum temperature (°C)		Mean minimum temperature (°C)		Mean annual average rainfall (mm)		Mean annual potential evapotranspiration (mm)		Mean wind speed (km/hr)	
	Arid	Semiarid	Arid	Semiarid	Arid	Semiarid	Arid	Semiarid	Arid	Semiarid
Andhra Pradesh	33.0	32.7	21.7	21.7	555	758	1786	1705	11.5	8.9
Gujarat	33.5	33.4	20.0	20.8	452	676	1865	1744	13.7	10.5
Haryana	32.8	31.4	17.4	17.6	455	573	1615	1481	7.0	[b]
Jammu and Kashmir	13.5	24.7	0.9	12.7	175	878	803	873	5.1	4.0
Karnataka	32.3	32.2	21.6	20.5	494	689	1765	1743	8.9	8.3
Madhya Pradesh	—[c]	31.9	—	18.3	—	770	—	1638	—	—
Maharashtra	33.6	33.2	19.6	19.7	473	718	1714	1759	9.9	7.9
Punjab	31.3	31.3	16.7	16.9	374	715	1362	1459	2.4	5.9
Rajasthan	33.4	31.9	18.9	18.5	343	650	1868	1594	10.7	9.9
Tamil Nadu	—	33.4	—	22.9	—	884	—	1725	—	[b]
Uttar Pradesh	—	31.2	—	18.7	—	737	—	1513	—	[b]

[a]From Krishnan (1969); Report of the National Commission on Agriculture (1976).
[b]Data not available.
[c]Not represented.

they have a greater magnitude of variability, both in space and time (Sharma, 1972). Droughts are frequent, but climatically none of the regions is a true desert (Stein, 1942). It is this marginal character that makes the conditions more critical and difficult. The arid regions are characterized by extremes of temperature (diurnal and annual) and erratic, irregular, and insufficient precipitation. In contrast, semiarid regions possess similar harsh conditions, yet the climatic parameters are not as erratic and unpredictable as they are in the arid regions. Some salient climatic features of arid and semiarid regions are summarized in the following paragraphs (see also Table II).

These regions experience higher temperatures (35–40°C in the northwest and 30–35°C in the peninsular regions) during May and June. The highest temperatures in arid regions range from 47.5 to 50°C and 40 to 47°C in the semiarid regions. However, the highest percentage frequency of fluctuation (to 50%) of maximum temperature in arid and semiarid regions is found for temperatures varying from 40 to 45°C. The diurnal variation in temperatures in the arid regions of Pakistan may be as much as 11–17°C (Pakistan, 1977–1978).

The mean minimum temperature varies throughout the year from 5 to 27.5°C in arid, and 5 to 22.5°C in semiarid, regions. The annual lowest minimum temperatures may reach −2.5 to 0°C in arid regions and 12.5 to 25°C in semiarid regions. During winter months, however, the gross surface minimum temperatures vary between 2.5 to 10°C in arid, and 2.5 to 17.5° in semiarid, regions. Frosts during the winter season are common, and up to 20 days of consecutive frosts have been recorded during some winters (Krishnan and Thanvi, 1973, 1977). Hot winds during the summer months are frequent in these regions.

Very high wind velocities have been recorded throughout the year, except during the post-monsoon season. In general, the wind speed through these regions decreases more in the north. The occurrence of dust haze and dust storms is a characteristic feature of these regions. In the arid regions of Rajasthan, an average of 8.7 dust storms per year has been recorded. Krishnan (1977a) reported that in arid regions of Pakistan and some parts of India, dust clouds reach heights of 6000 to 9000 m.

The mean rainfall normally ranges from 0 to 20 cm (30 cm at a few places) in arid regions, and 40 to 70 cm in semiarid regions. The main rainy season is from June to September, and it is during this period that 90–95% (10–20 cm in arid, 20–50 cm in semiarid) of the total annual rainfall is received under the influence of the southwest monsoon. In arid regions, especially in the northwestern part, the onset of this monsoon is from July 1 each year, and its withdrawal is completed by September 1. In the semiarid regions, however, the monsoon sets in from June 1 through June 30, and the season ends each year between September 15 and October 15 (*Agroclimatic Atlas of India,* 1978). On average, rainfall in the arid regions is received for not more than 20 days per year, and the maximum number of rainy days has been recorded from June through September. In the semiarid regions, on the other hand, the annual average of days of rainfall varies between 20 and 50. In addition to the main rainy season (June–September), during which rainfall may occur on a maximum number of days (15 to 40) in a year, precipitation is also recorded for relatively fewer days (1 to 20) during winter months (October–December). Occasional rainfalls can occur during the remaining part of the year (Spate and Learmonth, 1967). Average relative humidity values in both regions are more or less similar. Mean annual evaporation is 200–300 cm in arid, and 200–350 cm in semiarid, regions.

In the arid regions, the mean evaporation during summer exceeds 12 mm/day, although it is rarely more than 5 to 6 mm/day during winter. Also, in these regions, the potential evaporation during summer varies from 7 to 9 mm/day. The values for potential evaporation during the other two seasons (winter and monsoon), however, are lower than those recorded during summer. The potential evaporation in the northern areas is higher during monsoon season, while in the south it is higher during winter months. The annual potential evapotranspiration values are generally higher in arid regions than in semiarid regions (Tables II and III). Vapor pressure values during the monsoon exceed 25 mbars, but decrease during the post-monsoon season. During the winter months, however, vapor pressure is less than 10 mbars. The relative humidity ranges from 75 to 80% in the morning, and 50 to 60% in the afternoon (Krishnan, 1977a).

Krishnan (1977a,b), using an analysis of the frequency distribution of rainfall data from several stations located in the arid region, found that once in 10 years the annual rainfall of the entire arid zone of northwest India can be less than 200 mm, except for some stations (Jhunjhnu, Hissar, Sikar, and Pali) where it may be even less than 60 mm. The median value, represented by 50% frequency values, is invariably less than the mean values and thereby confirms the skewness of distribution of rainfall.

Table III

Seasonal Means of Potential Evapotranspiration (mm/day) for Some Locations in the States of Gujarat, Haryana, and Rajasthan, of the Indian Republic[a]

	Winter	Summer	Monsoon	Postmonsoon	Annual total (mm)
Bhuj	3.2	7.7	5.2	4.3	1397
Barmer	2.9	7.7	5.4	3.9	1858
Jodhpur	2.9	7.9	5.3	3.6	2063
Jaisalmer	2.5	8.8	5.3	3.8	2041
Phalodi	2.5	8.8	6.5	4.2	1771
Bikaner	2.0	7.3	6.3	3.2	1662
Ganganagar	1.8	6.9	6.4	2.9	1843
Hissar	1.9	6.8	5.6	3.1	1616

[a]From Krishnan (1977a).

Krishnan (1968b) noted an extremely poor moisture retention in the longitudinal range from 20°W to 70°E, and in the latitudinal belt between 30 and 40°N. In the longitudinal range 80 to 120°E, these values, however, are comparatively higher.

> Thus the comparatively little coverage of the arid zone in the belt 20° to 30°N and the occurrence of the arid zone right in the region of the lowest pressure during the main rainy season are peculiar to our longitudinal range. This may be mainly ascribed to the large circulation changes owing to the Asiatic monsoon. Further, the intertropical convergence zone lies only with[in] 3° to 10°N latitude in the Pacific Ocean, up to 18° latitude in Africa and up to 30°N latitude in India and South East Asia. In view of the above mentioned peculiarities in the circulation, the Indian arid zone falls in a peculiar category [Krishnan, 1973].

Further, although atmospheric humidity is usually fairly high, and is comparable with that of places with higher rainfall in the semiarid and subhumid regions, not much rainfall is received in these regions (Krishnan, 1968b).

C. Geology, Topography, and Edaphic Factors

1. GEOMORPHOLOGY

These regions present a rather desolate appearance with great areas covered by windblown sand and alluvium. Solid rock is practically concealed with the exception of hilly outcrops. Also, the substratum has evolved through long periods of subaerial degradation and aggradation, interrupted by short periods of tectonic activity (Ghosh *et al.*, 1977a,b). Furthermore, these areas have evolved, on different lithological formations, during the Quaternary period under different climatic phases. The present-day fluvial and aeolian landforms in the arid zones are thus polygenic in origin. Sand dunes and sand plains were created during the pre-Holocene and were stabilized during that epoch. However, later aeolian activity in the late Holocene created barchan dunes, shrub–coppice dunes, and sand sheet on the existing landforms. The geomorphology has been further modified due to biotic stresses (Ghosh *et al.*, 1977a).

2. TOPOGRAPHY

In general, the landscape is comprised of plains with sand dunes of old or deltaic alluvium, hills, and marshes. The residual hills and escarpments are in the form of domes, bosses, meas, scarps, knolls, and inselbergs. The physiography in these regions is uneven with no fixed pattern in variation (Gupta and Prakash, 1975b). By and large, the relief in both of these regions is constituted of different topographies. The aspect and degree of elevation varies in different zones. Thus, the regions in Pakistan, Rajasthan, Tamil Nadu, and Uttar Pradesh show an increase in the elevation from west to east. The gradient direction in cold and semiarid regions of Jammu and Kahmir, however, is from south to north. In

Table IV

Topography of Arid and Semiarid Regions of the Indian Subcontinent[a]

Country/state	Relief with respect to height above mean sea level (m)				Aspect
	Arid		Semiarid		
	Maximum	Minimum	Maximum	Minimum	
Andhra Pradesh	616.50	345.30	653.80	377.10	E to W
Gujarat	154.00	57.00	180.90	62.70	Not very distinct
Haryana	310.60	226.80	283.90	237.30	No distinct trend
Jammu and Kashmir	3000.00	1800.00	3000.00	1800.00	S to N
		1350.00	4500.00		
Karnataka	696.20	542.60	779.80	630.30	E to W
Madhya Pradesh	—[b]	—	577.00	355.00	E to W
Maharashtra	520.80	467.50	589.10	407.00	E to W
Punjab	210.20	201.30	275.80	246.10	Not distinct
Rajasthan	440.80	253.30	505.20	340.60	W to E
Tamil Nadu	—	—	473.90	132.10	W to E
Uttar Pradesh	—	—	199.70	181.80	W to E
Pakistan	150.00	50.00	—	—	W to E

[a]From Report of the National Commission on Agriculture (1976); Pakistan (1977–1978); Agroclimatic Atlas of India (1978).
[b]Not represented.

regions found in Andhra Pradesh, Madhya Pradesh, Maharashtra, and Karnataka, the increase in the elevation is from east to west. In the arid and semiarid regions of Gujarat, Punjab, and Haryana, the aspect is more or less plain (*Agroclimatic Atlas of India,* 1978; Table IV). Sandy plains are interspread through the landscape in Rajasthan, and in Sindh, Khairpur, and the Bhawalpur districts of Pakistan. These regions are flat. However, in some places hummoden topography can also be noticed (Spate and Learmonth, 1967).

3. DRAINAGE

Table V summarizes the details of the major river systems flowing through these regions. The drainage in the northern counterpart is internal except where open rivers and depressions occur. Rivers in the southern regions are more open but flow through deep valleys that have been cut through the peninsular rocks. In the arid region of Pakistan, drainage is from north to south and is carried on by the river Indus and its tributaries. On the Indian side, in the extreme east of the semiarid region, the Aravallis form the major watershed division. The drainage from the regions contiguous to the Aravalli system flows northeastward via the Chambal River into the river Yamuna. Drainage from the western and southern regions is aligned toward the Rann of Kutch. Banas, Mahi, Sabarmati, and Luni are the major drainage systems of Rajasthan and Gujarat. The river Luni has been found to have a flood cycle of 16 years (Sharma, 1972). Rivers in the peninsular region are perennial and have set annual and seasonal drainage patterns.

4. SOILS

Raychaudhuri *et al.* (1963), Roy *et al.* (1971), Kanzaria (1972), and Dhir (1977a) have studied the soil types of these regions and have suggested various classifications. The solid types have been described as desert soils and gray-brown soils by Raychaudhury *et al.* (1963), and as sierozems (less than 300 to 500 mm rainfall) and red desert soils (less than 200 to 400 mm of rainfall) by Mathur *et al.* (1972). According to the soil map of India (Fig. 3), however, the soils of arid and semiarid regions of India belong to 10 types: (1) forest and hill types, (2) gray and brown soils, (3) alluvial soils, (4) gangetic alluvium, (5) desert soils, (6) red and yellow soils, (7) deep black to medium black soils, (8) coastal alluvium, (9) red loam soils, and (10) red gravelly soils. Soils of the arid region of Pakistan come under the dry group, pedocals, which have a high $CaCO_3$ content and low organic matter (Pakistan, 1977–1978). These soils consist of old and new alluvium (Asghar and Hafiz, 1964) and are similar to desert soils (see Raychaudhury *et al.*, 1963). Also, they usually lack diagnostic horizons and possess a highly variable textural pattern, which is mostly due to characteristic aggradation. The upper horizons, when present, usually are light

Table V

Surface Water Availability of Arid and Semiarid Zones of the Indian Subcontinent[a]

River system	Total length (km)	Catchment area (km^2)	Average annual flow (ha m)	Utilizable flow (ha m)	Approximate utilization (1974) (ha m)	Remarks
Indus	2880	1,165,000 (87,900 in India)	7.7	4.6	3.70	Flows through arid tracts of Pakistan and Punjab (India); has an efficient system of canals in semiarid regions
Ganges	2525	861,404 in India	51.0	25.0	8.5	Flows through only a very small semiarid portion of Uttar Pradesh; however, small rivers and rivulets from semiarid regions of eastern Rajasthan and Madhya Pradesh drain their water in the Ganges
Luni	320	34,247				Main river in the arid regions of Rajasthan, drains its water in the Runn of Kutch; it is often flooded during the monsoon and this extra water potential can be utilized by building reservoirs, etc.; some noteworthy irrigation projects at present are Bisalpur

			2.5	2.0	0.5	Project, Gang Canal Project, Jawai River Project, and the Rajasthan canal project near Bilasa on Luni
Sabarmati and Mahi	300 533	21,674 34,842				Southern Rajasthan and Gujarat form the chief drainage basin; construction of many reservoirs like Gajod, Sanandro, Kankavati, Bhamban, and Bhimdad are under active consideration
Narmada and Tapti	1300 724	98,796 65,145	6.2	4.9	0.6	Drains water from the semiarid regions of Madhya Pradesh and southern Gujarat
Godavari	1465	312,812				Largest of the peninsular rivers; semiarid tracts of Andhra Pradesh, Karnataka, Madhya Pradesh, and Maharashtra form the drainage basin
Krishna	1400	258,943	22.5	19.0	7.3	Drains water from semiarid tracts in Andhra Pradesh, Karnataka, and Maharashtra; reservoirs have been constructed on several tributaries, especially Tungabhadra and Pennar
Kaveri	800	87,900				Drains areas of Kerala, Karnataka, and Tamil Nadu; Kadapa-Kurnool Canal irrigates semiarid regions of Tamil Nadu

[a]From Report of the National Commission on Agriculture (1970); Raza and Ahmed (1978).

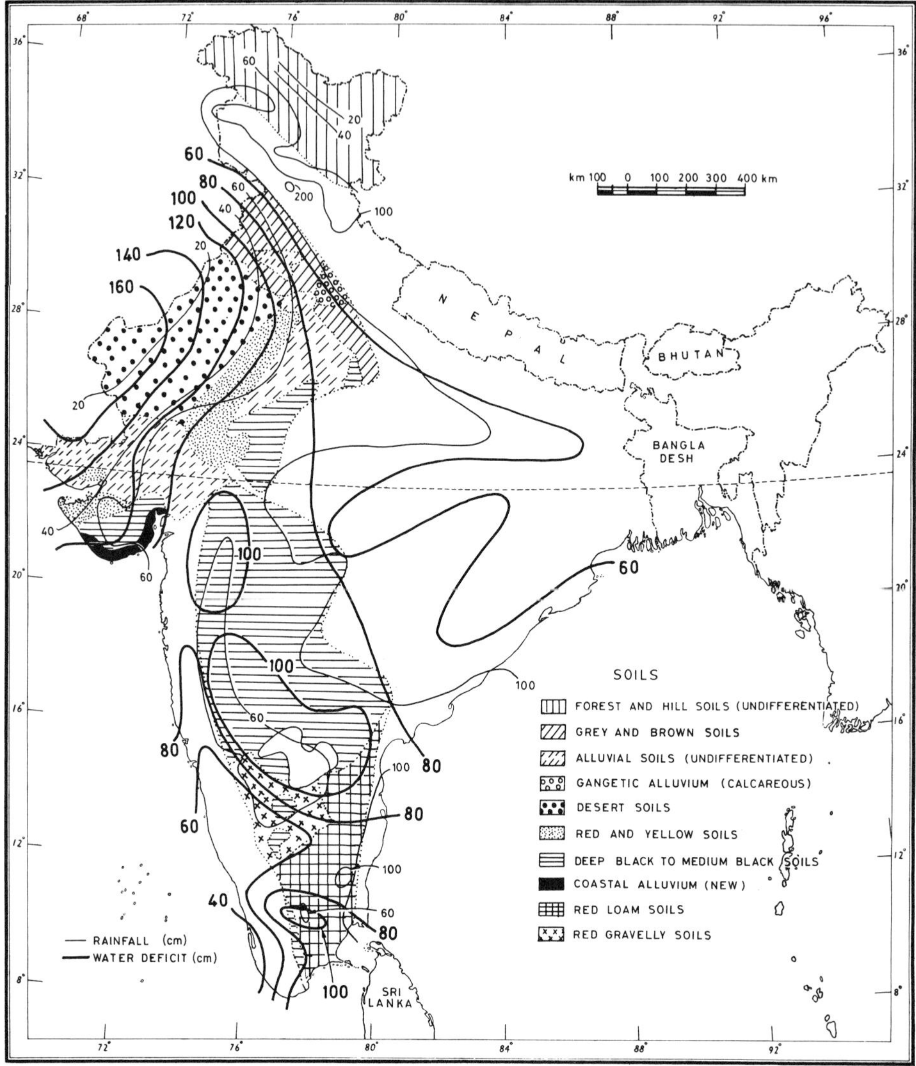

Fig. 3. Soils, annual average rainfall, and water deficit in arid and semiarid zones of the Indian Republic. Survey of India outline map. Copyright 1980 Government of India.

brown to dark yellowish brown in color. Some characteristic features of the soil types found in these regions are summarized in Table VI.

Further, in these regions, the overall total mineral wealth is limited. However, due to a broad range of geological formations, from ancient crystalline rocks to recent alluvium, a variety of minerals are available. In the Pakistan region,

Table VI

General Features of Soils of Arid and Semiarid Regions[a]

Soil no.	Soil type	Origin	Physical characteristics		pH	Chemical characteristics	Remarks
			Texture	Color			
1	Forest and hill soil	Alluvial	Loamy to sandy loam		Slightly above 7	P_2O_5 and K_2O supply moderate, N concentration varies from 0.04 to 0.08%	White incrustation, especially with and after flooding
2	Gray and brown soil	Alluvial; under hot and humid climate	Silty clay loam	Dark brown in upper horizons	Around 7	C/N ratio low; available C, N, P_2O_5, and total soluble salts, low to medium; K_2O supply adequate	Well drained; small, white concretions in top layers increase with depth
3	Alluvial	Alluvial river deposits				C/N ratio 7–14; P_2O_5, K_2O supply sufficient; N and organic matter poor (0.057–0.126% C); deficient in lime (except the old alluvium); base saturation poor	Most fertile; and less, "Kankary"
	(a) New alluvium	(i) Sandy	Light colored	7.0 to alkaline			
		(ii) Sandy to clayey loam	Light colored	7.0 to alkaline		In Kutch area, soil surface heavily cracked	
	(b) Old alluvium	Clayey		9.0			

(*continued*)

Table VI (*Continued*)

Soil no.	Soil type	Origin	Physical characteristics		pH	Chemical characteristics	Remarks
			Texture	Color			
4	Gangetic alluvium	Alluvial river deposits	Sandy loam	Light colored	>8	Calcareous; saline or alkaline, poor in P_2O_5, N, organic matter; K_2O supply moderate; lime content higher	Fertile; drainage imperfect
5	Desert soil	Derived partly from disintegration of subjacent rocks; largely contributed by the blown sand from coastal region and Indus Valley	Sandy to loamy–sand (75–90%)	Light, pale yellow to light yellow	7.2–9.2	With high percentage of soluble salts; P_2O_5 concentration varies from 0.5 to 0.1%; N supply low (0.02–0.07%), Ca varies from 1 to 1.5%; Ca content increases manyfold with depth	Dispersion mainly due to high Na–clay formation in the subsoil; water retention capacity appreciable
6	Red and yellow soils	Originated from ancient crystalline and metamorphic rocks through weathering	Light, porous, friable, and sandy	Pale yellow to reddish	5.5–8.8	P_2O_5 from 0.02 to 0.07%; poor in Ca and free CO_3^-; total salt, K, Ca, and Mg contents low; P fixation is more with increase in the acidity	Yellow color mainly due to hydration of ferric oxide in the red soils; poor retention of water
7	Deep black to	(a) Originated	Sandy to	Black	7.2–8.5	Generally deficient in	Sticky with high water

	medium black soils	from Deccan and Maharashtra Rajmahal trap system (b) Ferruginous gneiss and schists occurring in Madras State; shallow to deep	clayey; gravelly along the Ghats			P_2O_5, N and humus; Ca and K supply sufficient; good Ca exchange capacity (50 meq/100 g); Si/SiO_2 4.5%; high base status	retention; becomes easily saline or alkaline; porosity and water-holding capacity high
8	Coastal alluvium	Part of sea filled up by silt carried down by rivers	Sandy to clayey loam	—	—	Calcareous, poor in organic matter; nutrient status low	Water-holding capacity low (35%)
9	Red loam soil	*In situ* formation from red granite	Loamy to sandy loam; rich in clay	—	Acidic, 6.5–8	Poor in N, P_2O_5, and organic matter. Ca and K_2O supply adequate; poor cation exchange and base status	Color mainly due to ferric oxide; drainage free, with intermittent leaching; water-holding capacity better than red and yellow soils
10	Red, gravelly soil	Red granite rocks	Gravelly, mainly due to pisolitic iron gravels	—	Around 7	P_2O_5, N, organic matter and salt contents low; rich in total and available K_2O and total P_2O_5	Restricted movement of water in lateral and downward directions

[a]From Raychaudhuri *et al*. (1963); Raychaudhuri (1966); Mathur (1975).

limestone, gypsum, and fuller's earth are found (Pakistan, 1977–1978); in the Indian region, salt, gypsum, lignite, copper, fuller's earth, ceramic minerals, building stones, deposits of iron ore, lead, zinc, tungsten, manganese, bentonite, asbestos, graphite, feldspars, mica, and other minerals are available in appreciable quantities. The gypsum reserves in arid and semiarid regions of Rajasthan alone may be more than 1500 million tons (Sinha and Verma, 1978). However, exploitable gypsum is present all through the arid and semiarid regions of the subcontinent (Narayanaswami and Deshmukh, 1978).

At present, the salt resources of Rajasthan alone produce 400,000–450,000 tons of salt per year. The existing climatic and edaphic conditions, and their stability, provide for more harnessing of this resource potential. It is possible in the Rajasthan state of India to recover 25,000 tons, and 15,000–20,000 tons, of sodium sulfate at Sambhar and Didwana, respectively, if full utilization is made of the bitterns and brines that are available at these places (Bhatt *et al.*, 1977).

D. Hydrology and Resources

In the arid and semiarid regions, characterized by meager and erratic rainfall and high surface evaporation, groundwater is the main source of water. The quantity and quality of this groundwater, however, vary with the geological formation, which determines percolation rate, the mode of aquifer recharge, and the discharge capacity of wells. Thus, the discharge capacity of wells in Precambrian crystalline rocks varies from 1.36 to 21.76 m^3/hr (Mithal, 1969; Pathak and Prasad, 1973), 3 to 45 m^3/hr in Precambrian sedimentary rocks, 1 to 2 m^3/day in Middle and Upper Jurassic rocks, 40 to 60 m^3/hr in Tertiary sedimentary rocks, and 40 to 90 m^3/hr in Quaternary sediments (Rao, 1975).

In general, the groundwaters all through these zones are saline. Salinity in the arid region results primarily from sodium and chloride ions, whereas in semiarid regions it is due to bicarbonate, sulfate, and divalent cations (Paliwal, 1971; Gupta, 1979). The groundwaters found in sandstone formations are, by and large, less saline and relatively more suitable for diverse purposes. The regional hydrological characteristics, as summarized from Krishnamurthy (1970), Paliwal (1971, 1972, 1978), Choudhary (1973), Deshmukh and Karanth (1973), Pathak and Prasad (1973), Radhakrishna and Dusan Duba (1973), Sehgal and Gulati (1973), Singhal (1973), Dhir (1977b), Baweja (1979a,b), Gupta (1979), and Micheal *et al.* (1979) are as follows:

1. ANDHRA PRADESH

Hard rocks (Archean, Cuddapan, Karnool formations, and traps) are found in larger parts of the area, and the yield of groundwater from wells in these rock types ranges from 80 to 300 m^3/day. The remaining region is covered mostly by Gondwana formation and Tertiary and Recent alluvium. These types of wells vary in yield from 1200 to 3650 m^3/day. The water level fluctuates from 2 to 8 m in the former and 1 to 5 m in the latter. Overall quality of groundwater is potable except for that affected by canal irrigation.

The groundwaters contribute 20% toward irrigation. The arid districts have a higher electrical conductivity (EC) (>2000 μmho/cm) than the semiarid districts (755–2000 μmho/cm). In the higher EC range, Na^+ (2.2–7.8 meq/liter) is more abundant than Mg^{2+} or Ca^{2+}, whereas in the lower EC range waters, the divalent cations are more concentrated. Potassium concentrations are very low. Chloride concentration ranges from 0.5 to 5.5 meq/liter, and fluoride from 0.05 to 0.14 meq/liter. In both regions, the sodium absorption ratio (SAR) is less than 10. However, the total dissolved solids in these waters vary from 340 to 1035 ppm.

2. GUJARAT

Hard rocks of Archean, Precambrian and Mesozoic sandstones, and Deccan traps are the common types found in this region. However, Quaternary, Mesozoic, and Tertiary sandstones and Quaternary alluvium are associated with groundwaters. Considerable fluctuations in the water table (up to 30 m) have been recorded. The quality of ground water varies widely in this region.

Of the total irrigated area, 80% is irrigated by groundwaters. In some extreme arid regions, ECs up to 10,000 μmho/cm have been recorded. In general, however, in semiarid and arid regions, it varies between 750 and 2250 μmho/cm. Sodium is dominant (50–95%) among the cations. Potassium concentrations vary up to 1 meq/liter only. Magnesium concentrations are invariably higher than calcium. Groundwaters conform to the Na^+–Mg^+/Ca^+ cation type and the Cl^-–SO_4^-, Cl^-–HCO_3^-, or HCO_3^-–Cl^- anion type. CO_3^- concentrations are negligible. The concentrations of HCO_3^- and SO_4^- are less than those of chlorides. Residual NA_2CO_3 concentrations are small. The SAR usually varies from 10 to 30, and rarely ranges between 26 and 55.

3. HARYANA

Most of the region is covered by Indo–Gangetic alluvium. In the northern part of the state this alluvium is deeper, with a good potential for groundwater. The depth of the water table varies between 3 and 11 m, and the fluctuations in the water table usually coincide with the monsoon season. Groundwater quality varies widely through the region.

The groundwaters are used for irrigating 40% of the area. More than 60% of the groundwaters have an EC higher than 8000 μmho/cm, but EC values up to 25,000 and even 40,000 μmho/cm have been noted from the Bhiwani and Hissar regions. Sodium dominates among the cations, and the concentration of magnesium is higher than calcium. The calcium concentration is less than 10%. In arid regions the groundwaters are of Na^+/Cl^- type, and in semiarid regions they are of Na^+/HCO_3^- or CA^+/HCO_3^- type. Boron concentration varies from 0.68 to 2.03 ppm, and fluoride from 0.013 to 0.7 meq/liter. SAR ranges from 5 to 18, but values up to 100 have also been recorded.

4. KARNATAKA

Most of the region is underlaid by crystalline and metamorphic rocks—mainly granites, gneisses, limestones, sandstones, quartzites, and shales. The annual fluctuation of water level has been found to vary between 0.5 to 8.0 m, and the erratic rainfall from 1969 onward has caused a great decline in these levels.

Only 23% of the area is irrigated by groundwaters. Most of these waters have EC values ranging from 100 to 400 μmho/cm, but values up to 10,000 μmho/cm have also been reported from some semiarid districts. Sodium dominates among the cations, its concentration ranging from 0.13 to 2.0 meq/liter. Potassium concentrations are less than those of sodium (0.025–1.0 meq/liter). Chloride and fluoride values vary from 0.28 to 0.66 and 0.40 to 0.065 meq/liter, respectively. The SO_4^- concentration ranges from traces to 62.5 meq/liter, and CO_3^- and HCO_3^- concentrations are also very high. Residual NA_2CO_3 concentrations range from traces to 0.28 meq/liter. SAR values from 5 to 18 have been recorded.

5. MADHYA PRADESH

The geological formations include hard rocks belonging to Archean and Precambrian as well as Gondwana semiconsolidated formations. Groundwater is more often associated with Gondwana and alluvial types. The fluctuations in the water table are cyclic, although because of irrigation, many wells have shown a rise in the water level. The overall water quality is good except in areas affected by pollution and mineralization because of poor drainage.

Forty percent of the area is irrigated by groundwater. The EC values of these waters range from 75 to 2250 μmho/cm; most have lower values of up to 750 μmho/cm. Residual Na_2CO_3 concentrations are very low and only a few groundwater locations have values up to 2.5 meq/liter. SAR values are low, ranging from 1.2 to 2.6 and occasionally up to 14.6.

6. MAHARASHTRA

Precambrian hard rocks, laterites, and smaller patches of Gondwana formations are the major rock systems found in this region. Fluctuations in the water table mainly reflect the southwest monsoon cycle, although in recent years the water levels in wells have shown a general decline. Overall groundwater quality is good.

About 40% of the total irrigated area is covered by groundwater irrigation. The

EC varies from less than 3000 to 9000 μmho/cm. Magnesium concentration is greater than that of calcium. Also, SO_4^- concentration is greater than the chloride and bicarbonate concentrations. The SAR value ranges from 7 to 14.

7. PUNJAB

Indo–Gangetic alluvium forms the major cover in this region and includes efficient aquifer systems. In some parts a consistent rise in the water level of wells has been noted.

Groundwaters are used to irrigate 50% of the area. The EC of the groundwater in the arid regions varies from 200 to 15,000 μmho/cm, and in semiarid regions, from 300 to 6000 μmho/cm. In the arid regions, Na^+ values are higher than Ca^{2+} and Mg^{2+} values (4–8 meq/liter), but in semiarid regions, some waters have Ca^{2+} and Mg^{2+} values higher than that of Na^+. In general, however, in the semiarid region there is not much difference between Na^+ (3–6 meq/liter) and Ca^{2+} and Mg^{2+} values (3.5–5.5 meq/liter). The concentration of boron is less than 1 ppm. HCO_3^- and CO_3^- values (6−12 meq/liter) are higher than that of chlorine (12.2–13 meq/liter). Sulfate and chlorine concentrations are similar in both regions. The SAR is higher in arid (up to 9.1) than in semiarid (1–4.1) regions.

8. RAJASTHAN

In the southern half, mostly hard rocks—granites, gneisses, schists, and rhyolites—are found. In the northern half, Mesozoic to Quaternary sediments comprising Lathi and Palana sandstones are found. The alluvium as well as the blown-sand-covered regions support potable aquifers. The fluctuations in water level are less in the northern than in the southern part. In the southeastern region the water table is generally within 15 m of depth, but in the northwestern part it is deeper (30–100 m). Since the mid-1970s, not much change in the water level has been noted, especially in the canal-irrigated areas. A wide variety of groundwaters are found in this region.

More than 50% of the total irrigated area is sustained by groundwater. The EC of groundwaters of arid regions varies from 10,000 to 15,000 μmho/cm. In some places, however, values up to 30,000 μmho/cm have also been recorded. In the semiarid region, on the other hand, the EC is usually less than 10,000 μmho/cm. Sodium dominates (60–80%) among the cations, followed by magnesium (5.3–11.4 meq/liter) and calcium (3.5–8.1 meq/liter). Boron concentration varies from traces to 8 ppm. Anions vary, as chlorine is greater than HOC_3^-, which is greater than SO_4^-, in the lower salinity areas, and chlorine in greater than SO_4^-, which is greater than HCO_3^-, in higher salinity areas. Chlorides, mainly of sodium and magnesium, have been found. Residual Na_2CO_3 concentration is seldom greater than 250 ppm. The SAR ranges from 10 to 30.

9. TAMIL NADU

Hard rocks cover the majority of the area. Tertiary formations and Cuddapan stone of Miocene age contain productive aquifers. The water level fluctuates from 1.31 to 9.4 m in these rocks and is seasonal. Overall ground water is potable.

Only 35% of the area is covered by groundwater irrigation. In general, the EC varies from 120 to 5350 μmho/cm. Sodium concentrations up to 80 ppm are found, but values up to 475 ppm have also been recorded. Calcium concentrations are lower (up to 1.25 meq/liter) than sodium. HCO_3^- and chlorine range from 0 to 1.71 and 0 to 0.11 ppm, respectively. SAR values range up to 7; however, SAR values from 14 to 108 have also been reported.

10. UTTAR PRADESH

Mostly granite and gneisses, and to a lesser extent Vindhyan formations, from the water-bearing rocks. Water levels are deep (~20 m) and decline during summer months. The groundwater is mainly potable.

Sixty percent of the area is irrigated by groundwaters. The EC values range from 600 to 8000 μmho/cm in the semiarid regions. However, in the Agra and Mathura regions values up to 34,000 μmho/cm have also been found. Sodium is dominant among cations (3–6 meq/liter, up to 14 meq/liter in Agra). Magnesium concentrations are higher than calcium (3–10 meq/liter). Generally, HCO_3^- levels are higher (6–10 meq/liter) than the level of chlorides (1–4 meq/liter), except in the Agra region where chlorine levels up to 38 meq/liter have been recorded. SAR values range from 2.0 to 12.8.

11. PAKISTAN

Groundwaters are of medium quality, are usually deeper waters, and are unfit for consumption. The EC ranges from 4000 to 10,000 μmho/cm. Sodium is dominant among cations. Potassium, calcium, boron, and chlorine are low in concentration. Higher amounts of residual Na_2CO_3 are found. Total dissolved salts are higher (>3000 ppm), and the SAR varies from 20 to 50.

III. DEMOGRAPHY AND SOCIOECONOMIC FACTORS

The arid and semiarid regions of the Indian subcontinent have an average density of 48 persons/km^2 (260 persons/km^2 in Pakistan). These regions are the most thickly populated regions in the world, by arid-zone standards, as compared to 3 persons/km^2 in other similar regions of the world (Malhotra, 1977; Pakistan, 1977–1978). However, the arid regions are less populated than the semiarid regions (Table VII). In general, population density varies with the local climate, edaphic factors, and vegetation type. The availability of water, however, seems to be the most decisive factor for predicting the population density of a place, and it has been shown that density distribution and rainfall are positively correlated ($r = .6079$) (Malhotra, 1978; Sharma, 1972). Thus, in extreme arid conditions, only 4 persons/km^2 are found in comparison to 157 persons/km^2 in areas with more rainfall (Malhotra, 1977). The rate of population growth has been steady (Table VII). In Rajasthan, from 1901 to 1971, there was an increase of 186.78% in rural areas and 240.38% in urban areas. However, from 1961 to 1971, the rate of population growth in arid and semiarid regions of India showed no definite pattern (Table VII). The age structure pyramid is normal, and it is noted that people under 34 years of age constitute 75% of the total population; this trend has stabilized during the postindependence period (Table VII). Further, in Rajasthan, 75% of the people in the 0 to 35 age group are married (Sharma, 1972). Also, there are 4037 females in the 15 to 44 age group from every 10,000 women of all ages, and more women belong to the first half than those in the latter half of the reproductive period (Table VII). In rural areas, the female population is greater than that of males (Sharma, 1972; Malhotra, 1977; census of India, 1961–1971). To a larger extent, the rapidly growing population, scarcity of available potential for sustaining this growth (mainly pastoral life-style, joint-family systems, long-standing social customs, and food habits), and the consequent social structure are the peculiarities of the scenario of these regions. Further, each family possesses a small landholding for cultivation, which, in these prevailing harsh environments, is most uneconomical; often such a holding becomes a liability rather than an asset. The meager returns, both in cash and kind, from these small areas over the years have brought a decline (86%) in the number of cultivators on one hand; on the other hand, options for other jobs, especially for seasonal and casual labor workers, have increased. Additionally, since local employment potential for this newly emerging labor class is not available, these working hands migrate to other areas, often in large numbers (Malhotra, 1977; 1978). As more young people move out, leaving the older people to cultivate the limited areas, the yield is further reduced. Furthermore, production is meager and therefore everything (100%) is consumed locally; the income, if any, is invested in land, land rights, and ornaments. This leads to very stringent economic conditions. In fact, the majority of families in these regions are under a constant debt (Sharma, 1972; Malhotra, 1978).

The increase in population has affected the available cultivatable areas. In arid and semiarid regions of Rajasthan alone, such areas have declined from 13.72 ha per household in 1951, to 9.95 ha per household in 1971 (Sharma, 1972). This has led to the cultivation of marginal lands; since the mid-1970s there has been an increase of 9.47% in the net sown area. This practice has caused the shrinking of pasture areas by 6.95% and has created hardships for the cattle population.

Table VII

Demography of Arid and Semiarid Regions[a]

	Average population density (persons/km^2)		Population variation (1961–1971) (%)		0–14 years				15–34 years	
		Arid		Arid	General		Arid semiarid		General	
Regions	General	semiarid	General	semiarid	Male	Female	Male	Female	Male	Female
Andhra Pra-	157	107	20.9	23.0	20.5	20.3	22.3	20.2	15.2	15.5
desh		159		23.4			20.5	20.0		
Gujarat	136	58	29.4	28.9	22.4	20.6	22.8	21.0	16.1	15.2
		145		29.5			22.9	20.1		
Haryana	227	166	32.2	37.2	24.7	21.6	23.0	21.0	15.3	13.9
		264		30.0			4.0	21.0		
Jammu and	21	29.7	18.5	21.1	20.8	17.8	16.7	16.2	14.6	
Kashmir	1		31.0			21.2	19.0			
	114									
Karnataka	153	113	24.2	27.0	21.3	21.1	22.3	20.4	15.5	15.2
		139		21.9			21.8	20.6		
Madhya	94	—[b]	28.7	—	22.5	21.2	—	—	15.0	14.5
Pradesh		102		31.1			22.7	21.2		
Maharashtra	164	151	27.4	21.8	21.2	20.2	22.2	21.2	16.1	15.0
		133		25.5			21.9	19.8		
Punjab	269	188	21.7	19.7	22.0	19.3	22.3	20.2	16.7	14.8
		316		22.0			23.4	19.7		
Rajasthan	75	47	27.8	27.0	23.2	21.0	23.7	20.1	15.5	14.6
		119		26.6			22.4	19.3		
Tamil Nadu	317	—	22.3	—	19.1	18.7	—	—	16.0	16.3
		289		19.4			19.2	17.8		
Uttar Pra-	300	—	19.8	—	22.4	19.5	—	—	15.3	14.3
desh		438		22.0			22.6	19.4		
India	177	104	24.8	25.4	21.7	20.4	22.0	20.0	16.5	15.7
		202		25.7			22.1	19.8		

[a]From Census of India Reports (1967, 1971).
[b]Not represented.

These short-term welfare measures, in the form of land allotments to marginal and submarginal farmers, have caused further deterioration of the habitat. With low water supplies, scanty rainfall, and a lack of perennial streams, these areas cannot long be sustained economically and are abandoned sooner or later. Further, with the increase in population, use of wood and other plant parts has increased; the destruction of the natural vegetational cover of grasses, shrubs, and trees creates more wastelands. Such a pattern of life and the growing population lead to an unnatural use of resources and causes ecological imbalances, which at times are too severe to return to the original structure even under protected management. The net result is an increased stress on the already scarce resources that are available in these arid and semiarid environments.

Population sex and age structure (%)											
15–34 years		35–59 years				60 years and above				Sex ratio Females per 1000 males	
Arid semiarid		General		Arid semiarid		General		Arid semiarid		General	Arid semiarid
Male	Female	Male	Female	Male	Female	Male	Female	Male	Female		
16.0	15.6	11.7	10.7	10.4	11.1	3.2	3.2	2.7	2.3	978	946
16.8	16.0			10.8	10.8			3.0	2.5		957
15.6	15.3	10.6	9.8	10.0	10.5	2.6	2.7	1.8	3.2	934	994
15.6	16.3			9.9	9.8			2.6	2.3		961
15.2	14.8	10.1	8.6	12.0	9.0	3.5	2.5	3.0	2.0	866	886
15.8	14.8			10.0	8.8			3.2	2.4		887
16.3	16.0	11.6	9.2	13.0	12.8	3.3	2.2	3.6	3.6	878	965
17.3	16.0			12.2	8.8			4.0	1.4		826
16.4	13.3	11.2	9.6	10.6	9.8	3.1	3.0	2.9	2.2	957	916
15.7	14.9			11.5	10.4			2.8	2.3		931
—	—	11.2	9.9	—	—	2.8	3.0	—	—	941	—
16.0	15.0			10.5	9.7			2.8	2.1		923
17.3	15.4	11.7	10.1	10.3	8.6	2.9	2.9	2.7	2.3	930	905
16.3	15.2			11.5	9.8			3.3	2.2		837
17.2	15.3	10.5	9.3	10.4	8.4	4.4	3.1	4.0	2.2	365	355
16.6	14.8			10.4	8.6			4.6	1.9		818
16.6	14.7	10.7	9.5	10.6	8.5	2.9	2.6	3.4	2.4	911	842
16.8	15.0			11.8	9.5			3.4	2.3		852
—	—	12.6	11.6	—	—	2.9	2.8	—	—	978	—
15.5	16.5			13.5	12.3			3.0	2.2		953
—	—	11.8	10.0	—	—	3.7	3.0	—	—	679	—
16.2	14.9			10.9	10.4			4.0	1.6		862
16.3	15.3	10.9	9.6	10.9	9.8	3.7	2.5	3.0	2.5	930	914
16.2	15.5			11.2	9.9			3.3	3.1		398

Besides agriculture and labor, the major means of livelihood for the inhabitants of these regions have been the earnings generated through the sale of cattle, milk, milk products, bones, skin, wood, and other products [Central Arid Zone Research Institute (CAZRI), 1979]. However, due to social and ecological stresses, the economic gains per household are very meager. This is more so since business happens to be very localized. In recent years, however, with the increased facilities for transport and means of communication, improvements in the economic returns per household are evident (CAZRI, 1974; 1979).

IV. PLANT RESOURCES

The predominantly dry conditions and the extremes of the other climatic components that characterize these regions support a poor flora. However, within these regions the number of families, genera, and species varies in relation to the degree of dryness. Thus, the flora in arid regions of Pakistan, and the contiguous Thar Desert on the Indian side, consists of 82 families of phanerogams; the ratio of genera to species in this region is 1 to 1.8. In contrast, in relatively less dry semiarid regions, both in the northern (i.e., in Rajasthan, contiguous to the arid zone, and in Punjab, parts of Uttar Pradesh, and north Gujarat) and southern (Deccan Plateau, Coimbatore Plateau, located in the shadow of the Nilgiri and Palni hills, and the extreme southeast of Tamil Nadu) regions, the floras are comparatively richer. Thus, 497 genera and 969 species belonging to 104 families are represented in the northern region, and 353 genera, 1184 species, 109 families in the southern region. Further, the ratios of genera to species in northern and southern regions are 1 to 9, and 1 to 11, respectively (Biswas and Rao, 1953; Meher-Homji, 1977; Bhandari, 1978). Such an overall low genera to species ratio suggests that the taxa are occurring in habitats that do not favor rapid evolution at the species level. Interestingly, however, at the infraspecific level, as shown by Dakshini and Tandon (1981), each taxon seems to possess a tremendous potential for adaptation and microevolutionary adjustment to the dynamic environmental situations that are available in such habitat complexes. Taxa belonging to phanerogamic families such as Asclepiadaceae, Capparaceae, Euphorbiaceae, Malvaceae, and Poaceae have a large representation (Cooke, 1901–1908; Oza, 1961; Puri *et al.*, 1964; Jain, 1970; Bhandari, 1978; Sharma and Tyagi, 1979).

Additionally, from a floristic–phytogeographical analysis of these regions (Bharucha and Meher-Homji, 1965), the following characteristics can be noted: (1) endemism is very low; only 2.3% of the total species in the northern arid and semiarid areas, and only 1.7% in the southern arid and semiarid areas, are endemic; (2) in the southern region, the Indian element is 41% of the total species represented, while in the northern region it is only 23.4%; (3) the eastern Indo–Malayan element is higher (15%) in the southern region, as compared with 11.2% in the northern region; and (4) the North African element in the semiarid zone, located in the northern region, is 9.1%, but only 1.9% in the southern region. In the Thar Desert (Pakistan and India), however, the representation of the North African element in the flora is 16.2%. Thus, in general, the flora of arid and semiarid regions found in the northwest, and also applicable to Pakistan, is more varied than similar areas in the southern peninsular regions.

The diversities in the floristic composition of the different regions are, in a way, suggestive of the fact that the plant resource potentials have developed, evolved, been selected, and been stabilized under the different—either general or localized—habitat processes available in these zones. Interestingly, in the predominantly pastoral societies of these poorly industrialized regions, a social structure has evolved whereby the dependence of human beings and livestock on the scanty vegetational cover has become rather delicate and critical. The use of plants, or plant parts, controls or influences different facets of life (e.g., food, shelter, fodder, and religious, medicinal, psychological, and philosophical attitudes) of the inhabitants of these harsher environments. Gupta and Dutta (1967), Dakshini (1971), Bhandari (1974), Sen and Bansal (1979), Singh and Pandey (1979), and Tiwari (1979) have listed plants from these areas along with their diverse usages. In fact, each of the 1600 plant species (phanerogams) represented in these regions has some economic value. Since only native plants have been a permanent feature of the arid and semiarid environments, however, they have become an essential component of the life-style of the inhabitants of these regions. In the following pages only a few of these native plants of longstanding usage (as cited by Charak, 1975; Cooke, 1901–1908; Watt, 1908; Duthie, 1903–1929; Kirtikar and Basu, 1935; Chopra *et al.*, 1956; Oza, 1961; Gupta and Bhandari, 1965; Gupta and Dutta, 1967; Dakshini, 1971; CAZRI, 1972, 1974, 1976, 1979; Shah, 1978; Sharma and Tiagi, 1979) have been enumerated to highlight the economic significance of the plant resources available in these regions. The information for each species within a family is as follows: scientific name, sanskrit or local name, plant part used, and general remarks.

A. Food Plants

Anacardiaceae

Rhus myosorensis G. Don—*dansro*; ripe fruit; especially liked by nomad tribes and shepherds

Apocyanaceae

Carissa spinarum Linn.—*jangli karaunda*; fruits; used for pickles and chutneys

Burseraceae

Boswellia serrata Roxb. *ex* Colebr.—*shallaki, salai*; flowers, seeds; mixed with other vegetables

Cactaceae

Opuntia elatir Mill.—*thuar, vidara*; fruit pulp; eaten raw or fried

Curcubitaceae

Citrullus colocynthis (Linn.) Schrad.—*mahendra-varuni*; seeds; powdered seeds used for making a special bread locally called *sagra*

Citrullus lanatus (Thunb.) Matsumara and Nakai—*matiro*; fruit; unripe fruits also used as vegetable

Cucumis melo var. *momomordica* Cogn.—*ervaru*; fruits, seeds; used as vegetable

Cyperaceae

Cyperus rotundus Linn.—*musta*; tubers, roots; powdered tubers and roots used for making bread

Euphorbiacea

Securinega leucopyrus (Will.) Muell.—*svetakamboja*; leaves, berries; used only under scarcity conditions

Fabaceae (Leguminosae)

Acacia leucophloea (Roxb.) Willd.—*shveta-barbura*; ground bark, flowers; ground bark mixed with other seeds and vegetables; flowers used to make bread

Butea monosperma (Lamk.) Taub.—*palash*; fruits; eaten as vegetable

Indigofera cordifolia Heyne *ex* Roth.—*bekario*; seeds; powdered seeds used to make cakes

Prosopis cineraria (Linn.) Druce—*khejri*; fruits, seeds, young shoots, bark; young shoots and powdered bark eaten during periods of extreme food scarcity

Lamiaceae (Labiatae)

Leucas cephalotus (Roth.) Spreng.—*gomo*; leaves, seeds; tender leaves used as vegetable

Molluginaeae

Gisekia pharnacioides Linn.—*balo ka sag, morang*; leaves; used as vegetable

Moringaceae

Moringa oleifera Lamk.—*sohanjana*; leaves, fruit; used as vegetable

Poeceae (Gramineae)

Cenchrus biflorus Roxb.—*bukt*; seeds; very nutritious, with a high fat and protein content

Cenchrus ciliaris Linn.—*anjan, dhaman*; seeds; used for making porridge and bread

Cenchrus setigerus Vahl.—*anjan, dhaman*; seeds; used for making porridge

Lasiurus hirsutus (Forsk.)—*sevan*; seeds; ground seeds mixed with *Pennisetum typhoides* seeds to make flour

Polygonaceae

Calligonum polygonoides Linn.—*phog*; young buds; eaten with buttermilk or salt

Rhamnaceae

Zizyphus nummularia (Burm. f.) Wt.—*badari*; fruits; powdered dried fruits used for making bread

Rubiacea

Morinda citrifolia Linn.—*ashyuka, barachand*; fruit, used as pickles

Salvadoraceae

Salvadora pleoides Decne.—*pilu*; leaves, fruit; tender shoots and leaves eaten during times of scarcity

Salvadora persica Linn.—*brihat pilu*; leaves, fruits; tender shoots and leaves used as vegetable

Tiliaceae

Grewia abutilifolia Vent.—*gangetti*; leaves, fruits; used raw

Grewia tenax (Forsk.) Fiori; *gangan*; leaves, fruits; used raw

B. Forage Plants

Acanthaceae

Lepidagathis cristata Willd.—*aewel kangio*; tender twigs and leaves; plants relished by goats

Asteraceae (Compositae)

Pluchea lanceolata (DC.) C. B. Clarke—*rasna*; tender shoots; preferred under scarcity conditions, rich in nutrients

Bignonciaceae

Tecomella undulata (Sm.) Seem.—*rohiro*; tender leaves and flower buds; a general fodder plant, palatable

Boraginaceae

Sericostoma pauciflorum Stocks—*laggera*; tender shoots; relished by camels and goats

Cactaceae

Opuntia elatior Mill.—*thuar*; tender parts; rich in nitrogenous contents, usually supplemented with grasses or other fodder

Chenopodiaceae

Kochia indica Wight.—*bui*; tender portions of the plant; nutritious

Combrctaceac

Anogeissus pendual Edgew.—*kala dhau*; tender leaves; a palatable fodder

Fabaceae (Leguminosae)

Acacia nilotica (Linn.) Del.—*babul*; foliage and pods; a prolific tree produces 800 kg of pods per year; seeds germinate much better after passing through alimentary canal of cattle
Acacia planifrons Wt. and Arn.—*odai*; foliage; well suited to semiarid regions of South India
Acacia senegal (Linn.) Willd.—*kumbat*; foliage and pods; relished by camels and goats
Alhagi pseudalhagi (M. Bilo.) Desv.—*jawansa*; young leaves and pods; relished by camels and goats
Crotalaria burhia Buch.-Ham.—*shinio*; whole plant; relished by camels
Crotalaria medicagenia Lamk.—*gugaria*; whole plant; good for camels
Momosa hamata Willd.—*jinjani*; tender shoots; a general fodder
Prosopis cineraria (Linn.) Druce—*khejri*; tender shoots and pods; a valuable fodder
Rhynchosia aurea (Willd.) DC.—*batti*; foliage and young pods; a general fodder
Rhynchosia minima var. *laxiflora* (Camb.) Baker.—*chirimotio*; foliage and young pods; a climber with profuse growth

Menispermaceae

Anamitra cocculus (Linn.) W. and A.—*kakamari*; foliage and pods; a fodder under conditions of scarcity

Poaceae (Gramineae)

Aristida adscensionis Linn.—*lampro*; whole plant; tender shoots, palatable
Cenchrus ciliaris Linn.—*anjan, dhaman*; whole plant; usually yields 90 kg/ha of seed and 1239

kg/ha of foliage in one cutting, very palatable with a well-balanced ratio of calcium and potassium, and 11% protein

Cenchrus setigerus Vahl—*dhaman, anjan*; whole plant; relished by cattle, can stand cutting well, yields 112–157 kg/ha in the first year

Cymbopogon jwarancusa (Jones) Schult.—*lamajjaka*; whole plant, preferably younger parts; usually eaten when no other grass or fodder available

Dactyloctenium aegyptium (Linn.) P. Beauv.—*makro*; whole plant; a useful fodder

Dactyloctenium sindicum Boiss.—*tantia*; whole plant; a useful fodder

Desmostachya bipinnata (Linn.) Stapf.—*dab*; younger shoots; only buffalo prefer this grass

Eleusine compressa (Forsk.) Aschers. and Schweiuf. *ex* C. Christensen—*ghora dhab, chimber*; whole plant; usually found in sandy soils; compares well with *Cyanodon dactylon* as fodder

Elyonurus hirsutus (Vahl) Munro *ex* Benth.—*sewan*; whole plant; has a high carrying capacity, a protein content of 9%, and a well-balanced calcium/potassium ratio

Eragrostis ciliaris var. *brachystachya* Boiss.—under punch; whole plant; common on eroded soils, heavy seed producer

Lasiurus ecaudatus Saty. and Shanker—*shevan*; whole plant; a valuable fodder

Melanocenchris jacquemontii Jaub. and Spach.—*anjan*; whole plant; usually preferred only under fodder scarcity conditions

Panicum turgidum Forsk.—*munt*; whole plant; palatable

Sehima nervosum (Willd.) Stapf.—*seran, parna*; whole plant; relished, can withstand cutting well

Rhamnaceae

Zizyphus nummularia (Burm. f.) Wt.—*badari*; tender leaves; a general fodder

Salvadoraceae

Salvadora oleoides Decen.—*pilu*; foliage and tender shoots; a reliable fodder

Salvadora persica Linn.—*brihat pilu*; foliage and tender shoots; a reliable fodder

Simaroubaceae

Ailanthus excelsa Roxb.—*mahanimba, maharukh*; foliage; coppices well

Tiliaceae

Grewia tenax (Forsk.) Fiori—*gangan*; tender twigs and foliage; liked by camels and goats

C. Medicinal Plants

Acanthaceae

Barleria prionites Linn.—*karunta*; leaves, stem, root; used to treat catarrhal affections, boils, toothache, paste of leaves used in hot poultices, also used in steam baths to cure stiffness of limbs, sciatica, and enlargement of the scrotum

Blepharis indica T. Anders.—*shikhi*; seeds, whole plants; expectorant

Lepidagathis cristata Willd.—*aewal, kangio*; whole plant; used as a tonic to cure itchy affections of the skin

Apocynaceae

Wrightia tinctoria R. Br.—*svetakutaja*; bark, seeds, stem; used as a tonic and aphrodisiac

Asclepiadaceae

Sarcostemma acidum (Roxb.) Voigt—*soma, khirkhimp*; stem, root, whole plant; alternative cooling agent

Asteraceae (Compositae)

Dicoma tomentosa Cass.—*vajradanti*; root, plant; used as a febrifuge, for putrescent wounds, and roots used as toothbrush

Boraginaceae

Heliotropium strigosum Willd.—*kali Bui*; plant; used as a laxative, and for pain in limbs, sore eyes, boils, ulcers, and wounds

Burseraceae

Boswellia serrata Roxb. *ex* Colebr.—*shallaki*; bark, gum; diaphoretic, diuretic; for rheumatism, nervous disorders, and skin diseases

Commiphora wightii (Arnott) Bhandari—*gugala*; bark; astringent, antiseptic, appetite stimulant

Cactaceae

Opuntia elatior Mill.—*vidara, thuar*; fruit, leaves, plant; antibiotic, and used for whooping cough, as a purgative, and for ophthalmia

Capparaceae

Capparis decidua (Forsk.) Edgew.—*karira*; leaves, bark, fruit, root; used for boils, swellings, cough, asthma, cardiac troubles, rheumatism, and as a diaphoretic

Chenopodiaceae

Haloxylon recurvum (Moq.) Bunge *ex* Boiss.—*khar*; plant; used to treat intestinal ulcers

Kochia indica Wight—*bui*; plant; used as a cardiac stimulant

Convoluvulaceae

Cressa cretica Linn.—*rudanti*; leaves, roots, seeds; used as a tonic, aphrodisiac, expectorant, and antibilious medication

Cucurbitaceae

Citrullus colocynthis (Linn.) Schrad.—*mahendra varuni*; seeds, root; used as a purgative, and for ascites, jaundice, and rheumatism

Euphorbiaceae

Euphorbia caducifolia Haines—*thuar*; young branches; smoke from the burning of young twigs inhaled to cure asthma

Fabaceae (Leguminosae)

Acacia senegal (Linn.) Willd.—*svetakhadira*; plant gums; used as a demulcent and emulsifying agent to treat skin inflammations

Butea monosperma (Lamk.) Taub.—*palasha*; plant, flowers, seeds; used as an astringent, and an anthelmintic, for checking roundworms and tapeworms

Crotalaria burhia Buch.-Ham.—*shimio*; branches, leaves; used as a coolant

Crotalaria medicagenia Lamk.—*gugario*; whole plant; used as a tonic

Dichrostachys cinerea (Linn.) Wt. and Arn.—*viravriksha*; root, stem; used as an astringent, and to treat rheumatism and eye diseases

Mimosa rubicaulis Lamk.—*shiah-kanta*; leaves, roots, plant; used to treat piles, and applied to burns; root ash powder used to check vomiting due to weakness

Prosopis cineraria (Linn.) Druce—*shami, khejri*; flowers, fruits, bark, branches; used to treat rheumatism, and by women during pregnancy as a safeguard against miscarriage

Flacourtiaceae

Flacourtia sepiaria Roxb.—*pracinamaloka*; root, bark, leaves; used for snake bite, and as a liniment for gout and rheumatism

Securinega leucopyrus (Willd.) Muell.—*svetakamboja*; leaves, whole plant; used to destroy worms in sores, and as fish poison

Lamiaceae (Labiatae)

Leucas cephalotus (Roth.) Spreng.—*gomo*; whole plant; used to treat fever, and pain in joints

Leucas utricaefolia (Vahl) R. Br.—*paniharai*; whole plant; used to treat colds, fever, and gastrointestinal troubles

Menispermaceae

Anamitra cocculus (Linn.) Wt. and Arn.—*kakamari*; whole plant, fruit; used for the treatment of inflamed surfaces, and as fish poison; fruit used as an antidote for poisoning; occurrence of picrotoxin reported

Molluginaceae

Mollugo nudicaulis Lamk.—*rangatio khar*; leaves, whole plant; applied to boils to draw out pus, and use for whooping coughs and athrepy

Poaceae (Gramineae)

Chionachne koenigii (Spreng.) Thw.—*kanda*; whole plant; toxic, emetic; used to treat diseases of the blood, biliousness, and hemorrhagic diathesis

Cymbopogon jwarancusa (Jones) Schutl.—*lamajjaka*; root, whole plant; used to purify blood, as a tonic, and to treat cough and rheumatism

Desmostachya bipinnata (Linn.) Stapf.—*darbha*; plant; used to treat dysentery and menorrhagia, and as a stimulant

Rhamnaceae

Zizyphus nummularia (Burm. f.) Wt.—*badari*; leaves, fruits, roots; used for curing scabies, boils, bilious affections, old wounds, and ulcers

Rubiaceae

Morinda citrifolia Linn.—*ashyuka*; root, leaves, fruit; stimulant; used for fever, epilepsy, rheumatism, diseases of the liver and spleen, dysentery, and leukorrhea

Salvadoraceae

Salvadora oleoides Decne.—*pilu*; fruits, leaves, seeds; used as a purgative, aphrodisiac, and for treating rheumatism

Salvadora persica Linn.—*brihatpilu*; leaves, bark, fruits; used to treat asthma, cough, rheumatism, and as a stimulant and tonic

Simaroubaceae

Ailanthus excelsa Roxb.—*mahanimba*; bark; used as febrifuge and tonic for chronic bronchitis, diarrhea, and dysentery; ailantic acid present; decoction of leaves used in menstrual disorders

Balanites aegyptiaca (Linn.) Delile—*ingudi, hingot*; seeds, fruits, bark, whole plant; used to treat cough and colic, and as an anthelmintic and purgative; juice used as fish poison; whole plant used for snake bite; plant ash used for dermatosis, urinary diseases, and for rejuvenation; a source of diosgenin

Tamaricaceae

Tamarix troupii Hole—*jahuro*; bark, branches, fruits; used to treat dysentery, and as a laxative and expectorant

Tiliaceae

Grewia tenax (Forsk.) Fiori—*gangan*; wood, stem; used to treat cough, and pain in the sides

Verbenaceae

Clerodendrum phlomidis Linn. f.—*vataghni*; root, leaves, whole plant; source of a bitter tonic used for neglected syphilitic complaints

D. Industrial Plants

1. FIBER PLANTS

Asclepiadaceae

Calotropis gigantea (Linn.) R. Br.—*moto-aak*; stem fiber usually used for making ropes

Calotropis procera (Ait.) R. Br.—*aak*; cordage and ropes made from stem fibers; floss from seeds used for stuffing pillows and for textiles

Leptadenia pyrotechnica (Forsk.) Decne.—*khimp*; used for rope making

Cactaceae

Opuntia elatior Mill.—*hatha-thor, thuar*; coarse fiber obtained from stem

Ehretiaceae

Cordia dichotoma Forst. f.—*lassaura*; fiber obtained from bark

Fabaceae (Leguminosae)

Acacia leucophloea (Roxb.) Willd.—*shveta-barbura, urajio*; coarse cordage fibers obtained from bark

Butea monosperma (Lamk.) Taub.—*palash*; contains bark fibers up to 30 cm in length
Crotalaria burhia Buch.-Ham.—*shinio*; fiber obtained from stem

Malvaceae

Sida cordifolia Linn.—*bala*; undershrub yields excellent fiber

2. FUEL PLANTS

Amaranthaceae

Aerva persica (Burm. f.) Merrill—*buni*; low shrub

Ascleipiadaceae

Leptadenia pyrotechnica (Forsk.) Decne.—*khimp*; shrub; common in sandy, gravelly habitat
Sarcostemma acidum (Roxb.) Voigt—*khirkhimp, soma*; shrub; good source of fuel

Burseraceae

Boswallia serrata (Roxb.) and Colebr.—*shallaki, salai*; medium-sized tree

Capparaceae

Capparis decidua (Forsk.) Edgew.—*karira*; many-branched shrub to small tree

Chenopodiaceae

Haloxylon recurvum (Moq.) Bunge *ex* Boiss.—*khar*; small tree, woody near the base, gregarious

Euphorbiaceae

Euphorbia caducifolia Haines—*thuar*; small tree
Securinega leucopyrus (Willd.) Muell.—*ghat-bor*; shrub

Fabaceae (Leguminosae)

Acacia jacquemontii Benth.—*bu-banwali*; medium-sized tree
Acacia nilotica (Linn.) Del.—*babul*; medium-sized tree
Acacia planifrons Wt. and Arn.—*odai*; medium-sized tree; excellent fuel
Acacia senegal (Linn.) Willd.—*kumbat, kuntia*; medium-sized tree
Prosopis cineraria (Linn.) Druce—*khejru, Vidara*; medium-sized tree, coppices well; good fuel

Moringaceae

Moringa oleifera Lamk.—*sohanjana*; tree, easy to propagate by cuttings

Polygonaceae

Calligonum polygonoides Linn.—*phog*; shrub

Rhamnaceae

Zizyphus nummularia (Burm. f.) Wt.—*badri*; common small shrub

Rubiaceae

Mitragyna parviflora Korth.—*kain*; tree

Simaroubaceae

Balanties aegyptiaca (Linn.) Delile—*hingota*; shrubby tree, propagates by root, gregarious

Tiliaceae

Grewia tenax (Forsk.) Fiori—*gangan*; small shrubby tree, coppices well

3. TIMBER PLANTS

Apocynaceae

Wrightia tinctonia R. Br.—*dudhi*; used for turnery and making lacquered toys

Bignoniaceae

Tecomella undulata (Sm.) Seem.—*rohiro, reodana*; source of strong, durable timber highly prized for making furniture, carving work, and agricultural implements

Burseraceae

Boswellia serrata (Roxb.) and Colebr.—*shallaki, salai*; source of industrial timber, but not very strong; used for small poles; seasons rapidly, decays rapidly in the open

Combretaceae

Anogeissus acuminata (Roxb.) Wall. *ex* Beb.—*chakwa*; used to produce inferior furniture; tree regenerates well through seeds

Anogeissus latifolia Wall.—*dhau*; used for ax handles, poles, cart axles, furniture, and ship building; recommended for studs; supposed to be fire resistant

Anogeissus pendula Edgew.—*kala dhau*; similar to *A. latifolia* in use; coppices well

Anogeissus rotundifolia Blatt. and Hall.—*dhauro*; a beautiful tree; used for furniture

Ehretiaceae

Crodia dichotoma Forst. f.—*lasaura*; a soft wood but fairly strong; seasons well; used for boat building, well curbs, gun stocks, agricultural implements, building purposes, and light packing cases

Cordia gharaf (Forsk.) Ehrenb. and Aschers—*goohdni*; used in the manufacture of agricultural implements

Fabaceae (Leguminosae)

Acacia nilotica (Linn.) Del—*babul*; not very good for timber purposes; used for small poles and inferior furniture

Acacia planiforns Wt. and Arn.—*odai*; hardwood; used for agricultural implements

Acacia senegal (Linn.) Willd.—*kumhat, kuntia*; used for small poles and weaving shuttles

Prosopis cineraria (Linn.) Druce—*jhand, khejri, vidara*; a relatively strong wood, used in locomotives

Flacourtiaceae

Flacourtia sepiaria Robx.—*shrawani, fracinamaloka*; not a very good timber

Moringaceae

Moringa oleifera Lamk.—*sohanjana*; a very soft wood, used for making floats and toys; recommended for use in the textile industry

Polygonaceae

Calligonum polygonoides Linn.—*phog*; used for building huts and well curbs

Rubiaceae

Mitragyna parviflora Korth.—*kadam, kain*; used for industrial purposes
Morinda citrifolia Linn.—*ashyuka, barachand*; used to make general furniture

Simaroubaceae

Ailanthus excelsa Roxb.—*mahanimba, mahrukh*; small poles good for light packing cases and match splints; wood easily perishable in the open

4. DYE AND TANNIN PLANTS

Apocynaceae

Wrightia tinctonia R. Br.—*kerno*; leaves yield a blue dye

Chenopodiaceae

Kochia indica Wt.—*bui*; herb; whole plant used in dyeing

Combretaceae

Anogeissus pendula Edgew.—*dhau*; dark green extract of leaves used in dyeing

Fabaceae (Leguminosae)

Acacia nilotica (Linn.) Del. subsp. *indica* (Benth.) Brenan—*babul*; moderate-sized evergreen tree, wood used for tanning purposes
Butea monosperma (Lamk.) Laub.—*palash*; fresh and dried flowers yield a light yellowish dye used for dyeing cloth and other materials; the water extract of flowers also used during the color festival
Cassia auriculata Linn.—*anwal*; an important tanning material

Rubiaceae

Morinda citrifolia Linn.—*ashyuka*; stems and roots yield a bluish dye, it is commonly believed that white ants do not destroy anything dyed with it.

5. GUM, WAX, AND RESIN PLANTS

Asclepiadaceae

Sarcostemma acidum (Roxb.)—*khir-khimp*; leafless, straggling shrub

Burseraceae

Boswellia serrata Roxb. *ex* Colebr.—*salai*; deciduous medium-sized tree; fragrant gum–resin obtained from bark

Commiphora wightii (Arnott) Bhandari—*guggal*; many-branched shrub; pale brown or dull green gum–resin obtained from bark during the cold season; source of ''Indian Bedellium''

Combretaceae

Anogeissus pendula Edgew.—*dhau*; moderate-sized tree; yields inferior quality gum; used for cloth printing

Fabaceae (Leguminosae)

Acacia jacquemontii Benth.—*bu-banvali*; moderate-sized tree; source of the best edible gum

Acacia nilotica (Linn.) Del. subsp. *indica* (Benth.) Brenan—*babul*; moderate-sized tree; gum used as gum arabic, and for calico printing

Acacia senegal (Linn.) Willd.—*kumbat*; small prickly tree; source of true gum arabic

Butea monosperma (Lamk.) Taub.—*palash*; small tree; ''Ruby gum'' obtained from bark

Prosopis cineraria (Linn.) Druce—*khejri, vidara*; large, many-branched shrub to a small tree; inferior quality gum

Moringaceae

Moringa oleifera Lamk.—*sohanjan*; medium-sized tree; white-gray to mahogany colored gum, seeds yield an oil used in perfume

Salvadoraceae

Salvadora oleoides Decne—*pilu*; many-branched evergreen tree

Salvadora persica Linn.—*brihat-pilu*; small evergreen tree

Simaroubaceae

Ailanthus excelsa Roxb.—*maharukh*; large tree; yields superior quality gum

E. Other Plants

Dried branches of *Alhagi pseudoalhagi* and *Calligonum polygonoides* are used to form screens that, after wetting with water, help ward off dessicating hot winds.

Dried plants of *Calligonum polygonoides, Leptadenia pyrotechnica,* and Zizyphus nummularia are often used for fencing fields.

Dried powder of the leaves of *Balanites aegyptiaca, Haloxylon recurvum,* and *Suaeda maritima* are often used for washing clothes.

Wooly empty perianths of *Aerva tomentose* are used for stuffing pillows.

Gums from *Boswellia serrata* and *Commiphora wifhtii* are used as incense, especially during religious ceremonies and for general fragrance.

Dried twigs of *Acacia planifrons, Calotropis gigantia, C. procera, Capparis decidua, C. spinosa,* and *Prosopis cineraria* are often used to thatch roofs.

The empty woody fruits of *Balanites aegyptiaca* are filled with gunpowder and used as fireworks.

Various plant parts and products of the following plants are used for different religious ceremonies: *Acacia senegal, Barleria prionites, Butea monosperma, Capparis decidua, Crataeva nurvala, Cucumis sativus, Desmostachya bipinnata, Prosopis cineraria,* and others.

V. SUMMARY

A. Ecology

The vegetation of arid and semiarid regions is scanty, xerophytic, and widely dispersed (Dakshini, 1971). Cold climate steppes and hot climate thorn forests are characteristic of the subdeserts and semiarid regions (Meher-Homji, 1977). An analysis of the biological spectrum of these regions reveals the dominance of therophytes in arid regions of India and Pakistan. The semiarid regions of both northern and southern parts, on the other hand, support a thero–chaemophytic vegetation (Table VIII) (Meher-Homji, 1962). Plant communities in such regions are less varied in structure and diversity in terms of plant species. Vegetation types of these regions, in the Indian subcontinent, have been described by Sarup (1952), Biswas and Rao (1953), Mathur (1960), Legris (1963), Bharucha and Meher-Homji (1965), and Meher-Homji (1977). Champion and Seth (1963), in a detailed treatment classify the vegetation of these regions under two major types: (1) dry deciduous type and (2) tropical thorn forests. Each of these types has been discussed by Champion and Seth in the separate context of northern and southern regions. These types have been treated as "edaphic climaxes." However, transitory successional stages are also indicated (see also Dakshini, 1971).

Gaussen *et al.* (1968, 1972) and Meher-Homji (1977) classified the vegetation of these regions under the following six types (the major criterion followed in this classification is the decreasing order of aridity): (1) *Calligonum polygonoides,* (2) *Prosopis cineraria, Capparis decidua, Zizyphus, Salvadora,* (3) *Acacia, Capparis decidua,* (4) *Acacia senegal, Anogeissus pendula,* (5) *Acacia catechu, Anogeissus pendula,* and (6) marginal sub-dry vegetation type comprising *Anogeissus pendula, A. latifolia, Albizzia amara,* and *Hardwickia binata.* The first two types, as indicated by *Calligonum polygonoides* and *Prosopis cineraria, Capparis decidua, Zizyphus,* and *Salvadora,* occur in the arid regions of India and Pakistan, and the other types are found in the semiarid regions of the Indian subcontinent. The salient characteristics of these types are summarized in Table IX.

A critical assessment of these types, their distribution, and habitat features clearly brings out a delimitation of the regions under various subzones depending on the repetition of the factors and the vegetation types. Due to such a correlative

Table VIII

Biological Spectrum of the Arid and Semiarid Regions of the Indian Republic[a]

Region	No. of dry months	Percentage representation								
		Ph	N	Ch	H	G	HH	Th	L	E
World		28.0	15.0	9.0	26.0	4.0	2.0	13.0	—	3.0
West India	11–12	9.1	4.6	18.9	15.5	3.4	—[b]	40.0	7.8	—
North India (semiarid)	9–10	11.0	12.0	18.0	10.0	5.0	3.0	33.0	6.0	—
South India (semiarid)	8–9	15.0	15.0	12.0	11.0	—	4.0	28.0	—	—

[a]From Meher-Homji (1962). Ph, Phanerophytes; N, Nannophanerophytes; Ch, Chamaephytes; H, Hemicryptophytes; G, Geophytes; HH, Hydrophytes; Th, Therophytes; L, Lianas; E, Epiphytes.

[b]Not represented.

Table IX

Vegetation Types of Arid and Semiarid Regions of the Indian Subcontinent[a]

Vegetation type	Regions	Habitat characteristics	Characteristic association
1. *Calligonum polygonoides*	Arid (India and Pakistan)	Annual average rainfall less than 200 mm; dry season exceeds nine months; dune slopes	*Acacia jacquemontii* (on low dunes), *Cyperus areuarius, Dipterygium glaucum, Haloxylon salicornicum* (interdunal areas), *Leptadenia pyrotechnica* (gravelly substratum), *Rhynchosia arenaria*
2. *Prosopis cineraria*	Arid and semiarid (Barmer, Bikaner, Jaisalmer, Jodhpur, Nagaur, Ja lni, Pali, Kutch, etc.)	Annual average rainfall 150–400 mm; 9–11 dry months; on old alluvial, sandy soils	*Acacia nilotica, Acacia senegal* (on rocky substratum), *Aerva persica, A. pseudotomentosa* (sand with lime or gypsum), *Balanites aegyptiaca, Calotropis procera, Cenchrus ciliaris, Cymbopogon jwarancusa* and other grasses, *Ephedra foliata, Lasiurus indicus, Maytenus emarginata, Salvadora persica*
3. *Acacia–Capparis*	Semiarid (Sirohi Jalore, Pali, Broach, Baroda, Kaira, Ahmedabad, Kathiawar region, Mehsana, Dhulia to Miraj, Bijapur, Gulburga)	Annual average rainfall 400–700 mm; 8–9 dry months; sandy soil; trap to gneissic complex, shale, sandstone quartzite, chlorite schists; several soil types; black-red alluvial; calcimorphic	*Acacia leucophloea, A. nilotica, A. planifrons* (in Kallinaurai, Tamil Nadu, Bellary), *Balanites aegyptiaca, Capparis decidua, Cassia auriculata, Commiphora mukul, Cordia gharaf, Euphorbia caducifolia, Flacourtia indica, Grewia tenax, Lycium europaeum, Taverniera num-*

Table IX *(Continued)*

Vegetation type	Regions	Habitat characteristics	Characteristic association
			mularia, Tecomella undulata, Xeromphis spinosa, Zizyphus spp.
4 and 5. *Acacia senegal, Anogiesus pendula* and *Acacia catechu, Anogeissus pendula*	Semiarid (on Aravallis, Sirohi, Pali, Ajmer, Jalni Barmer, Alwar, Kota, Chittorgarh, Udaipur)	Annual average rainfall 400–700 mm; 8½–10 dry months; Soil brown-black; (*A. senegal* in nonhumid regions)	*Aegle marmelos, Balanitis aegyptiaca, Bauhinia racemosa, Boswellia serrata, Butea monosperma, Commiphora mukul, Cordia gharaf, Crataeva religiosa, Dendrocalamus strictus, Dichrostachys cinerea, Diospyros melanoxylon, Dyerophytum indicum, Euphorbia caducifolia, Flacourtia indica, Grewia flavescens, Holoptelea integrifolia, Holarrhena antidysenterica, Maytenus emarginata, Mimosa rubicaulis, Mitrgyna parviflora, Moringa concanensis, Rhus mysorensis, Sericostoma pauciflorum, Soymida febrifuga, Zizyphus mauritiana*
6. Marginal subdry vegetation	Semiarid (Circar plains of Tamil Nadu and Andhra Pradesh, Mysore, part of Udaipur, Sirohi)	Annual average rainfall 600–1500 mm; 6–8 dry months; Ferruginous to clayish black soils	*Albizzia amara, Hardwickia binata*

[a]From Gaussen *et al.* (1968, 1972) and Meher-Homji (1977).

evolution and development, the habitats have become so distinctive that each location acquires its own characters. This is amply illustrated when the different landforms, hills, different pediment types, older and younger alluvial plains, different types of sand dunes, intradunal plains, and shallow saline depressions that have been created over the years are considered along with their ecological potentials, especially edaphic factors and available water resources (Ghosh *et al.*, 1977a). Each one of these landforms provides a specific substratum for both evolution and adjustments of the vegetational components. These landforms are ecologically related. Further, like any other ecological situation, the arid and semiarid habitats (landforms) are also dynamic in terms of composition, structure, and behavior of edaphic and biological components. The successions of habitat features and vegetational components that have taken place or are taking place in these landforms have been controlled both by natural causes and the indigenous biotic stresses. In fact, through these processes the vegetation is able to utilize the inherent but cryptic potentials of each of the habitats (Tandon, 1977). An understanding of these phenomena are a must for an economical utilization of the untapped resources that are available in these regions, which are under severe ecological stress. Interesting data, collected by the Central Arid Zone Research Institute (CAZRI) in Jodhpur, have shown that minor habitat changes may lead to many responses by plants. Thus, it was found (CAZRI, 1976) that the density and biomass of *Capparis decidua* in an undisturbed *Prospis–Capparis* stand changed with slight variation in the composition of the vegetational cover under the influence of the diverse dynamic processes of this habitat; changes were from 10 to 15 ha and 7.6 to 14.4 ha, respectively, to 302 plants ha and 55.1 plants/ha, respectively. Similarly, with degradation of the habitat the grass cover of *Eleusine compressa* and *Aristida funiculata* is replaced by *Tephrosea purpurea* and *Oropetium thomaem;* during this succession the biomass and vegetative cover values changed from 2.3% and 15 plants/ha to 0.2 to 0.7% and 0.35 plants/ha, respectively (CAZRI, 1976). Dakshini (1971) has summarized data pertaining to total acid number and other chemical constituents that change with variation of the habitat. Also, variation of the hyposcyamine content of *Datura innoxia,* collected from different habitats in Rajasthan, (Table X) has been reported (CAZRI, 1976).

Furthermore, in addition to the role of extremes and erratic behavior of the climatic components in these regions, the ecological implications of the data, similar to those summarized in Tables XI and XII in modulating the habitat process will have to be appreciated for a profitable harnessing of the resources available in these regions.

Additionally, the role of biota—livestock, in particular—in changing or controlling the ecology of these areas is far more significant and requires immediate attention. Wild animals in the present unfavorable environment are rather few and may not be of any consequence in drastically changing the ecology. Howev-

Table X

Hyoscyamine Content in *Datura innoxia*[a]

Location	Percentage of active principle calculated as hyoscyamine
Bikaner	0.0975
Kolayat	0.208
Jodhpur	0.30

[a]Time of collection of *Datura*, June 1976. From CAZRI (1972).

Table XI

Available Nutrient Contents in Field with Light-Textured Soils and Fence Line Depositions[a]

		Available nutrients (means and standard deviations)		
Site	Number of samples	Organic C (%)	P/kg/ha (average)	K/kg/ha (average)
Field proper	42	0.15 ± 0.05	13.00 ± 2.43	138.0 ± 6.2
Fence line depositions	46	0.10 ± 0.06	16.70 ± 7.60	176.0 ± 73.0

[a]From CAZRI (1972).

Table XII

Ionic Components of Rainwater at Some Sites in Western Rajasthan (Means and Ranges)[a]

Site	pH	Conductivity (μ mhos/cm^2)	Sodium (ppm)	Potassium (ppm)
Jodhpur	7.5 (7.3–7.7)	33 (20–76)	4.5 (0.75–15.00)	1.20 (0.50–4.0)
Pali	8.2 (7.7–9.2)	183 (6.5–383)	52.5 (11.5–160)	2.29 (0.5–29)
Jaisalmer	7.4 (6.6–7.8)	90 (40–227)	11.7 (2.5–31.75)	4.28 (1.0–9.5)
Bikaner	7.8 (7.5–8.1)	47 (28–109)	22.87 (7.25–56.75)	5.85 (2.0–16.2)
Palsana	8.9 (7.4–8.4)	83 (21–276)	11.5 (1.5–26.75)	4.85 (1.0–14.5)

[a]From CAZRI (1972).

er, the density of livestock varies throughout these regions. Raheja and Sen (1964) found a positive correlation between the degree of aridity, grass cover, and sheep and goat populations. Thus, to a larger extent, livestock control can modify the stability of the productive potential of these regions and indirectly influence the pattern of life of their inhabitants.

Livestock population far exceeds human population, and in arid and semiarid regions of Rajasthan (India) alone it has increased from 9.4 million in 1951 to 28 million in 1977–1978 (Sharma, 1972). During this period, however, the goat and sheep population has increased from 57.1 to 68.3%, while the overall cattle population has decreased by ~11%. Interestingly, through this period, although the degree of drought increased, on a relative basis the goat population showed a maximum increase of 34% (Sharma, 1972). Such a rise in the sheep and goat population brings about the hazards of erosion and an increase in unpalatable plant species such as *Crotalaria burhia, Cyperus* spp., *Indigofera cordifolia, Tephrosia purpurea, Tribulus terrestris,* and shrubby species like *Acacia leucophloea, Balanites aegyptiaca, Flacourtia indica, Lycium europaeum,* and *Mimosa hamata.* It is likely that in other regions, in a similar way, the increase of cattle and other large animals brings about a biotic elimination of desirable plants by controlling their regeneration. Incidentally, with this vegetational change and the destruction of useful and edible plant species the ecological equilibrium of the rodent population is disturbed so much that the depleted resources in their own habitats compel these rodents to extend and even change their area of movement. This enhances biotic pressure on new areas. Further, because of the scarcity of food in the neighborhood, these rodents collect in very large numbers the seeds and other propagules of desired plants to the extent that the regeneration of natural vegetation is drastically affected. The potential damage to the ecology of the habitat by these rodents may be appreciated by the following observations by Mann (1977b) on *Gerbillus gleadowi,* which is abundant in sand-dune habitats: "It is estimated that these gerbils alone excavate 61,500 kg of soil/day/cm^2 in cultivated fields during summer and 1,043,800 kg/day/km^2 in uncultivated areas."

In addition to the above, a perusal of Table XIII provides valuable information on the general ecological conditions of the available areas due to different land utilization practices, as well as the magnitude of the different ecological problems that exist under these conditions. In fact, a thorough understanding of the available resources is required for their proper management and optimum use.

B. Projections

Acharya *et al.* (1977) noted, "In the arid region of Rajasthan the estimated forage (dry matter) from the range land, top feed, crop residues, etc., is about 15.41 million [metric] tonnes against the total requirement of 17.46 million

Table XIII

Land Utilization Statistics (1969–1970) for Arid and Semiarid Regions (Percentage of Reported Area)[a]

State	Forest		Area under nonagricultural uses		Permanent pastures and other grazing lands		Land under miscellaneous trees, crops, and groves		Cultivable waste land		Fallow land		Net sown area	
	Arid	Semi-arid	Arid	Semi-arid	Arid	Semi-arid	Arid	Semi-arid	Arid	Semi-arid	Arid	Semi-arid	Arid	Semi-arid
Andhra Pradesh	15.67	20.67	6.00	7.67	1.77	5.67	0.70	1.70	5.00	4.00	19.19	19.85	51.67	40.44
Gujarat	5.93	5.18	2.80	3.53	6.25	8.32	0.15	0.14	3.08	2.30	33.94	10.44	47.85	70.04
Haryana	1.18	3.25	6.72	15.15	0.27	2.10	0.00	0.30	2.27	1.15	5.05	2.50	84.51	75.55
Jammu and Kashmir	0.10	17.30	50.50	19.18	1.60	5.73	2.00	3.05	15.30	7.80	1.20	6.58	29.30	40.36
Karnataka	7.40	5.10	5.40	5.90	5.90	9.30	0.90	1.40	3.50	3.30	17.60	12.60	59.30	62.40
Madhya Pradesh	—[b]	27.70	—	11.60	—	7.80	—	0.00	—	3.50	—	1.80	—	47.60
Maharashtra	2.16	10.26	3.56	6.00	2.89	5.63	0.53	0.22	0.33	3.30	8.33	2.09	82.77	71.87
Punjab	1.50	3.48	7.00	14.64	0.00	0.12	0.00	0.06	2.00	1.61	2.61	2.02	36.89	78.06
Rajasthan	1.36	5.65	3.28	5.14	4.43	7.59	0.00	0.05	14.33	10.53	31.01	23.73	45.54	57.26
Tamil Nadu	—	12.56	—	11.66	—	1.34	—	1.36	—	4.16	—	17.63	—	51.29
Utter Pradesh	—	1.81	—	7.69	—	0.34	—	1.20	—	4.07	—	3.08	—	77.41

[a] From Report of the National Commission on Agriculture (1976).

[b] Not represented.

tonnes, showing a shortfall of 2.05 million tonnes.'' Similarly, agricultural produce is also very limited—so much so that everything produced is consumed locally in the region and nothing is available to be exported from the area. The existing social customs, joint-family system, social values for more children, and the subsistence-oriented agricultural practices further augment the scarcity conditions. Further, the economic conditions worsen not only because of the higher growth rate of human and livestock populations, but also because of changing preferences for professions. Regional disparities in the distribution of agricultural products and livestock bring about degradation of the habitat and a multitude of other related problems. All such phenomena lower the productivity of these regions and make the habitat or substratum more unstable, thus adding to the existing hostile environment. These interrelated ecological, social, and economic problems call for proper planning and development of the following: (1) fodder, (2) agriculture, (3) horticulture, (4) social forestry and medicinal plants, (5) identification of plant indicators of geological, soil, and water resources, (6) sedentarization of nomadic tribes, and (7) stabilization of sand dunes. Scientists at the Central Arid Zone Research Institute in Jodhpur have collected valuable data during the past 10 years on some of the aspects mentioned above (CAZRI publications). However, similar data must be collected from different arid and semiarid regions in order to harness the vast available resources. Based on these earlier observations, the following comments are made in order to draw attention to the need and potentials available in these regions.

1. FODDER

Because the pastoral system is the life-style of the majority, fodder is one of the most important requirements. Dairy products, livestock, hides, wool, and other products are the major sources of income to a household (Sharma, 1972). Development of pastures, elimination of the destruction of existing native vegetation, and selection of some rapidly growing species must be planned. Trees and shrubs provide not onlyshade on the range lands, but fodder as well during the lean periods of the year. The nutrient qualities of top feed vary (Table XIV), however, and this, in addition to the palatability and acceptance of the feed by livestock, must be kept in mind while selecting and multiplying particular plant species. Grass cover is mostly perennial in these areas and therefore provides a tremendous potential for development. The average forage yields of various grass covers occurring in different substrata are 473 kg/ha (sandy to gravelly substratum with an annual rainfall below 200 mm), 478 kg/ha (gravelly substratum with 300 to 400 mm rainfall), 925 kg/ha (deep sandy areas with more than 400 mm rainfall), and 1630 kg/ha (in shallow to heavy rocky substratum in areas with more than 400 mm rainfall). It was further noted that the average yield of forage under protected grazing conditions increases by about 198.3, 91.9, and

Table XIV

Nutrient Status (Percent of Dry Matter) of Plants Inhabiting Arid and Semiarid Regions in Rajasthan[a]

	Crude protein	Fiber	Nitrogen free extract	Ash	P	Ca^{2+}	Mg^{2+}
Acacia niltica	13.9	9.2	69.8	7.1	0.1	2.6	0.4
Acacia senegal	10.3	9.5	65.7	16.4	0.05	6.9	0.6
Anogeissus pendula	7.6	19.0	65.3	8.1	0.1	3.5	0.3
Aristida funiculata	2.39	33.2	53.31	15.05	0.15	2.3[b]	—
Calligonum polygonoides	7.4	20.7	64.2	7.6	2.1	2.4	0.8
Cenchrus ciliaris	6.50	35.19	45.73	12.69	0.41	0.44	—
Dactyloctenium aegyptium	7.25	33.74	45.32	8.65	0.21	0.65	—
Prosopis cineraria	13.9	20.3	59.2	6.5	0.2	1.9	0.5
Salvadora oleoides	9.6	9.3	40.2	40.8	0.1	11.9	0.7
Salvadora persica	14.2	9.4	44.9	31.4	0.2	3.3	0.7
Tecomella undulata	11.7	16.0	65.0	7.2	0.2	1.6	0.3
Zizyphus nummularia	11.7	16.6	65.0	7.2	0.2	1.6	0.3

[a] From Ganguli *et al.* (1964).
[b] Not available.

116.3% in poor, fair, and good conditions of range lands, respectively (Ahuja, 1977, 1978). *Cenchrus ciliaris, C. setigerus, Dichanthium annulatum,* and *Lasirus sindicus* are the grasses found in abundance on different landforms. The nutritive values of these grasses have been shown to vary with the habitat type (Gupta, 1975; Ahuja, 1977). Some of these grasses are also good for hay production. Interestingly, the optimum forage production of these grasses under reseeding practices increases from 45.4 to 107.3% (CAZRI, 1979). The success of reseeding and the occurrence of specific plants or groups of plants in diverse landforms suggest the acceptability of these plants for each habitat. However, this relationship has evolved through the years and its delicate balance must be clearly understood before any development of fodder resources is planned. The management, selection, maintenance, and improvement of fodder resources will succeed only if these measures go hand in hand with the specialized ecology of the landforms present in these regions.

2. AGRICULTURE

Local varieties of conventional crops such as *Cyamopsis tetragonoloba, Penninsetum typhoides, Phaseolus aconitifolius, Sessamum indicum,* and *Vigna radiata,* though adapted to low fertility and moisture status, are very low yielding. The cropping patterns are traditional and subsistence oriented. Further, because of social customs, the local farmers insist on using these varieties, and the choice of crops and varieties is very restricted. Additionally, low cropping intensity and inefficient inputs further decrease the yield. Furthermore, the extension of cultivation to marginal and submarginal tracts brings about deterioration of the cultivated area and loss of nutrients and water, leading to soil erosion and low productivity. The basic need, therefore, is to preserve the ecological balance of the area and increase the yield. This may be achieved by introducing diversification, utilizing intercropping with castor, guar, *Cenchrus ciliaris,* and other crops, and recycling runoff water. Some success has been achieved in this direction at the Central Arid Zone Research Institute. However, more work has to be done along these lines.

3. HORTICULTURE

There are no organized orchards in these regions. However, *Capparis decidua, Citrus* spp., *Cordia myxa, Grewia* spp., *Morus alba, Phoenix dactylifera,* and *Zizyphus rotundifolia* have been grown successfully for their fruits in some areas of Rajasthan. It is therefore likely that other similar regions can be developed for profitable orchards. Further, it has been noted that rocky and gravelly piedmonts and flat, aggraded, older alluvial plains are good landforms in terms of available groundwater; these can be utilized for growing valuable fruit trees.

4. SOCIAL FORESTRY AND MEDICINAL PLANTS

This is another area where immediate attention is required. Even though production forestry is required, social forestry—which generates social benefits such as supplies of fuelwood, small timber, fodder, protection of agricultural fields against winds, and recreational needs—is most urgently needed to alleviate the living conditions of the inhabitants of these regions. Social forestry is an integral part of the Gandhian philosophy of economic growth and community development (Srivastava and Pant, 1979), and the social benefits arising from this growth and development and the self-reliance achieved as a consequence would go far in bringing about a stabilized community. It should, however, be emphasized that in the selection of plant species for such purposes, priority must be given to local species rather than depending on exotic species. The exotic species may perform much better in a new environment in the beginning, but their successful survival after this initial period is rather limited; during this short period these introductions can bring about tremendous damage to the native species and the environment. For example, it has been noted that the rapidly growing roots of *Acacia tortilis,* an exotic reputed to be very successful in terms of the yield of timber, foliage, and such, compete with the native crops and drain the fields of nutrients. Species such as *Acacia nilotica, A. planifrons, Ailanthes excelsa, Calligonum polygonoides, Commiphora mukul, Cordia myxa, Ehretia laevis, Prosopis cineraria,* and *Zizyphus nummularia,* which have evolved within these environments, might be preferred (see also Patil and Pathak, 1977). However, much has to be done in this direction.

5. PLANT INDICATORS

The association of *Calotropis procera* with gypsum in the substrata has often been found in Rajasthan. Dakshini (1971) has shown the value of plants for indicator purposes. Figures 4 and 5 show the realtionships of some plant associations to groundwater, depth of water table, mineral content of the substratum, and geomorphic environment. However, data on these aspects are meager; a detailed study to establish similar relationships for other plants should be invaluable in any attempt to economically utilize the diverse resources available in these regions.

6. SEDENTARIZATION OF NOMAD TRIBES

Nomad tribes were at one time the most important lifeline in these areas, especially because they supplied the day-to-day articles used by inhabitants in very remote areas. Nomads played an important role in earlier days when communication was almost nonexistent and life in these areas was dependent on the

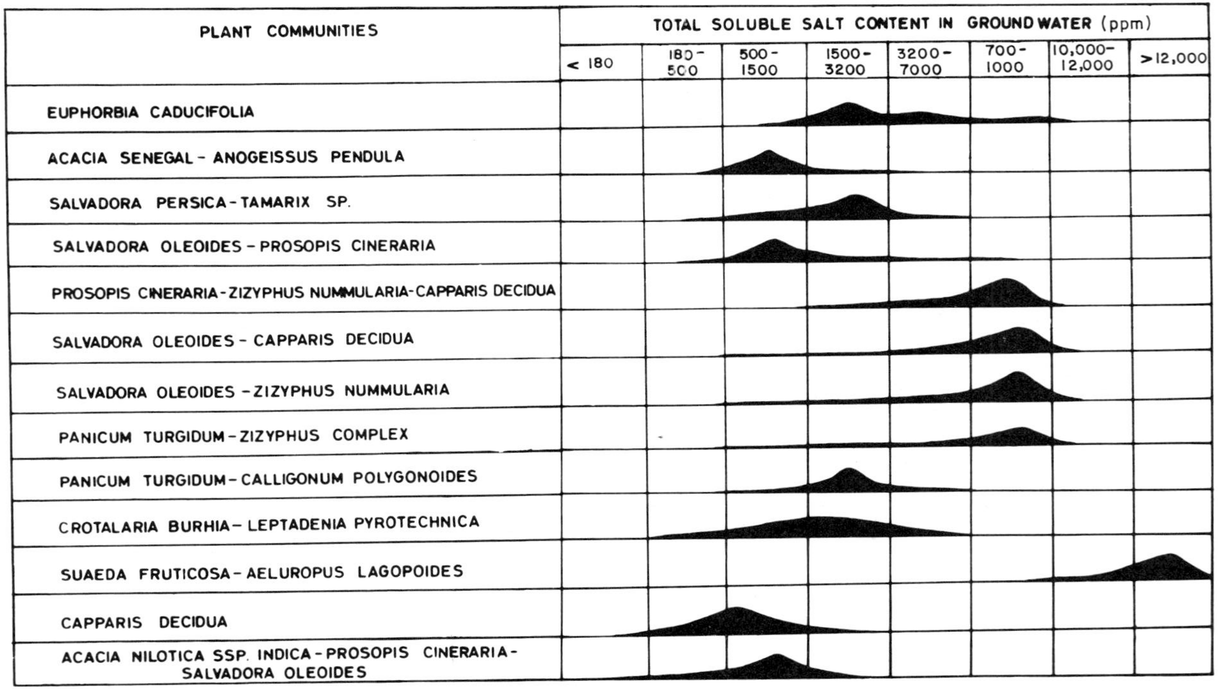

PLANT COMMUNITIES	TOTAL SOLUBLE SALT CONTENT IN GROUNDWATER (ppm)							
	< 180	180–500	500–1500	1500–3200	3200–7000	700–1000	10,000–12,000	> 12,000
EUPHORBIA CADUCIFOLIA								
ACACIA SENEGAL – ANOGEISSUS PENDULA								
SALVADORA PERSICA – TAMARIX SP.								
SALVADORA OLEOIDES – PROSOPIS CINERARIA								
PROSOPIS CINERARIA – ZIZYPHUS NUMMULARIA – CAPPARIS DECIDUA								
SALVADORA OLEOIDES – CAPPARIS DECIDUA								
SALVADORA OLEOIDES – ZIZYPHUS NUMMULARIA								
PANICUM TURGIDUM – ZIZYPHUS COMPLEX								
PANICUM TURGIDUM – CALLIGONUM POLYGONOIDES								
CROTALARIA BURHIA – LEPTADENIA PYROTECHNICA								
SUAEDA FRUTICOSA – AELUROPUS LAGOPOIDES								
CAPPARIS DECIDUA								
ACACIA NILOTICA SSP. INDICA – PROSOPIS CINERARIA – SALVADORA OLEOIDES								

Fig. 4. Plant communities and total soluble salt content in groundwater. After Chatterji and Gupta (1969).

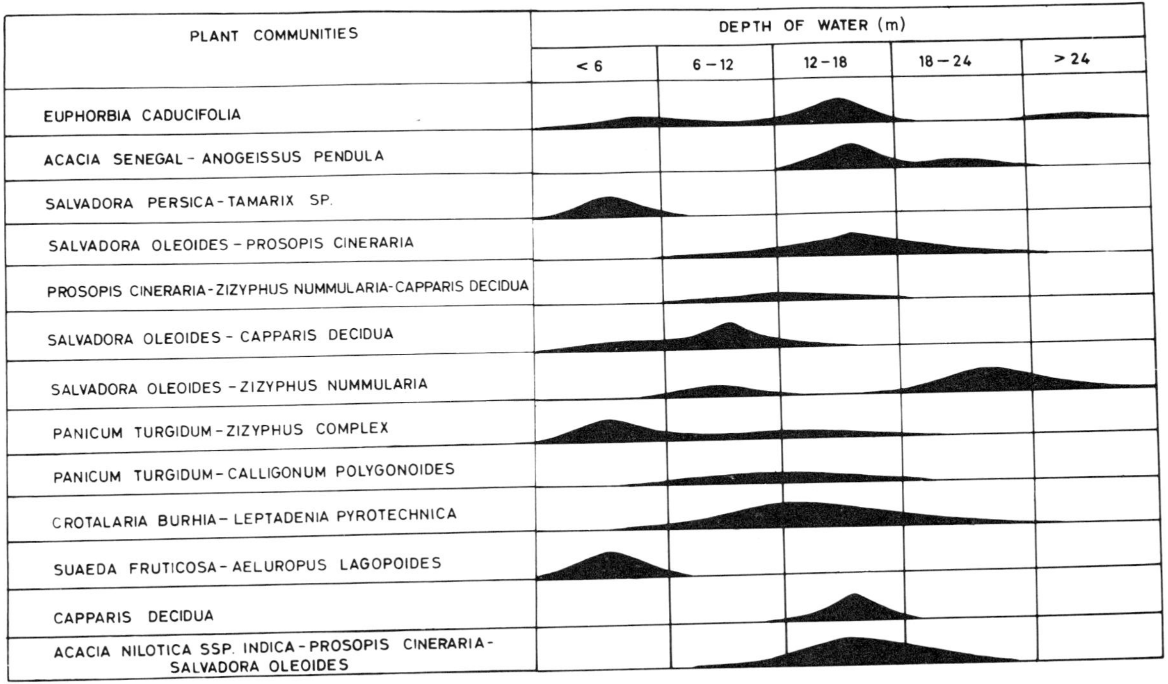

PLANT COMMUNITIES

DEPTH OF WATER (m)

< 6 | 6 – 12 | 12 – 18 | 18 – 24 | > 24

EUPHORBIA CADUCIFOLIA

ACACIA SENEGAL - ANOGEISSUS PENDULA

SALVADORA PERSICA - TAMARIX SP.

SALVADORA OLEOIDES - PROSOPIS CINERARIA

PROSOPIS CINERARIA - ZIZYPHUS NUMMULARIA - CAPPARIS DECIDUA

SALVADORA OLEOIDES - CAPPARIS DECIDUA

SALVADORA OLEOIDES - ZIZYPHUS NUMMULARIA

PANICUM TURGIDUM - ZIZYPHUS COMPLEX

PANICUM TURGIDUM - CALLIGONUM POLYGONOIDES

CROTALARIA BURHIA - LEPTADENIA PYROTECHNICA

SUAEDA FRUTICOSA - AELUROPUS LAGOPOIDES

CAPPARIS DECIDUA

ACACIA NILOTICA SSP. INDICA - PROSOPIS CINERARIA - SALVADORA OLEOIDES

Fig. 5. Plant communities and depth of groundwater. After Chatterji and Gupta (1969).

activities of these tribes. However, the changes brought by the development of roads and railway lines have tremendously improved the supply position; in fact, these tribes now contribute only minimally to the economic life in these regions. Instead, their large cattle herds bring about considerable destruction to the already poor and scanty vegetation (Sharma, 1972). Also, the tribes are becoming a social liability rather than an asset. It is therefore necessary that these tribes be encouraged to stabilize and develop their own communities. Such a step would not only help in controlling the destruction of vegetational cover, but bring about an improvement in their standard of living.

7. STABILIZATION OF MOVING SAND DUNES

It is needless to mention the significance of such a phenomenon for improving the habitat. Further, different types of sand dunes have the capacity to retain sufficient groundwater; such a potential can be used for agriculture and the development of grasslands.

C. Recommendations

Diversification of economic efforts is necessary for the evolution of any society and more so for those in relatively harsher environments. Further, with the postindependence growth rate of other regions, extension of the "urban syndrome" to the arid and semiarid regions would be more desirable. For any such approach, however, the totality of the scenario cannot and should not be neglected. From the data presented in the preceding pages, it is evident that the regions' overall bioclimatic pattern conforms to the total cultural and natural environment. Also, the stability and peculiarities of the arid and semiarid environments in these regions have left such an impressive mark on the substratum that to a large extent the landscape seems to have become mummified. This monotony of the landscape is magnified by a decrease in the rate of change and by rapid adjustments to any change in the habitat. Additionally, in such a habitat, which does not provide any opportunity for plants to modify their environment—either due to physiological stress or sparseness in distribution—the multitude of problems related to growth and development varies from plant to plant. In fact, each individual inhabiting these regions has its own characterisic evolutionary history. Thus, in reality, a tremendous diversity in the germplasm is maintained under these hostile conditions. Should this invaluable diversity be destroyed by changing the habitat or the plant cover? Also, can these stable environments sustain drastic changes? Increases in plant diseases and nematodes and the resultant yield losses in many introduced crops are being gradually realized in the arid regions of Rajasthan despite 10–15 years of extended canal irrigation. Any improvement or development in any of the spheres—agriculture, pasture, industry, urbanization, recreation, or forestry—in these regions will have to be planned in an integrated way to obtain the best advantage of the opportunities that these arid and semiarid lands provide.

> In Ladakh, extreme aridity combined with low temperature limits the possibility of growing crops about five months out of each year. Hence the strategy for agricultural development in Ladakh has to depend largely on the cultivation of quick growing cereals, oil seeds and fodder crops, and the rearing of goats, giving Pashmina wool. The hot regions, in contrast, have an abundance of sunshine, land and soil capable of responding to management, well-adapted grasses and trees, excellent breeds of sheep, goat and cattle and considerable reserves of water [Swaminathan, 1973].

Further, the conclusions drawn from the study of floristics and vegetation lead to the realization that the natural landscape is more suitable for forestry and range management than it is for cropping. In fact, a neglect of this vital observation is the cause for increased desertification within these regions. Afforestation and overall range management must be the focal points for harnessing these resources for the maximum betterment of the inhabitants of these regions. The existing ecological conditions cannot be drastically changed, however, and therefore overall improvement can be achieved only through proper management of these regions' immense available resources such as landforms, groundwater, soil types, and plant diversity. Further, it may be pointed out that since native plants in particular have a special ecological sanction for successful survival, proper management of this potential will go a long way in improving the living conditions of the local people. Although each one of the plants found growing in these regions has potential, the following plants have exhibited relatively more promise; therefore it is suggested that priority be given to the following plants for establishment, multiplication, and management in the arid and semiarid regions of the Indian subcontinent:

Acacia leucopholea (Roxb.) Willd. is a very fast growing tree that is successful on piedmont plains.

Acacia nilotica (Linn.) Del. is a tree with a moderate growth rate and a lifespan of 20 to 30 years; harvesting can be done profitably 6–10 years after planting; seed germination is slow, but viability is very high, and it can be successfully planted in diverse habitats, including semirocky areas and highly saline conditions. It is a good plant for shelterbelts.

Acacia planifrons Wight and Arn. is a plant the multiplication of which should be encouraged; growth potentials are fairly high. It also should prove valuable for landscaping purposes.

Ailanthus excelsa Roxb. can be successfully used for afforestation in arid and semiarid regions and even in shifting sand dune areas: coppicing is very pronounced and the growth rate is very high. It can be grown on diverse landforms.

Albizzia lebbek (Linn.) Benth.; is a large, unarmed, deciduous tree with a fairly rapid growth. It is very adaptive to local soil conditions and can be successfully planted on shallow sandy-loam soils overlying hard calcareous pan and on sandstone, rocky, and semirocky sites as well; crude protein content of the foliage has been found to be as high as 29.2% of the dry matter (Ganguli *et al.*, 1964); seed set is profuse and the seedling mortality rate under nursery conditions is low (3.7%); seedlings show a higher survival rate in metallic containers (CAZRI, 1979). It can be used as an avenue or windbreak tree.

Anogeissus pendula Edgew. is a moderate-sized and gregarious tree suitable for diverse landforms, especially the rhyolite residual hills. Easy to maintain, it can be used on avenues or as a windbreak.

Balanites aegyptiaca (Linn.) Delile is a gregarious shrub or small tree that propagates by root suckers and grows well in sandy, undulating, hummocky plains.

Barleria prionites Linn. is a prickly undershrub suitable for diverse landforms, especially hard gravelly substratum. Free flowering with quite a high seed set, it is a promising plant for horticultural purposes.

Boswallia serrata Roxb. *ex* Colebr. is a valuable tree suitable for northern and southern regions; can be used for afforestation in undulating to plain terrain, sandy loam, stabilized sands, or semirocky moderate slopes. It is a highly exploited tree and hence multiplication is most desirable.

Butea monosperma (Lamk.) Taub. is a small tree with a high seed set and germination percentage. It is suitable for diverse landforms and requires little care for maintenance; lopping and harvesting can be done after 10 years of growth. It is promising plant for horticultural purposes.

Calotropis procera (Ait.) R. Br. is an erect many-branched shrub, 2–3 m high, with a good regeneration capacity; seed set and germination are very high. It is suitable for arid zones, especially sand dunes, and also on loose sand overlying lime or gypsum beds; requires no special maintenance and can be used as a perennial ornamental plant.

Cassia fistula Linn. has good prospects for ornamental horticulture; should be planted in groups to have a mass effect of showy yellow flowers in pendulous racemes.

Cenchrus ciliaris Linn. is a perennial grass with a net primary productivity ranging from 662 to 977 g/m^2 (CAZRI, 1979). It is most suitable for arid and semiarid regions and for diverse landforms. Biomass, seed set, and germination potentials are very high. It is a very good and palatable grass for range and pasture development.

Cenchrus setigerus Vahl. is a valuable pasture grass that is successful on rocks, gravel, or sand. Often gregarious, it grows in tufts. Intercropping with legumes has shown a 29–30% increase in yield (CAZRI, 1979); seeding of 2 to 3 g/ha during monsoon months gives maximum germination; under average conditions of growth, production has been found to vary from 0.95 to 9.95 g/m^2/day, and net primary productivity from 647 to 837 g/m^2 (CAZRI, 1979). It has a tremendous potential for adaptation to and rehabilitation of overexploited desert grasslands.

Citrullus colocynthis (Linn.) Schrad. is a perennial trailing scabrid herb. It is a promising soil binder, especially in shifting sand dunes; cultivation should be encouraged. Seed set is profuse and germination and growth rate are quite high.

Cordia rothii Roem. and Schult. is a hardy tree that is successful on dry, gravelly substratum as well as in saline tracts. Seed set and viability of seeds is quite high. Seedlings are raised more successfully in the nursery but transplantation and maintenance pose no problems.

Crotalaria burhia Buch.-Ham. is a good soil binder; grows well in sandy and sandy-loam soils, as well as in saline and alkaline tracts, which are rich in carbonates. It can be easily multiplied using seeds.

Dichrostachys cinerea (Linn.) Wt. and Arn. is a small, compact tree that should be successful on different landforms. Growth is initially slow, but later regeneration of foliage is quite satisfactory. The bicolored pink and yellow pendulous racemes should make this point acceptable for horticultural purposes. Multiplication and maintenance should not be a problem in arid and semiarid environments.

Flacourtia sepiaria Roxb. has a grayish white stem and small leaves that change color with age, making this plant most suitable for landscaping. Seed set is poor, and hence multiplication through the use of cuttings should be tried. Growth rate is average and the plant can sustain heavy pruning.

Hardwickia binata Roxb. is successful on hard gravelly to deep, black, clayey soils. Growth and regeneration under irrigation is fairly fast. It should prove effective for semiarid regions.

Lasiurus sindicus Henr. is a perennial grass that is very palatable and provides a good potential for grassland development in arid tracts. Seeding at 5 kg/ha during the monsoon (July, August) yields well (CAZRI, 1979). It remains productive under proper care and management for more than 10 years; net primary productivity has been found to range from 630 to 652 g/m^2 (CAZRI, 1979).

Mimosa hamata Willd. is a large straggling shrub with straw-colored hooked prickles and an attractive pink globose inflorescence. It has very good prospects for ornamental horticulture. Seed germination and growth rate are fairly high.

Momordica dioca Roxb. *ex* Willd. is a tuberous-rooted plant that should be valuable as a vegetable and for medicinal purposes.

Moringa oleifera Lamk. is a very fast-growing tree that can grow successfully on diverse landforms under moderate irrigation. Multiplication by cutting is very successful.

Prosopis cineraria (Linn.) Druce is a successful tree under arid and semiarid conditions on shallow to deep piedmonts as well as on other landforms. Multiplication through seeds and coppicing is fairly easy and the growth rate is fairly fast. It is a reliable plant for diverse purposes.

Salvadora oleoides Decne. shows a potential for use in the afforestation of shifting sand dune areas. On the basis of nutrient contents and palatability ratings of the leaves, it is second only to *Acacia nilotica* (CAZRI, 1979).

Sarcostemma acidum (Roxb.) Voigt. is a useful shrub that is suitable for sand and gravelly substratum. Its initial establishment is difficult, but later maintenance is no problem even under very dry conditions. Its potentials should be exploited.

Tecomella undulata (Sm.) Seem. is a much exploited tree. Its multiplication, which is very easy, should be given priority. Its growth rate is slow but foliage regeneration is fairly fast. It has a good potential for ornamental horticulture due to its showy yellow flowers.

Wrightia tinctoria R. Br. is a tree suitable for gravelly substratum. Its seed set and germination are fairly high. The aromatic cream-colored flowers and pendulous dipyrenes should prove to be of great ornamental value.

Zizyphus nummularia (Burm. f.) Wt. is a straggling shrub that needs very little maintenance and is very suitable for afforesting shallow sandy-loam soils overlying calcareous pan. Its cultivation should be encouraged, especially on shallow sandy soils and piedmont plains.

ACKNOWLEDGMENTS

I am most grateful to Shri S. K. Gupta, J. K. Soni, and Km E. Roshini Nayar for their help in preparing the manuscript.

REFERENCES

Acharya, R. M., Patnayak, B. C., and Ahuja, L. D. (1977). Livestock production—problems and prospects. *In* "Desertification and Its Control" (P. L. Jaiswal, ed.), pp. 275–280. Indian Cent. Agric. Res., New Delhi.

Agroclimatic Atlas of India (1978). India Meteorological Dep., Div. Agric. Meteorology, pp. 1–91. Pune, India.

Ahuja, L. D. (1977). Improving rangeland productivity. *In* "Desertification and Its Control" (P. L. Jaiswal, ed.), pp. 203–214. Indian Cent. Agric. Res., New Delhi.

Ahuja, L. D. (1978). Range management, utilization and production in arid zones. *Proc. Indian Natl. Sci. Acad.* **44B**(5), 252–265.

Asghar, A. C., and Hafiz, M. A. (1964). Integrated surveys in the Indus plains, W. Pakistan. Toulouse Conf.

Baweja, B. K. (1979a). "Summarized Results—National Hydrograph Observation Well Stations (1969–1980)," Vol. 1. Central Ground Water Board, Gov. India.

Baweja, B. K. (1979b). "Summarized results—National Hydrograph Observation Well Stations (1969–1980)," Vol. 2. Central Ground Water Board, Gov. India.

Bhandari, M. M. (1974). Native resources used as a famine food in Rajasthan. *Econ. Bot.* **28,** 73–81.

Bhandari, M. M. (1978). "Flora of the Indian Desert." Scientific Publ., Jodhpur.

Bharucha, F. R. (1955). Structural and physiological features of Rajasthan desert. *Proc. Montpellier Symp. Pl. Ecol.,* 34–37.

Bharucha, F. R., and Meher-Homji, V. M. (1965). On the floral elements of the semi-arid zones of India and their ecological significance. *New Phytol.* **64**(2), 330–342.

Bhatt, M. P., Gadre, G. T., and Mehta, D. J. (1977). Salt and byproduct resources of Gujarat and

Rajasthan. *In* "Desertification and Its Control" (P. L. Jaiswal, ed.), pp. 281–190. Indian Cent. Agric. Res., New Delhi.

Biswas, K., and Rao, R. S. (1953). Rajputana desert vegetation. *Proc. Natl. Inst. Sci. India* **19,** 411–421.

CAZRI (1972). "Annual Report." Cent. Arid Zone Res. Inst., Jodhpur.

CAZRI (1974). "Basic Resources of Bikaner District (Rajasthan)." Cent. Arid Zone Res. Inst., Jodhpur.

CAZRI (1976). "Annual Report." Cent. Arid Zone Res. Inst., Jodhpur.

CAZRI (1979). "Twenty-five years of Arid Zone Research (1952–1977)." Cent. Arid Zone Res. Inst., Jodhpur.

Census of India (1961). Registrar General and Census Commissioner, Gov. India.

Census of India (1971). Registrar General and Census Commissioner, Gov. of India.

Charak (Maharshi) (1975). "Charakasamhita" (J. D. Vidyalankar, ed.), Parts 1 and 2. Motilal Banarsi Das, Delhi.

Chopra, R. N., Nayar, S. L., and Chopra, I. C. (1956). "Glossary of Indian Medicinal Plants." Council of Scientific and Industrial Research, New Delhi.

Choudhary, K. K. (1973). Ground water provinces of Madhya Pradesh State, India. *Proc. Int. Symp. Dev. Ground Water Resour., Madras* **3**(V), 123–132.

Cooke, T. (1901–1908). "The Flora of the Presidency of Bombay," Vols. 1–2. London [Repr. 1958], Vols. 1–3, Calcutta.

Dakshini, K. M. M. (1971). Indian subcontinent. *In* "Wildland Shrubs—Their Biology and Utilization" (C. M. McKell, J. P. Blaisdell, and J. R. Goodin, eds.), pp. 3–15. Utah State Univ., Logan.

Dakshini, K. M. M., and Tandon, R. K. (1981). The mechanism of adaptation in *Oropetium thomaem* (Linn. f.) Trin. (In press).

Deshmukh, D. S., and Karanth, K. R. (1973). Ground water resources and their exploitation in arid zones of India. *Proc. Int. Symp. Dev. Ground Water Resour., Madras* **3**(VI), 1–8.

Dhir, R. P. (1977a). Western Rajasthan soils: their characteristics and properties. *In* "Desertification and Its Control" (P. L. Jaiswal, ed.), pp. 102–115. Indian Cent. Agric. Res., New Delhi.

Dhir, R. P. (1977b). Saline waters—their potentiality as a source of irrigation. *In* "Desertification and Its Control" (P. L. Jaiswal, ed.), pp. 130–148. Indian Cent. Agric. Res., New Delhi.

Duthie, J. F. (1903–1929). "Flora of Upper Gangetaic Plain and of the Adjacent Siwalick and Sub-Himalayan Tracts," Vol. 3 (Vol. 3, Pt. 3 by R. N. Parker; index, 1952). Calcutta. [Repr. 1960].

Gamble, J. S. (1972). "A manual of Indian Timbers," pp. 1–868. Reprinted by Bishen Singh Mahendra Pal Singh, Dehr Dun.

Ganguli, B. N., Kaul, R. N., and Nambiar, K. N. T (1964). Preliminary studies on a few desert top feed species. *Ann. Arid Zone* **3,** 33–37.

Gaussen, H., Legris, P., Blasco, F., Meher-Homji, V. M., and Troy, J. P. (1968). Sheet Kathiawar, *In* "International Map of Vegetation and Environmental Conditions." Indian Cent. Agric. Res., New Delhi, and *Trav. Sect. Sci. Tech. Inst. Fr. Pondichery,* Series No. 9.

Gaussen, H., Meher-Homji, V. M., Legris, P., Blasco, F., Delacout, A., Gupta, R. K., and Troy, J. P. (1972). Sheet Rajasthan, *In* "International Map of Vegetation and Environmental Conditions." Indian Cent. Agric. Res., New Delhi, and Trav. Sect. Sci. Tech. *Inst. Fr. Pondichery,* Series No. 12.

Ghosh, A. (1952). The Rajputana Desert; its archaeological aspect. *In* "Symposium on Rajputana Desert" (S. D. Hora, ed.). *Bull. Natl. Inst. Sci. India* 37–42.

Ghosh, B., Singh, S., and Kar, A. (1977a). Geomorphology of Rajasthan desert. *In* "Desertification and Its Control" (P. L. Jaiswal, ed.), pp. 69–76. Indian Cent. Agric. Res., New Delhi.

Ghosh, B., Singh, S., and Kar, A. (1977b). Desertification around the Thar—A geomorphological interpretation. *Ann. Arid Zone* **16**(3), 190–301.

Gupta, I. C. (1979). "Use of Saline Water in Agriculture in Arid and Semi-Arid Zones of India." Oxford and IBH Publ., New Delhi.
Gupta, R. K. (1975). Plant life in the Thar Desert. *In* "Environmental Analysis of the Thar Desert" (R. K. Gupta and I. Prakash, eds.), pp. 202–236. Engl. Book Depot, Dehra Dun.
Gupta, R. K., and Bhandari, M. M. (1965). Flora of Rajasthan—a review. *Ann. Arid Zone* **4,** 236–238.
Gupta, R. K., and Dutta, B. K. (1967). Vernacular names of the useful plants of northwest Indian arid regions. *J. Agric. Bot. Appl.* **14,** 401–453.
Gupta, R. K., and Prakash, I. (1975a). The landscape and land forms. *In* "Environmental Analysis of the Thar Desert" (R. K. Gupta and I. Prakash, eds.), pp. 1–21. Engl. Book Depot, Dehra Dun.
Gupta, R. K., and Prakash, I. (1975b). Origin and geomorphic evolution. *In* "Environmental Analysis of the Thar Desert" (R. K. Gupta and I. Prakash, eds.), pp. 22–37. Engl. Book Depot, Dehra Dun.
Jagannathan, P, Bhalve, H. N., and Rakhlecha, P. R. (1978). Fluctuation and climate over Rajasthan. *Proc. Indian Natl. Sci. Acad.* **44B**(6), 364–380.
Jain, S. K. (1970). Floral composition of Rajasthan—a review. *Bull. Bot. Surv. India* **12**(1–4), 176–187.
Kanzaria, M. V. (1972). Soils of Gujarat. *Soil Ind. Fert. Assoc. India.*
Kirtikar, K. R., and Basu, B. D. (1935). "Indian Medicinal Plants," 4 Vols. L. M. Basu, Allahabad.
Krishnamurthy, R. (1970). A study of the quality of well water in Kovilpatti taluk. *Madras Agric. J.* **57,** 143–146.
Krishnan, A. (1968a). Distribution of arid areas in India. *Proc. Symp. Arid Zone, 21st Int. Geog. Congr., Jodhpur* 11–19.
Krishnan, A. (1968b). Delineation of different climatic zones in Rajasthan and their variability. *Indian J. Geog.* **3,** 33–40.
Krishnan, A. (1969). Some aspects of water management for crop production in arid and semi-arid zones of India. *Ann. Arid Zone* **8,** 1–17.
Krishnan, A. (1973). Why does it not rain in the Indian desert? *Indian Fmg.* **23**(3), 21–24.
Krishnan, A. (1977a). A climatic analysis of the arid zone of north-western India. *In* "Desertification and Its Control" (P. L. Jaiswal, ed.), pp. 4257. Indian Cent. Agric. Res., New Delhi.
Krishnan, A. (1977b). Climatic changes related to desertification in the arid zone of north west India. *Ann. Arid Zone* **16**(3), 302–309.
Krishnan, A., and Thanvi, K. P. (1973). "Agroclimatological Report of Bikaner District. Report of Division IV." Cent. Arid Zone Res. Inst., Jodhpur.
Krishnan, A., and Thanvi, K. P. (1977). Quantification of rainfall in different regions of western Rajasthan. *Ann. Arid Zone* **16**(2), 185–194.
Krishnan, M. D. (1952). Geological history of Rajasthan and its relation to present day conditions. *Proc. Symp. Rajputana Desert, Natl. Inst. Sci.* pp. 19–31.
Lakhanpal, R. N., and Bose, M. N. (1951). Some tertiary leaves and fruits of Guttiferae from Rajasthan. *J. Indian Bot. Soc.* **30,** 131–139.
Malhotra, S. P. (1977). Socio-demographic factors and nomadism in the arid zone. *In* "Desertification and Its Control" (P. L. Jaiswal, ed.), pp. 310–323. Indian Cent. Agric. Res., New Delhi.
Malhotra, S. P. (1978). Socio-economic situation in arid zone of Rajasthan. *Proc. Indian Natl. Sci. Acad.* **44B**(6), 423–430.
Mann, H. S. (1977a). The desertification process—an overview. *Ann. Arid Zone* **16**(3), 279–280.
Mann, H. S. (1977b). Introduction. *In* "Desertification and Its Control" (P. L. Jaiswal, ed.), pp. 1–5. Indian Cent. Agric. Res., New Delhi.
Mathur, C. M., Ganu, S. N., Moghe, V. B., and Jain, S. V. (1972). "Soils of Rajasthan." Soil Survey Org. Dept. Agric., Rajasthan.

Meher-Homji, V. M. (1977). The arid zones of India: bio-climatic and vegetational aspects. *In* "Desertification and Its Control" (P. L. Jaiswal, ed.), pp. 160–175. Indian Cent. Agric. Res., New Delhi.

Micheal, A. M., Sharma, R. B. S., Ashok Raj, P. S., and Chowdhary, R. R. (1979). "Resources Inventory Mahbubnagar Dist. Andhra Pradesh." Indian Cent. Agric. Res., New Delhi.

Mithal, R. S. (1969). Proc. Symposium on underground water studies in arid and semi-arid regions. Univ. of Roorkee.

Narayanaswami, S., and Deshmukh, G. P. (1978). Development of mineral resources of western Rajasthan. *Proc. Indian Natl. Sci. Acad.* **44B**(6), 418–422.

NDRI (1977). "Annual Report." Natl. Dairy Res. Inst., Karnal.

Oza, G. M. (1961). Flora of Pavagarh. Ph.D. Thesis.

Pakistan (1977–1978). "An Official Handbook." Dep. Films Publ., Minist. Inf. Broadcasting, Gov. Pakistan, Islamabad.

Paliwal, K. V. (1971). Quality of irrigation water and its effect on soil properties in Rajasthan. *Ann. Arid Zone* **10,** 226.

Paliwal, K. V. (1972). "Irrigation with Saline Water." Water Technol. Cent. Indian Cent. Agric. Res., New Delhi.

Paliwal, K. V. (1978). Quality distribution of ground waters and their utilization in western Rajasthan and Delhi regions. *Proc. Symp. Land Water Manage. Indus Basin (India), Ludhiana* **1,** 241–251.

Pathak, B. D., and Prasad, K. K. (1973). Ground water resources and their exploitation in arid zones of India. *Proc. Int. Symp. Dev. Ground Water Resour., Madras* **3**(V), 1–17.

Patil, B. D., and Pathak, P. S. (1977). Energy plantation and silvipastoral systems for rural areas. *Invention Intelligence* **12**(182), 79–88.

Pramanik, S. K., Hariharan, S. P., and Ghosh, S. K. (1952). Analysis of the climate of Rajasthan desert and its extension. *Indian J. Met. Geophys.* **3,** 131–140.

Puri, G. S., Jain, S. K., Mukerjee, S. K., Sarup, S., and Kotwal, N. N. (1964). Flora of Rajasthan. *Rec. Bot. Surv. India* **19,** 1–159.

Radhakrishna, B. P., and Dusan Duba (1973). Ground water resources evaluation in Mysore state. *Proc. Int. Symp. Dev. Ground Water Resour., Madras* **3**(V), 63–73.

Raheja, P. C., and Sen, A. K. (1964). "Resources in Perspective." Souvenir Volume. Gov. Press, Nasik.

Rao, K. N. (1958). Some studies on rainfall of Rajasthan with particular reference to trends. *Indian J. Met. Geophys.* **9**(2), 97–116.

Rao, K. N. (1963). Climatic changes in India. *In* "Ranges of Climate." *Proc. Rome Symp. Arid Zone Res.* **20.** UNESCO, Paris.

Rao, K. N. (1975). Unpublished.

Raychaudhuri, S. P., Aggarwal, R. R., Dutta Biswas, N. R., Gupta, S. P., and Thomas, P. K. (1963). "Soils of India." Indian Cent. Agric. Res., New Delhi.

Raza, M., and Ahmed, A. (1978). "General Geography of India." NCERT, New Delhi.

Report of the National Commission on Agriculture (1976). Minist. Agric. Irrig., Gov. India, New Delhi.

Roy, B. B., Gupta, R. K., and Pandey, S. (1971). Integrated approach for development of natural resources in Santalpur block of Gujarat State. *Ann. Arid Zone* **10**(1).

Sehgal, S. R., and Gulati, A. D. (1973). Hydrological consequences of irrigation and drainage projects in the Punjab. *Proc. Int. Symp. Dev. Ground Water Resour., Madras* **3**(V), 133–147.

Sen, D. N., and Bansal, R. P. (1979). Food plant resources of the Indian desert. *In* "Arid Land Plant Resources" (J. R. Goodin and D. K. Northington, eds.), pp. 357–370. Texas Tech Univ., Lubbock.

Sen, A. K., and Mann, H. S. (1977). A geographical appraisal of the expansion and deterioration of the Indian desert. *Ann. Arid Zone* **16**(3), 281–289.

Shah, G. L. (1978). "Flora of Gujarat State," Vol. I, pp. 1–602; II, pp. 603–1074. Sardar Patel Univ., Vallabh Vidyanagar.

Sharma, R. C. (1972). "Settlement Geography of the Indian Desert." Kumar Brothers, New Delhi.

Singh, G. (1977). Climatic changes in the Indian desert. *In* "Desertification and Its Control" (P. L. Jaiswal, ed.), pp. 25–30. Indian Cent. Agric. Res., New Delhi.

Singh, V., and Pandey, R. P. (1979). Medicinal plant-lore of the tribals of eastern Rajasthan [Abstract]. *J. Ind. Bot. Soc.,* Second Botanical Conference (Dec. 29–31). pp. 1–115.

Singhal, B. B. S. (1973). Some observations on the occurrence, utilization, and management of ground water in the Deccan Trap areas of the central Maharashtra, India. *Proc. Int. Symp. Dev. Ground Water Resour., Madras* **3**(V), 75–81.

Sinha, A. K., and Verma, T. P. (1978). Gypsum industry in Rajasthan. *Proc. Indian Natl. Sci. Acad.* **44B**(6), 411–417.

Spate, O. H. K., and Learmonth, A. T. A. (1967). "India and Pakistan: A General and Regional Geography," 3rd ed. Methuen, London.

Srivastava, B. P., and Pant, M. M. (1979). Social forestry on a cost–benefit analysis framework. *Indian For.* **105**(1), 2–35.

Stein, A. (1942). A study of the ancient sites along the 'lost' Saraswati River. *Geogr. J.* **99,** 173–182.

Swaminathan, M. S. (1973). "Extracts from the Sardar Patel Memorial Lectures of the All India Radio. Science and Agriculture" (S. Ramanumam, E. A. Siddiqi, V. L. Chopra, S. K. Sinha, eds.). [Marketed by Assoc. Publ., New Delhi].

Tandon, R.K. (1977). "Ecological Investigations on *Oropetium thomaem* (Linn. f.) Trin." Ph.D. Thesis, Univ. Delhi, India.

Thornthwaite, C. W. (1948). An approach towards a rational classification of climates. *Geogr. Rev.* **38,** 35–94.

Tiwari, M. N. (1979). The distribution of medicinal plants in the arid and semi-arid regions of Rajasthan—Thar Desert. *In* "Arid Land Plant Resources" (J. R. Goodin and D. K. Northington, eds.), pp. 186–194. Texas Tech. Univ., Lubbock.

Vishnu-Mittre (1977). Origin and history of the Rajasthan desert—paleobotanical evidence. *In* "Desertification and Its Control" (P. L. Jaiswal, ed.), pp. 6–9. Indian Cent. Agric. Res., New Delhi.

Wadia, D. N. (1960). The post glaciation dessication of central Asia. *Natl. Inst. Sci. India, Monogr.* **10,** 1–25.

Watt, Sir George. (1908). The commercial products of India. John Murray, London.

Wealth of India (1948–1976). "A Dictionary of Indian Raw Materials and Industrial Products," Vols. I–XI. CSIR, New Delhi.

Wheeler, Sir Mortimer (1960). "Civilization of the Indus Valley and Beyond." Thames and Hudson, London.

4

THE MIDDLE EAST

Loutfy Boulos
National Research Center
Cairo, Egypt

I. INTRODUCTION

Since time immemorial the Middle East, with its vast arid and semiarid lands, has provided mankind with food and other useful plants. These include wheat (*Triticum* spp.) and barley (*Hordeum* spp.) among the cereals; chickpea (*Cicer arietinum* Linn.), broadbean (*Faba vulgaris* Moench), lentil (*Lens culinaris* Medik.), and fenugreek (*Trigonella foenum-graecum* Linn.) among the pulses; flax (*Linum usitatissimum* Linn.), the antique textile fiber and oil plant; carob tree (*Ceratonia siliqua* Linn.), which yields fruits used as a fodder and is sometimes edible; common fig (*Ficus carica* Linn.); pistachio nuts (*Pistacia vera* Linn.); and many medicinal herbs, such as rue (*Ruta graveolens* Linn.), a narcotic and stimulant, and opium poppy (*Papaver somniferum* Linn.), which yields opium from the latex of unripe capsules, a raw material for many other important drugs.

II. PHYSIOGRAPHY

A. Size and Location

The Middle East comprises 17 countries belonging to three continents, but most occur in Asia. Two countries bridge more than one continent: Egypt is mainly African and partly Asiatic (Sinai), and Turkey is mainly Asiatic (Anatolia or Asia Minor) and partly European. The countries of the Middle East with their areas in square kilometers and the percentage of arid and semiarid regions within each country are given in Table I.

The percentage of arid lands within each of the above countries varies both in quantity and degree of aridity. The distribution of arid zones within the Middle East is shown in Fig. 1. Additional meteorological data are given in Table II. Three categories are recognized: (1) hyperarid, or extremely arid, usually with a mean annual rainfall below 50 to 250 mm, and (2) semiarid, with a mean annual

Table I

Area of Countries in Middle East and Proportion of Each That is Arid and/or Semiarid

Country	Area (km^2)	Percentage	
		Arid[a]	Semiarid
Egypt (Arab Republic of Egypt)	1,000,000	100	—
Israel (State of Israel)	21,000	35	40
Jordan (Hashemite Kingdom of Jordan)	102,798	70	25
Syria (Syrian Arab Republic)	184,479	70	20
Lebanon (Republic of Lebanon)	10,400	—	10
Turkey (Republic of Turkey)	780,576	—	50
Saudi Arabia (Kingdom of Saudi Arabia)	2,253,300	95	5
Kuwait (State of Kuwait)	15,540	100	—
Bahrain (State of Bahrain)	598	100	—
Qatar (State of Qatar)	10,600	100	—
United Arab Emirates (U.A.E.)	83,600	100	—
Oman (Sultanate of Oman)	212,380	100	—
Northern Yemen (Yemen Arab Republic)	195,000	50	45
Southern Yemen (People's Democratic Republic of Yemen)	287,684	100	—
Iraq (Republic of Iraq)	448,742	75	20
Iran (Islamic Republic of Iran)	1,648,000	65	25
Afghanistan (Republic of Afghanistan)	657,500	40	40
	7,912,197		

[a]The percentages of arid and semiarid areas are quoted after Paylore and Greenwell (1979).

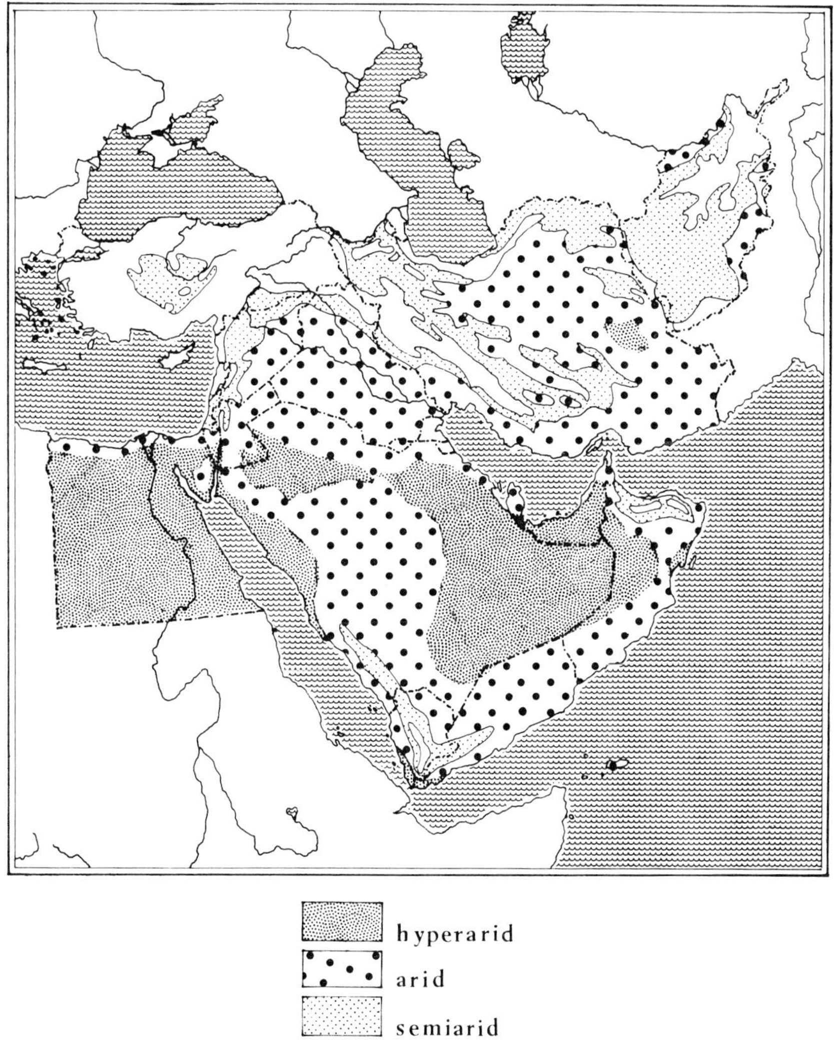

Fig. 1. Distribution of arid zones in the Middle East. Redrawn from UNESCO map (1977).

rainfall from 250 to 450 mm. Whereas the majority of Egypt is hyperarid, only a narrow northern coastal strip and a small mountainous area in Sinai are arid. Another hyperarid area is located within the Arabian Peninsula, and relatively small areas occur in Iran, Jordan, and Israel. The remaining area is either arid or semiarid. Turkey possesses a central large semiarid region, but Lebanon has relatively few arid lands. The majority of Israel, Jordan, Syria, Iraq, Iran, and Afghanistan is arid or semiarid. The amount of precipitation available to plants,

Table II

Meteorological Data, Coordinates, and Elevations for Selected Stations within Hyperarid, Arid, and Semiarid Regions of the Middle East[a]

Station	Latitude north	Longitude east	Elevation (m)	Mean annual rainfall (mm)	Mean minimum air temperature, January (°C)	Mean maximum air temperature, July (°C)	Prevailing surface winds, January	Prevailing surface winds, July
Egypt								
Sollum	31°31′	25°11′	4	114	9.2	30.9	S,SW,W	NW,W
Mersa Matruh	31°20′	27°18′	28	144	8.3	29.2	SW,S	NW,W
Alexandria	31°12′	29°53′	32	184	9.3	29.7	SW,NW	NW
Port Said	31°17′	32°14′	8	93	11.4	30.4	SW,W	NW,N
El-Arish	31°07′	33°45′	15	97	8.5	30.6	SW,S	W,NW
Almaza (Cairo)	30°06′	31°22′	74	24	8.7	35.4	SW,NE	N,NE
Suez	29°56′	32°33′	8	21	9.4	36.3	SW,S	NW,N
Tor	28°14′	33°37′	3	13	8.8	34.3	NW,N	NW,N
Asiut	27°11′	31°06′	71	0.3	6.8	36.8	W,NW,SW	NW,W
Luxor	25°40′	32°42′	95	1	5.6	40.5	NW,S	NW,W
Aswan	24°02′	32°53′	108	3	9.3	41.8	N,NE,NW	N,NW
Dakhla	25°29′	29°00′	110	1	4.1	38.5	NW	NW
Kharga	25°26′	30°34′	70	0	5.9	39.4	N,NW	N,NW
Quseir	26°08′	34°18′	5	4	13.9	33.0	NW,N	NW,N
Gaza Zone								
Gaza	31°38′	34°27′	16	379	9.4	38.6	SE,SW,S	NW,SE,W
Israel								
Tiberias	32°49′	35°36′	−212	436				
Beersheba	31°14′	34°47′	275	202	5.4	34.7		
Eilat	29°33′	34°57′	34	27	8.0	40.0	N,S	N,NW
Jordan								
Wadi Ziglab	32°31′	35°36′	−150	390	9.8	35.5		

Deir Alla	32°12′	35°37′	−224	291	10.1	38.3		
Shouneh South	31°54′	35°37′	−230	158	11.5	38.1		
Aqaba Airport	29°33′	35°00′	8	35	10.0	40.0	N,S	N,S
Amman Airport	31°59′	35°59′	766	272	3.7	32.1	W,SW,E	W,NW
Shoubak	30°30′	35°32′	1365	325	0.3	27.5		
H-4	32°30′	38°12′	686	83	2.9	38.1	W,S,SW	W,NW,SW
Mafraq	32°17′	36°14′		150	2.2	32.8	SE,W,S	W,NW
Syria								
Aleppo	36°11′	37°13′	395	323	1.8	36.4	E,NE,N	W,SW
Damascus	33°29′	36°14′	729	213	2.5	35.8	W,SW,NW	NW,W,S
Kamishly	37°03′	41°33′	455	492	2.8	40.3	N,E,NE	W,N,NW
Hassakeh	36°30′	40°45′	295	291	1.3	40.3	W,E,NW	W,NW
Deir-ez-Zor	34°20′	40°09′	212	149	2.5	40.4	W,NW,E	W,NW
Palmyra	34°33′	38°18′	404	127	2.6	38.0	W,NE,E	W,NW
Abu Kamal	34°25′	40°55′	182	108	2.3	40.4	W,SE,E	W,NW
Lebanon								
Beirut	33°49′	35°29′	16	743	9.1	29.1	E,S,SE,SW	SW,S,W
Rayak	33°52′	36°00′	921	618	0.4	32.3	SW,NE,S	SW,W,S
Fakeha	34°15′	36°25′	1060	210	2.9	31.9		
Turkey								
Erzurum	39°55′	41°16′	1829	471	−12.9	25.8	SW,W,SE	NE,E,SW
Sivas	39°45′	37°01′	1285	411	−8.1	27.2	SW,S,NW	N,NW,NE
Ankara	39°57′	32°53′	902	359	−3.7	30.1		
Diyarbakir	37°53′	40°11′	677	481	−2.6	38.2		
Saudi Arabia								
Al-Wajh	26°13′	36°27′	20	38	9.8	33.8		
Tabouk	28°24′	36°35′	774	76	−1.1	42.7		
Jeddah	21°30′	39°12′	11	63	15.8	39.5	N,NW	NW,N
Jizan	14°54′	42°30′	0	37	19.8	39.4		
Hail	27°31′	41°44′	914	77	3.9	38.4		
Taif	21°29′	40°32′	1395	166	4.2	38.2	SW,W	NW,N

(continued)

Table II *(Continued)*

Station	Latitude north	Longitude east	Elevation (m)	Mean annual rainfall (mm)	Mean minimum air temperature, January (°C)	Mean maximum air temperature, July (°C)	Prevailing surface winds, January	Prevailing surface winds, July
Riyadh	24°42′	45°43′	594	105	3.0	44.8	SE,NW	N,NW
Dhahran	26°17′	50°09′	72	85	5.9	45.2	NW	
Kuwait								
Kuwait	29°14′	47°59′	55	97	8.4	44.6	NW,SE,N	NW,N
Bahrain								
Manama	26°12′	50°30′	6	81	13.9	37.8		
Qatar								
Doha	25°15′	51°32′	11	78	12.7	41.2		
United Arab Emirates								
Dubai	24°10′	55°20′	5	112	17.1	55.2		
Sharjah	25°20′	55°24′	6	107	12.2	37.8		
Oman								
Muscat	23°37′	58°35′	4	106	18.9	36.1		
Masirah Island	20°41′	58°54′	16	37	23.9	31.1		
Salalah	17°03′	54°06′	10	96	17.8	27.8		
Northern Yemen								
Hodaieda	14°50′	43°00′	5	100	20.3	36.7		
Sanaa	15°27′	44°12′	2350	600	3.0	27.8		
Taez	13°35′	43°57′	1100	610	14.9	32.6		
Southern Yemen								
Riyan (Mukala)	14°39′	49°23′	25	60	19.6	33.4		
Khormaksar (Aden)	12°50′	45°01′	6	38	22.2	36.4		
Beihan	14°53′	45°42′	1097		8.3	37.4		

Dhala	13°42′	44°44′	1395	377	10.6	33.1		
Mukeiras	14°00′	45°42′	2041	240	3.4	25.7		
Iraq								
Mosul	36°19′	43°09′	222	389	2.5	43.4	E,SE,NW	W,N,NW
Kirkuk	35°28′	44°24′	330	373	4.5	42.8	S,SE,NW	W,NW,SW
Khanaqin	34°18′	45°26′	201	311	4.6	43.7	W,S,NW	W,NW
Rutbah	33°02′	40°17′	615	121	1.7	38.5	W,NW,S	W,NW
Habbaniya	33°22′	43°34′	43	112	4.4	43.4	NW,N,S	NW,W
Baghdad	33°20′	44°24′	34	145	4.3	43.4	NW,N,SE	NW,W,N
Kut-El-Hai	32°30′	45°45′	14	134	5.5	43.5	NW,W,E	W,NW
Diwaniya	31°59′	44°59′	20	119	4.1	42.7	W,NW	W,NW
Nasiriya	31°01′	46°14′	3	110	5.9	42.8	NW,W,N	NW,W
Basrah	30°34′	47°47′	2	140	7.0	40.5	NW,W,SE	NW,W
Iran								
Tabriz	38°08′	46°15′	1362	350	−4.9	32.3	E,W,NE	E,SE
Mashhad	36°16′	59°38′	985	199	−4.3	33.8	S,NW,N	NE,E,SE
Tehran	35°41′	51°19′	1191	165	−0.3	36.6	N,W,NW	SE,S,W
Abadan	30°22′	48°15′	3	135	7.5	44.1	NW,W,SE	NW,W
Bushehr	28°59′	50°50′	14	197	10.4	36.9	NW,N	W,SW
Kerman	30°15′	56°58′	1749	101	−2.7	35.5	W,SW,S	NE,N,NW
Esfahan	32°37′	51°40′	1598	301	−1.8	36.8	W,S,SW	E,NE,W
Shiraz	29°36′	52°32′	1491	296	−0.1	37.4	NW,W,N	NW,W,E
Afghanistan								
Kandahar	31°36′	64°40′	1050	178	−0.5	38.9		
Kabul	34°30′	69°13′	1800	338	−7.8	33.3		

[a] Stations with rainfall over 450 mm are given for comparison.

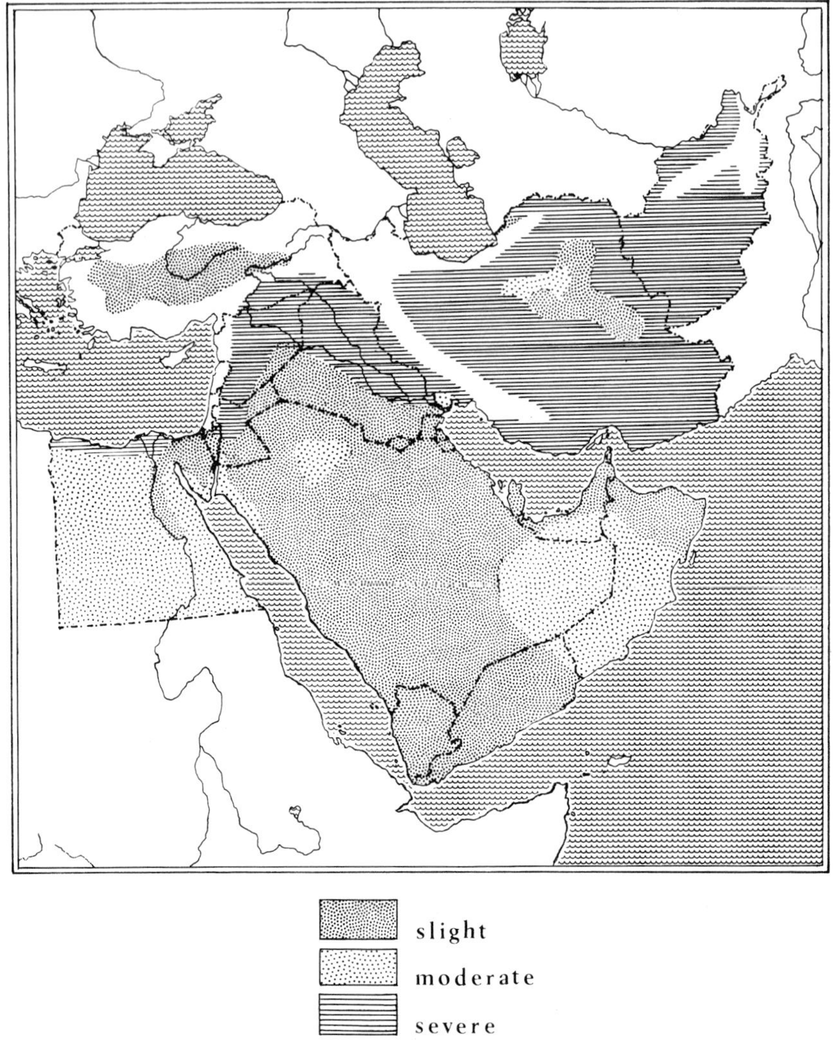

Fig. 2. Status of desertification in the Middle East. Redrawn from UNEP/FAO/UNESCO/WMO map (1977).

together with temperature, influences where the line is drawn between the above categories representing the degree of aridity in any of the above regions. Soil variation is less a determining factor in vegetation composition than either moisture or temperature. Therefore, moisture remains the main factor used in drawing the lines between the above categories. In this connection, Zohary (1973) stated "there are tremendous extents of abiotic desert in Arabia, Egypt, and Iran, resulting from moisture deficiency."

The majority of the Middle East arid zones are liable to desertification, though

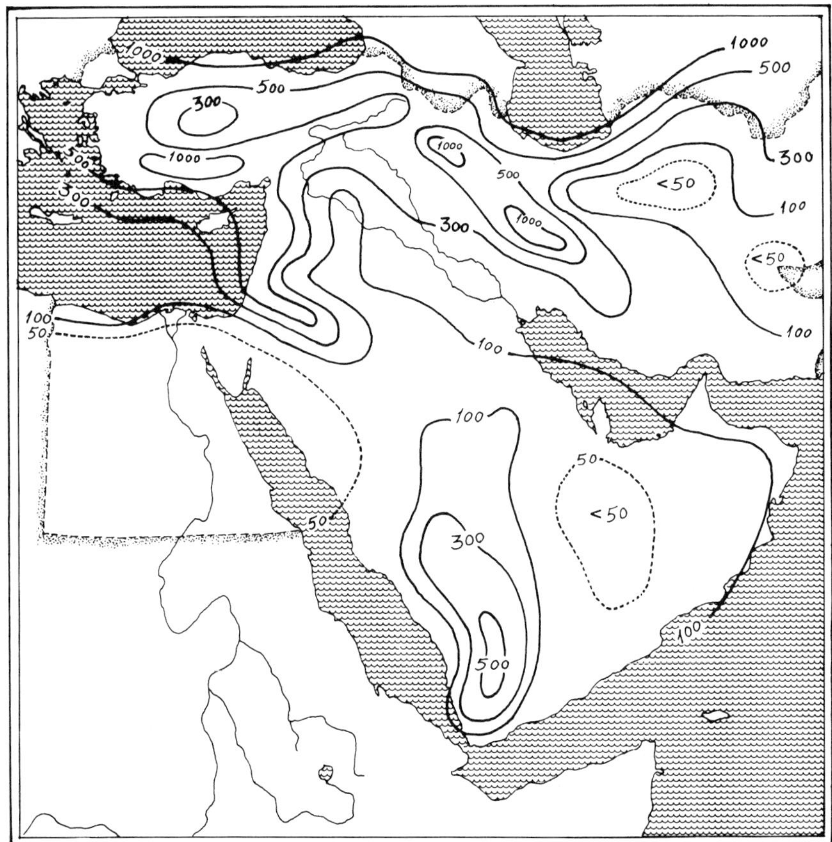

Fig. 3. Mean annual rainfall (mm) in the Middle East. Courtesy of Dr. Kamel Hanna Soliman.

in different degrees (Fig. 2). Consequently, the lines between the three categories of arid zones, besides being arbitrary, remain subject to change, especially where desertification is very high or may become so in the future.

The Middle East is characterized, in general, by its prevailing Mediterranean climate: long, dry, and hot summers and rather mild, rainy winters, with the exception of some mountainous regions (partly arid) in the southwestern Arabian Peninsula and southern Iran where summer rains can occur (Fig. 3). In the highlands of Yemen, for example, according to Fisher (1963), there is a well-marked and major rainy season from July to September due to the southwest monsoon. Soliman (1978) pointed out that the prevailing winds in winter are mainly easterlies, except over Egypt and the southeast Mediterranean where southwesterly to westerly winds prevail. The Mediterranean depressions, however, move eastward to Iraq and Iran, temporarily disturbing the high-pressure pattern. These depressions are associated with clouds and rain (and snow on high ground), and affect the area north of 25°N. Surface winds become variable in direction, are occasionally strong, and cause sandstorms in sandy regions. In summer the pattern is reversed and a huge thermal low covers south Asia, extending its effect to the east Mediterranean and Arabia. The prevailing winds in summer are northerly; the frequency of strong winds is less than in winter, except for Iraq, where strong northwesterly winds may cause dust storms (Soliman, 1978).

Autumn and spring are not definite seasons, but rather transitional. Autumn is usually more stable, with pleasant warm days, than spring.

Summer days are practically cloudless with long hours of sunshine; this results in a low relative humidity and high evaporation rate, especially inland away from the maritime effect that is enjoyed in the coastal regions. Therefore, the maximum day temperatures are much higher than those at the equator. The highest maximum daily temperatures ever recorded in the world are known from the interior of the African Sahara, Arabia, Iraq, and Iran, and may exceed 50°C. Temperatures as high as 42 to 48°C are normally experienced in some of these regions during summertime (see Fig. 4). The mean maximum July air temperature in Dhahran, Saudi Arabia, for instance, is 45.2°C.

Almaza, a suburb of Cairo ~200 km south of the Mediterranean Sea coast, is an example of an inland arid station. The difference between the mean daily maximum and minimum temperature is highly pronounced (see Table III), which gives warm days and cold nights during winter and hot days and cool nights during summer. The evaporation rate is high during the hot, dry summer and low during the warm, rainy winter (see Table III). In contrast to the very hot summer, winter is usually mild along the coasts, but snow can fall inland as far south as Yemen at altitudes of more than 3000 m. In general, in the mountainous regions winter can be severe and snow can cover some areas for periods ranging from a few days to several weeks, depending on latitude and altitude.

The rainy season is from October to May, but most of the rain falls from December to February; June to September is dry. Rainfall is scanty and erratic; the mean annual rainfall is generally below 400 mm (see Table II) and is not regularly distributed throughout the rainy season. The mathematical mean is misleading because several successive seasons can have rainfall amounts much below the average (Table IV). This can, in turn, intensify the aridity and causes degeneration of the natural vegetation, failure of crops in regions fully dependent on dry farming, and deficiency of fresh water for use in urban zones. As was mentioned previously, most of the rain for a given rainy season can fall during a very short period, leaving the rest of the season receiving very little rain. For example, although the mean annual rainfall in Aqaba Port, Jordan is 42 mm, 65.7 mm was recorded in 24 hr in February 1975; but only 10 mm was recorded for the remaining months from September 1974 through September 1975 (Boulos and Lahham, 1977). Thus, rainfall is sporadic and is erratic in terms of quantity,

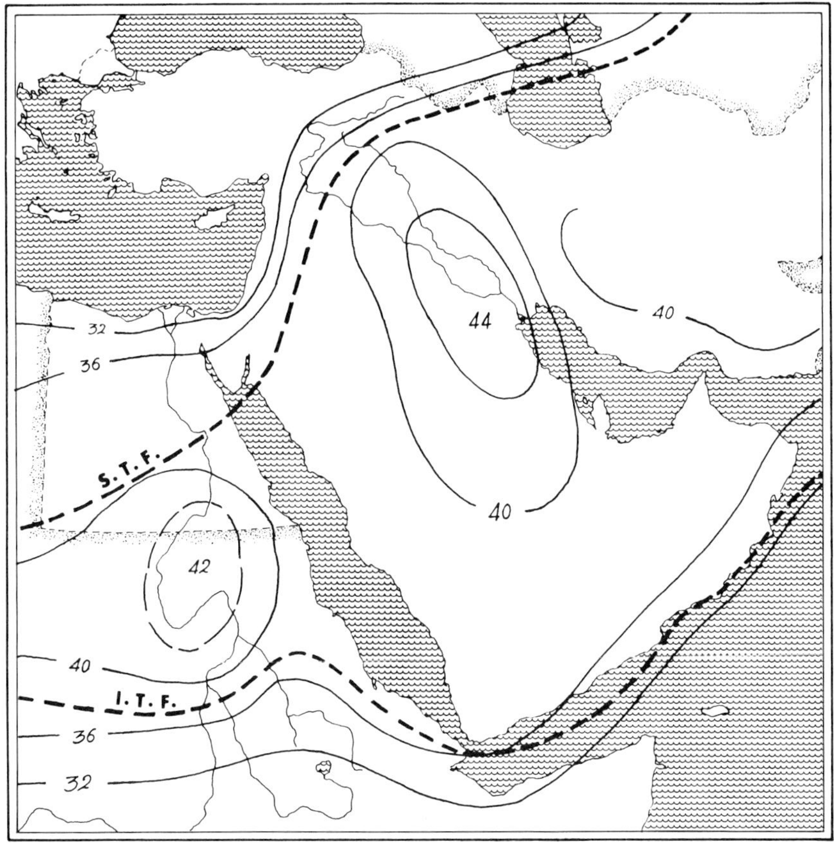

Fig. 4. Mean maximum air temperature in July (°C) in the Middle East. S.T.F., Subtropical front; I.T.F., intertropical front. Courtesy of Dr. Kamel Hanna Soliman.

number of rainy days, and intervals between them, whether during the same season or from one season to another. Table V illustrates the contrasts between the highest and lowest monthly rainfall at eight stations in Egypt during different months of the year; the maximum amount of rainfall in 24 hr for each month is also given. If Alexandria, with a mean annual rainfall of 184 mm, is taken as an example of an arid coastal station, the highest rainfall in December is 191.5 mm, the lowest is 4.7 mm, and the maximum in 24 hr during the same month is 54.3. mm. If Cairo, with a mean annual rainfall of 24 mm, is taken as an example of an extremely arid or hyperarid inland station, the highest rainfall in December is 67.1 mm, the lowest is 0 mm, and the maximum in 24 hr during the same month is 50 mm.

According to Koller (1959), in order to compensate for the erratic nature of

Table III

Climatic Particulars of Almaza, Calculated as Averages of 24 Years of Records (1931–1954)[a]

	Temperature (°C)			Relative humidity			
	Mean daily maximum	Mean daily minimum	Mean of day	Mean	12 hr	Rainfall (mm)	Evaporation piche (mm/day)
January	19.5	7.9	13.9	58	45	1.9	5.4
February	21.0	8.8	14.8	54	38	4.0	6.4
March	24.1	16.8	17.2	52	35	3.1	7.9
April	28.3	13.5	20.9	46	28	0.8	9.4
May	32.2	17.1	25.3	42	24	1.3	11.9
June	35.1	20.0	27.6	47	28	0.0	11.5
July	36.2	21.6	28.8	54	33	0.0	10.8
August	35.3	21.7	28.4	55	33	0.0	9.3
September	32.6	19.8	26.2	59	40	0.0	7.7
October	30.7	17.6	24.0	58	39	2.1	7.2
November	25.6	13.9	20.2	62	44	1.1	5.8
December	20.9	9.5	15.2	64	48	6.4	4.8
Year	28.4	15.7	21.9	54	36	20.7	8.2

[a] After Kassas and Iman (1957).

rain, seeds of many desert annuals can fail to germinate unless enough moisture is available to guarantee both germination and plant growth to the stage of seed production, although plants can mature in a reduced form. According to several authors and our own field observations, seeds of desert plants retain their viability for many years and remain in the soil awaiting the next effective shower, which permits both successful germination and seed production.

Perennial plants respond to irregularity of rain in a different fashion: according to Zohary (1973), during rainy years perennials produce seeds in abundance, whereas in dry years growth is reduced to a minimum and plants may enter a phase of dormancy lasting as long as the drought threatens the plants. In extreme cases, a perennial plant may become reduced to the extent of keeping its body restricted to lignified tissues.

B. Topography, Geology, and Edaphic Factors

The Middle East embraces a diverse surface topography ranging from the high peaks of the Taurus and Zagros Mountains to the lowest point in the world—the Dead Sea—and from the alluvial fertile lowlands of Iraq and Egypt to the vast "sterile" desert of Rub Al-Khali and the eastern part of the Sahara with its high

Table IV

Annual Rainfall Records from Six Stations within the Cairo Area and Yearly Deviation from a 50-Year Mean[a]

Year	Rain (mm)	Yearly deviation from 50-year mean (mm)	5-year mean (mm)	Year	Rain (mm)	Yearly deviation from 50-year mean (mm)	5-year mean (mm)
1906	21	−6		1931	11	−16	
1907	51	+24		1932	8	−19	
1908	66	+39		1933	12	−15	
1909	39	+12		1934	18	−9	
1910	14	−13	38.2	1935	2	−25	10.0
1911	21	−6		1936	9	−16	
1912	17	−10		1937	38	+11	
1913	24	−3		1938	14	−17	
1914	13	−14		1939	20	−7	
1915	9	−18	16.8	1940	15	−12	19.2
1916	44	+17		1941	19	−8	
1917	38	+11		1942	15	−12	
1918	36	+9		1943	34	+7	
1919	44	+17		1944	45	+18	
1920	55	+28	43.4	1945	36	+9	29.8
1921	87	+60		1946	17	−10	
1922	13	−14		1947	22	−5	
1923	44	+17		1948	23	−4	
1924	42	+15		1949	37	+10	
1925	23	−4	41.8	1950	14	−13	22.6
1926	21	−6		1951	62	+35	
1927	17	−10		1952	34	+7	
1928	11	−16		1953	10	−17	
1929	31	+4		1954	24	−3	
1930	32	+5	22.4	1955	19	−8	29.8

[a] Mean, 27 mm/year; standard deviation, 15.2. Also included are 5-year means partly after Kassas and Inam (1957).

plateaus and green oases. The geology of the area is unique as well, housing the major oil fields of the world, the rift structures of the Red Sea system, and many others.

1. ZONES

a. The Arabo–Nubian Massif. This massif consists of a platform of ancient Precambrian rocks, some of the most ancient in the world, and represents a northerly offshoot of the old African shield. It comprises, according to Zohary

Table V

Contrasts Between the Highest and Lowest Monthly Rainfall at 8 Stations in Egypt during Different Months of the Year[a]

	Jan	Feb	Mar	Apr	May	Jun	Jul	Aug	Sep	Oct	Nov	Dec
Highest monthly rainfall (mm)												
Sollum	55.2	46.0	58.5	34.2	31.7	8.7	Tr	0.0	21.2	73.1	271.7	133.9
Mersa Matruh	102.4	59.3	78.7	13.4	22.5	59.8	0.0	17.3	10.0	55.5	101.2	77.3
Alexandria	158.3	90.7	47.8	35.7	9.8	0.3	Tr	8.8	23.2	41.6	117.5	191.5
Port Said	50.9	34.9	32.5	22.8	20.2	Tr	0.0	Tr	5.4	81.9	36.1	78.4
El-Arish	114.3	69.5	37.7	42.3	61.0	0.1	0.0	5.7	16.0	54.0	96.0	105.0
Cairo	28.3	13.5	14.6	5.4	6.2	3.6	0.0	Tr	0.1	14.4	20.8	67.1
Tor	11.8	10.8	36.0	3.2	5.8	Tr	0.0	0.0	0.0	13.0	38.3	34.0
Quseir	0.8	1.5	9.0	2.0	1.5	Tr	0.0	0.0	Tr	15.8	34.0	15.0
Maximum rainfall in 24 hr (mm)												
Sollum	38.4	44.4	30.0	27.0	17.1	2.7	Tr	3.5	9.4	53.9	97.6	33.6
Mersa Matruh	35.1	26.2	12.2	8.0	20.1	47.2	0.0	15.0	9.0	55.5	75.5	53.5
Alexandria	47.9	28.0	21.1	13.0	8.8	0.3	Tr	8.8	22.6	39.0	64.6	54.3
Port Said	11.6	15.3	10.9	22.8	19.5	0.6	0.0	Tr	5.4	39.6	18.0	47.7
El-Arish	40.5	31.0	24.5	42.3	59.3	0.1	0.0	5.7	16.0	52.0	39.2	35.0
Cairo	9.6	10.4	10.0	3.8	6.0	3.6	0.0	Tr	0.1	13.8	18.5	50.0
Tor	9.5	10.0	22.0	3.2	5.6	Tr	0.0	0.0	0.0	13.0	37.4	22.0
Quseir	0.8	0.8	9.0	2.0	1.5	Tr	0.0	0.0	Tr	11.0	34.0	4.3
Lowest monthly rainfall (mm)												
Sollum	Tr	Tr	0.0	0.0	0.0	0.0	0.0	0.0	0.0	0.0	Tr	Tr
Mersa Matruh	0.9	1.1	0.0	0.0	0.0	0.0	0.0	0.0	0.0	0.0	0.2	0.4
Alexandria	0.9	5.6	Tr	0.0	0.0	0.0	0.0	0.0	0.0	0.0	0.0	4.7
Port Said	Tr	Tr	0.0	0.0	0.0	0.0	0.0	0.0	0.0	Tr	0.2	0.0
El-Arish	0.0	0.0	0.0	0.0	0.0	0.0	0.0	0.0	0.0	0.0	0.0	Tr
Cairo	0.0	0.0	0.0	0.0	0.0	0.0	0.0	0.0	0.0	0.0	0.0	0.0
Tor	0.0	0.0	0.0	0.0	0.0	0.0	0.0	0.0	0.0	0.0	0.0	0.0
Quseir	0.0	0.0	0.0	0.0	0.0	0.0	0.0	0.0	0.0	0.0	0.0	0.0

[a] Partly after Soliman (1978). The maximum amount of rainfall in 24 hr for each month is also given.
[b] Tr, Trace.

(1973), most of the Arabian Peninsula, southern Palestine (Jordan and Israel), the plateau of southern Sinai, and the Red Sea hilly country in eastern Egypt.

According to Said (1962), Abd-El-Gawad (1969), Picard (1970), and others, a table of sediments belonging to the Phanerozoic overlies this massif on both sides in its Asian and African parts. Within this massif lies one of the most significant tectonic features of the world—the Red Sea Basin. Its northern part, represented by the Gulf of Suez, dates back to the early Paleozoic, while the southern prolongation is much younger, of Oligocene–Miocene age.

Since the Precambrian, the Arabo–Nubian massif has been nearly stable, subject only to gentle epeirogenic movement. On this rigid, partly peneplained landmass, a thick sequence of continental and shallow-water shelf sediments, along with deep marine rocks, was deposited during the Phanerozoic.

The Paleozoic generally is represented by clastic sediments with a few carbonate interbeds, developed during the Upper Paleozoic (Carboniferous and Permian) in Egypt and the Arabian Peninsula. The Mesozoic section is a mixture of both continental clastics and marine carbonates; the latter was developed more during the close of this era. On the other hand, Jurassic evaporites are known from the Arabian Peninsula. In places on the Arabo–Nubian massive, coarse clastics are common from the Cretaceous (Lexique Stratigraphique, 1968).

The Tertiary rocks show a wider and more complex variation on both sides of the older shield. Carbonates with clay interbeds, from the Paleocene and the Eocene, are well developed to the west, whereas carbonates and evaporites are known from east of the massive. The Miocene clastics and evaporites distinguish the western outcrops, whereas sandy limestone and sandstone characterize the eastern side of the massive. Although changes in the facies often alter the character of the rock types during different epochs, this general pattern holds true for most of the Arabo–Nubian massif and its overlying sediments, both outcropping and subsurface sections.

Volcanic rocks are common and cover a wide span of time, from Paleozoic to Quaternary. El-Shazly *et al.* (1974) recorded volcanics as recent as the Quaternary from St. John Island in the Red Sea.

b. The Irano–Anatolian Folded Zone. According to Zohary (1973), this zone consists of sedimentary strata comprising the Pontas, Taurus, Elburz, and Zagros systems, as well as the presumably younger east Mediterranean system comprising the Aegean and Syro–Palestine ranges. Whereas the folded system in the eastern part of the area consists of symmetrical, orderly folds with a definite trend, the western system was heavily disturbed by accessory movements and became broken into asymmetrical arcs and garlands. The eastern and western parts of this zone have some similarities, however, such as the local basins with Tertiary deposits and the fertile valleys between the mountains of Iran and Turkey.

c. The Intermediate Zone. This zone, according to Zohary (1973), lies between the two above-described zones and comprises a belt with comparatively simpler landforms and a less disturbed geological structure. It is outstanding for its alluvial plains and rather flat lands, such as the Syrian Desert and the Mesopotamian lowland. One of the most spectacular features of this zone is the Syro–African Rift Valley, a Tertiary formation that resulted from faulting. It extends from the Amanus Mountains in the north to East Africa and includes, in the Middle East, the valley of the Orontes River, Beqa'a in Lebanon, the Litani and Jordan Rivers, and the Dead Sea Depression to the Aqaba Gulf southward.

Faulting was accompanied by volcanic eruptions, particularly along the eastern side of the Rift Valley. This resulted in the formation of enormous volcanic cones and produced the highest peaks in the Middle East—Mount Ararat (5156 m) in northeastern Turkey, and Mount Damavend (5766 m) in northern Iran.

Rock strata on the southern flanks of the fold mountains are tilted into great domes; the vast deposits of petroleum that make the Middle East one of the world's leading oil provinces have accumulated in these domes.

2. SOIL TYPES

The basic types of soil that can be encountered in the arid and semiarid regions of the Middle East are the following, summarized mainly from Clawson *et al.* (1971).

a. Alluvial Soils. These are formed from geologically young materials, are mainly calcareous, of different textures, and are deposited by running water in stream valleys, deltas, or at the foot of slopes, such as in the Nile Valley and Delta in Egypt.

b. Barren Desert Soils. These soils have little or no vegetation cover and occupy vast areas. They consist of gravels, limestone, sandstone, sand over limestone, or sand dunes of different shapes and sizes, and occur in almost rainless regions.

c. Chestnut Soils. Medium textured, these soils develop under vegetation on calcareous materials where rainfall is moderate, as in the hilly terrain north of Damascus, Syria.

d. Desert Soils. These develop in areas with low rainfall and support shrub and annual desert vegetation. These soils are variable in texture, depth, and mother rocks. Desert soils are common on level or undulating plains over chalk, sandy plains, and terraces along rivers and basins.

e. Lithosol Soils. Lithosols are thin, with a high proportion of bare limestone, sandstone, lava beds, basic igneous rocks, or differentiated sedimentary rocks. These soils occur on rough desert terrain, hilly terrain, level or undulating windswept plains, and dissected plateaus with wadis having calcareous and saline alluvial soils. These soils are of general occurrence in all arid zones of the Middle East.

f. Noncalcic Brown Soils. Nonacid soils, these do not contain free calcium carbonate. Examples are the sandy coastal plains of Israel and Lebanon.

g. Reddish Brown Soils. These develop either from limestone or other sedimentary rocks on level, sloping, or undulating plains with clayey or loamy textures, or on hilly upland terrain from basic volcanic rocks, or on slopes of volcanic cones. They also may occur on basic igneous rocks on hilly terrain. These soils are known from northern Syria, southern Turkey, northern Iraq, and western Iran.

h. Red Desert Soils. Gravelly or sandy, these soils formed on level to undulating terrain, or differentiated on plains and basins of oases. These soils usually are associated with different types of lithosols. The red desert soils are known in the deserts of Egypt, Jordan, Israel, Syria, Iraq, Iran, and Afghanistan.

i. Sand Dunes. These occur as shifting dunes of the barren desert, dunes adjacent to seacoasts as part of a sand dune–beach–marsh complex, or dunes with solonchak. They are known from Egypt, Israel, Lebanon, Syria, and southwest of the Euphrates in Iraq.

j. Solonchak Soils. Salty soils, these occur within desert depressions where evaporation of runoff waters has left salt accumulations, such as on the broad Qattara depression in Egypt or along beaches.

k. Salt Marshes. These are formed in low-lying areas along coasts.

l. Sierozen Soils. Developed in semiarid regions, these soils originate from chalk, limestone, sand, or basalt, on level, undulating, or rolling plains or hilly terrain. Examples include large areas in northern Syria between arid and humid regions toward the northern mountains.

C. Hydrology and Water Resources

Information presented in this section has been summarized mainly after Gischler (1979).

The amount of fresh water available within a given habitat determines the potential for vegetative production, and hence for the framework of various life forms within the habitat. Freshwater resources may be in the form of precipitation (rainfall, snow, dew), surface flow (runoff, ephemeral streams, freshwater lakes, rivers), and underground water. Additionally, inland saline and brackish waters represent a nonconventional, albeit much less productive, source of water.

Rainfall, although erratic in most arid areas of the Middle East, remains the only reliable source of water for some isolated areas. Some mountain chains act as precipitation collectors—for example, the Red Sea hills in Egypt, Asir Mountains in Saudi Arabia, and Djebel Akhdar (green mountain) in Oman. Surface runoff from the semiarid mountains feeds the intermittent and perennial rivers and supplies the infiltration waters for the recharge of acuifers. Examples are the eastern Sahara aquifers of Nubian sandstone and those of the Arabian Peninsula.

Water input from rain, snow, river floods, discharge, and such is variable from year to year, and from season to season, and although different in origin and nature, represents a renewable resource. Fresh water of some lakes may, however, from irrational manipulation, become brackish. Two such examples are what happened to Lake Mariut several centuries ago and, more recently, to Lake Qarun in Egypt.

In arid and semiarid regions, evaporation always exceeds rainfall if the whole year is taken into consideration. Therefore, evaporation is a major problem for surface storage of water, especially in sandy soils, because of rapid percolation. In some cases, evaporation results in large scale precipitation of salts and allows formation of *sebkhas* (salt marshes), which are usually situated along coastal regions. In semiarid mountainous regions with sparse vegetation, the capacity of the vegetational cover for absorbing water is poor and the runoff coefficient correspondingly high. As a result, this lack of a root system (which normally keeps the soil *in situ*) allows soil to be easily carried away by runoff waters; deep erosion gullies are cut into the mountain slopes.

Man-made lakes, such as Lake Nasser in Egypt and Sudan and Lake Tharthar in Iraq, are constructed mainly to serve as water storage reservoirs. These and other large water masses stop the horizontal river flow and force deposition of sediment load.

The impoundment of the Nile at Aswan by the construction of the Aswan High Dam, and the resulting formation of Lake Nasser, have introduced radical changes in the ecology of Lake Nasser as well as in the Nile Valley. Nile water is mainly used for agriculture, urban, and industrial activities; only drainage water reaches the Mediterranean Sea. Silt is deposited at the entrance of the storage reservoir, where a new delta is being formed, mainly between 420 and and 300 km upstream from the Aswan High Dam. Consequently, silt-free water being released from the dam is eroding the river bed. Moreover, the ~50 million metric tons of fertile Nile silt, which in the past was spread yearly over the land by river floods, must be replaced by fertilizers. Again, because of the disturbance of the dynamic equilibrium between the supply to, and the losses of sediments from, the Mediterranean shores, erosion of the delta coast is taking place.

Projects of agricultural expansion based on the stored Nile water were implemented in the late 1960s, mostly at the Nubariya area west of Alexandria (136,000 ha). Unfortunately, the scheme was ill planned and has shown gross defects. Irrigation water is brought into a high-level canal by a series of pumps and then used in the fields by gravity flow. Much of the water has escaped underground and then reappeared in ~22,800 ha as stagnant saline lakes. Underground water, at a depth of 40 m at the beginning, has since risen at the rate of 4 m per year.

III. DEMOGRAPHY

Population Size, Distribution, and Densities

Population size, distribution, and density in the Middle East are tremendously variable, not only from country to country (see Table VI), but also from one part of a country to another. Saudi Arabia, with its national territory of 2,253,300 km^2 has only 8.2 million inhabitants, whereas the 21,000 km^2 of Israel is inhabited by 3.9 million. The 42.1 million population of Egypt is concentrated in the narrow cultivated strip around the Nile, which is ~40,000 km^2 or 4% of the entire country; the estimated population figure for the year 2000 is 64.9 million.

There is a substantial and uncontrolled migration from the desert communities to the villages. However, the rate is much more pronounced from the village to the towns because the population size in all desert regions of Egypt represents only a small fraction of a million. Similar migration problems are taking place in almost all countries of the Middle East.

Immigration from one country to another is taking place on a small scale and may be illustrated by the few thousand Egyptian farmers in Iraq who are becoming settlers in the newly reclaimed agricultural lands. Oil producing countries, short in manpower, import labor from the neighboring overpopulated countries, such as Egypt and Pakistan.

IV. SOCIOECONOMIC FACTORS

At present, oil is the primary economic resource in several of the Middle Eastern countries, such as Saudi Arabia, Iran, Kuwait, Iraq, United Arab Emirates, Qatar, and Oman. Egypt's oil exports recently have become, and seemingly will continue in the near future to constitute, a major part of the country's national income because of the increase of oil prices. Syria, Lebanon, and Jordan, although producing little or no oil, receive some income from the pipelines passing through their territories. Iran and Iraq depend to a large extent on their income from oil. The economy of these two countries is also dependent on agricultural and industrial products; the majority of their people are engaged in agriculture.

Table VI

Population Data for Some Middle East Arid Countries[a]

Country	Annual birth rate (per 1000 pop.)	Annual death rate (per 1000 pop.)	No. years to double population	Infant mortality rate (per 1000 pop.)	Percentage of population under age 15	Life expectancy at birth	Per capita GNP (U.S. dollars)	Population estimation mid–1980 (millions)	Population projected for the year 2000 (millions)
Egypt	38	10	26	90	40	55	400	42.1	64.9
Israel	25	7	38	15	33	73	4,120	3.9	5.5
Jordan	46	13	21	97	48	56	1,050	3.2	5.9
Syria	45	13	21	114	49	57	930	8.6	16.2
Saudia Arabia	49	18	23	150	45	48	8,040	8.2	15.5
Kuwait	42	5	19	39	44	69	14,890	1.3	3.1
Qatar	44	14	23	138	45	48	12,740	0.2	0.4
United Arab Emirates	44	14	23	138	34	48	14,230	0.8	1.6
Oman	49	19	23	142	45	47	2,570	0.9	1.7
Northern Yemen	48	25	30	160	47	45	580	5.6	9.5
Southern Yemen	48	21	26	155	49	45	420	1.9	3.4
Iraq	47	13	20	104	48	55	1,860	13.2	24.5
Iran	44	14	23	112	44	58	—	38.5	66.1
Afghanistan	48	21	26	226	45	37	240	15.9	26.4

The main agricultural products of the Middle Eastern countries, summarized after Clawson *et al.* (1971), are as follows:

Afghanistan: cereals (mainly wheat, maize, barley, and rice), cotton, sugar beets, sugarcane, vegetables, oilseeds, fruits (mainly melons and pomegranates), nuts

Bahrain: dates, alfalfa, vegetables

Egypt: cereals (mainly wheat, maize, rice, and barley), cotton, sugarcane, pulses, vegetables, fruits, forage plants (mainly berseem: *Trifolium alexandrinum* L.)

Iran: cereals (mainly wheat, barley, and rice), sugar beets, cotton, tea, tobacco, timber, pulses, sesame, flax, fruits (a great variety), nuts, vegetables

Iraq: cereals (mainly wheat, barley, and rice), fruits (mainly dates), cotton, tobacco, vegetables, timber

Israel: cereals (mainly wheat and barley), fruits (mainly citrus, olives, and grapes), vegetables, sugar beets, tobacco, forage crops

Jordan: cereals (mainly wheat and barley), tobacco, ''vegetables'' (mainly tomatoes and cucumbers), fruits (mainly olives and grapes)

Kuwait: little grain is grown due to lack of water; most food is imported

Lebanon: cereals (mainly wheat), fruits, vegetables, tobacco

Northern Yemen: coffee, cereals (mainly millet, maize, oats, sorghum, barley, and rice), fruits (mainly dates, almonds, grapes, and oranges), cotton, tobacco, forage plants, qat (*Catha edulis*)

Oman: cereals (mainly wheat and sorghum), fruits (mainly dates, pomegranates, and citrus), vegetables, pulses, cotton

Qatar: vegetables, fruits

Saudi Arabia: cereals (mainly wheat, millet, and maize), fruits (mainly dates), alfalfa

Southern Yemen: cereals (mainly sorghum, millet, wheat, and barley), sesame, fruits, vegetables, cotton

Syria: cereals (mainly wheat, barley, and millet), cotton, tobacco, vegetables (mainly onions), pulses, olives, fruits (mainly grapes, figs, and tomatoes), sugar beets

Turkey: cereals (mainly wheat), tobacco, cotton, fruits, nuts, tea, oilseeds, forage crops, timber, opium, medicinal plants

United Arab Emirates: fruits (mainly dates), vegetables

In the past, several among the above countries produced enough food to meet the needs of their own people, and in some cases, extra products were exported. In recent years, however, the population has grown at a higher rate than has food production, which has made it necessary for some countries to import major amounts of wheat, rice, sugar, fruit, vegetable oils, etc., depending on the production of each country. Table VII illustrates the annual increase or decrease

Table VII

Annual Increase or Decrease in Cereal Grain Yields (%)

Country	1960–1969	1970–1978	Net increase or decrease
Egypt	3.4	−0.03	(−)
Israel	2.6	4.6	(+)
Jordan	5.6	0.0	(−)
Syria	7.0	11.3	(+)
Saudi Arabia	0.08	6.6	(+)
Northern Yemen	0.0	1.9	(+)
Southern Yemen	−4.5	4.8	(+)
Iraq	6.3	3.0	(−)
Iran	2.5	1.2	(−)
Afghanistan	2.0	1.7	(−)

in the percentage of cereal grain production in some countries of the Middle East, calculated from the weighted average of coarse grain, wheat, and rice yield figures for the 1960s and 1970s (after Paylore and Greenwell, 1980).

V. PLANT RESOURCES

In addition to the famous fertile regions of the Middle East, which are heavily cultivated with traditional crops, the arid and extremely arid parts of the region support numerous useful native plants that have the potential of becoming future new crops that may provide extra resources for food, forage, medicinals, fibers, oils, ornamentals, etc.

The following is a synopsis of selected species listed according to their uses. The habit of the plant is given, as well as its geographical distribution within the Middle East, that is, if a species is cosmopolitan, its distribution within the Middle East is mentioned without going into any other details regarding its full distribution. Similarly, "Mediterranean" denotes the Mediterranean part of the Middle East, which in fact is the eastern Mediterranean, usually from Egypt to Turkey.

All distribution records have been compiled by the author unless a reference is given. References for the use or uses of plants are given, otherwise the information is heretofore unpublished work of the author. If the habit and distribution is not mentioned for a certain species, this denotes that the same sepcies has already been treated under a previous group, such as food plants or forage plants, where the habit and distribution are first mentioned.

A. Food Plants

Kunkel (1984) enumerated numerous plants from the Middle East that are suitable for human consumption. The following are some examples cited by other authors:

Aizoaceae

Aizoon canariense Linn.—annual herb; desert of the Middle East; leaves eaten as salad (Zohary, 1966)

Amaranthaceae

Digera muricata (Linn.) Mart.—annual herb; tropical regions of the Middle East; eaten as a potherb (Zohary, 1966)

Araceae

Arisarum vulgare Targ.-Tozz.—perennial herb with corms; Mediterranean; leaves eaten as a potherb

Eminium spiculatum (Blume) Kuntze—perennial herb with corms; Mediterranean; leaves boiled to obtain a soup (Boulos, 1959)

Asclepiadaceae

Glossonema edule N. E. Br.—perennial herb; Arabia; fresh fruits edible (Boulos, 1978)

Capparaceae

Capparis spinosa Linn.—shrub; Mediterranean to Arabia; flower buds used as pickles

Chenopodiaceae

Chenopodium album Linn.—annual herb; throughout the Middle East; seeds highly nutritive because of high vitamin C content; used as a salad plant; formerly cultivated as bread plant (Zohary, 1966)

Chenopodium murale Linn.—annual herb; throughout the Middle East; salad herb (Zohary, 1966)

Chenopodium opulifolium Koch and Ziz—annual herb; Mediterranean to Iran; used as a potherb (Zohary, 1966)

Compositae

Cichorium pumilum Linn.—annual herb; Mediterranean to Afghanistan; leaves consumed as a salad

Gundelia tournefortii Linn.—perennial herb; central Turkey, Mediterranean to Iran; achenes edible, young heads eaten as artichokes (Feinbrun-Dothan, 1978) (I have eaten them in Jordan, raw and cooked, both are delicious!)

Helminthotheca echioides (Linn.) Holub—annual herb; Mediterranean to Iran; Potherb and salad plant (Zohary, 1973)

Silybum marianum (Linn.) Gaertn.—annual spiny herb; Mediterranean to Iran; young shoots used as a raw salad

Sonchus oleraceus Linn.—annual herb; throughout the Middle East; young shoots and leaves consumed as a raw salad (I have seen bundles of this plant for sale in a vegetable market in a small village in the Nile Delta, Egypt)

Cruciferae

Diplotaxis acris (Forsk.) Boiss.—annual herb; Egypt, Israel, Jordan, Arabia, Iraq; leaves eaten as a fresh salad

Nasturtium officinale R. Br.—perennial aquatic herb; throughout the Middle East; leaves and young shoots used as a salad

Geraniaceae

Erodium cicutarium (Linn.) L'Hérit.—annual herb; throughout the Middle East; leaves used as a potherb (Zohary, 1972)

Erodium hirtum Willd.—perennial herb with tuberous roots; hot arid regions of the Middle East; tubers and green parts used as potherbs (Zohary, 1972)

Erodium malacoides (Linn.) L'Hérit.—annual herb; throughout the Middle East; young branches and leaves used as potherbs (Zohary, 1972)

Erodium moschatum (Linn.) L'Hérit.—annual herb; throughout the Middle East; leaves used as a potherb (Zohary, 1972)

Gramineae

Cymbopogon olivieri (Boiss.) Bor—perennial aromatic grass; Iraq, Iran, Afghanistan; hummocky tufts smell strongly of lemon; used in the desert to flavor tea (Bor and Guest, 1968)

Echinochloa colonum (Linn.) Link—annual tufted grass; throughout the Middle East; grains eaten by poor classes in times of scarcity (Chakravarty, 1978) [Grains were found in the intestines of mummies dating back 6000 years, which suggests that the plant was cultivated as a cereal grain by the ancient Egyptians (Täckholm and Drar, 1941).]

Echinochloa crus-galli (Linn.) Beauv.—annual herb; throughout the Middle East; grains consdiered to be an inferior millet, used in times of scarcity (Chakravarty, 1976)

Panicum turgidum Forsk.—perennial grass; deserts of the Middle East [In the Sahara, where the plant grows on dunes, the grain is said to be collected and used for human consumption (Bor and Guest, 1968).]

Leguminosae

Alhagi graecorum Fisch. (= *A. maurorum, A mannifera*)—perennial shrub; throughout the Middle East; sugary excretion (known as manna) obtained from plant by shaking twigs over a dry cloth (in some districts of Iran) [It was also known as a source of manna in Egypt and Syria (Townsend and Guest, 1974).]

Astragalus carduchorum Boiss. and Hausskn.—subshrub forming low cushions; Iraq, Iran; one of the manna plants; gum eaten by people in Iran, but not sold (Townsend and Guest, 1974)

Ceratonia siliqua Linn.—evergreen tree; Mediterranean; widely used food source for cattle and man, large pods rich in protein and sugars; layer of gum in pod is polysacharide of widespread application (Zohary, 1973); refreshing drink made (in Egypt) by soaking ground fruits in water

Vicia angustifolia Linn., *V. canescens* Lab., *V. peregrina* Linn., *V. sativa* Linn.—fairly well distributed annual herbs in the Middle East; seeds could be used as pulses

Malvaceae

Malva parviflora Linn.—annual herb; throughout the Middle East; used as a potted herb, and often cultivated (in Egypt) and sold in vegetable markets

Menispermaceae

Cocculus pendulus (Forst.) Diels—liana on cliffs or trees; Egypt, Israel, Jordan, Arabia; leaves edible (Zohary, 1966)

Moraceae

Ficus sycomorus Linn.—large tree; Egypt, Israel, Jordan, Arabia; fruits edible

Palmae

Hyphaene thebaica (Linn.) Mart.—branched palm with palmate leaves; Egypt, Israel, Arabia, and probably Jordan; fruit edible, also used for making molasses (Täckholm and Drar, 1950)

Polygonaceae

Rheum palaestinum Feinbrun—perennial herb; endemic to Israel and Jordan; young inflorescence and thick upper portion of root used as fresh salad or potherb

Rumex conglomeratus Murr.—biennial or perennial herb; throughout the Middle East; leaves used as potherb (Zohary, 1966)

Rumex crispus Linn.—perennial herb; throughout the Middle East; wild salad plant and potherb with high vitamin C content (Zohary, 1966)

Rumex vesicarius Linn.—annual herb; Egypt, Israel, Jordan, Syria, Arabia; leaves eaten both green and cooked (Zohary, 1966)

Portulacaceae

Portulaca oleracea Linn.—succulent annual; throughout the Middle East; used as a vegetable and salad; often cultivated or collected from the wild in Egypt and other Mediterranean countries and sold in vegetable markets

Rhamnaceae

Ziziphus lotus (Linn.) Lam.—deciduous thorny shrub; Egypt, Israel, Jordan, Syria, Turkey, Arabia; fruits edible but less tasty than those of *Z. spina-christi* (Zohary, 1972)

Ziziphus spina-christi (Linn.) Desf.—evergreen spiny tree; hot and warm regions of the Middle East; fruits edible; often cultivated in Egypt, Israel, Jordan, Syria, Arabia, etc., for shade and edible fruit (Zohary, 1972)

Rosaceae

Crataegus aronia (Linn.) DC.—tree or shrub with thorny twigs; Israel, Jordan, Syria, Iraq, Iran; fruits edible, sold in markets

Crataegus azarolus Linn.—tree with thorny twigs; Mediterranean; fruits edible, often sold in markets

Eriolobus trilobatus (Poir.) Roem.—deciduous tree or shrub; Turkey, Lebanon, northern Israel; fruits edible

Pyrus elaeagnifolia Pallas—spinescent tree; Turkey; fruits edible; plant used as stock on which other cultivated varieties are grafted (Zohary, 1973)

Rubus sanctus Schreber—evergreen shrub; Mediterranean to Afghanistan; berries edible

Salvadoraceae

Salvadora persica Linn.—small tree or shrub; hot deserts of the Middle East; fruits edible, and young shoots and leaves eaten as a salad (Zohary, 1972)

Sapindaceae

Dodonaea viscosa (Linn.) Jacq.—south Egypt, South Arabia, Iraq, Iran; seeds said to be edible, and fruits once used as substitute for true hops (*Humulus lupulus* Linn.) in making yeasts and beer (Chakravarty, 1976)

Tiliaceae

Corchorus olitorius Linn.—herbaceaous annual; hot regions of the Middle East; a popular summer potherb in Egypt, Jordan, Israel, and Syria

Umbelliferae

Crithmum maritimum Linn.—perennial shrubby herb; rocky Mediterranean coasts; fleshy leaves eaten as a salad and pickled in vinegar (Zohary, 1972)

Foeniculum vulgare Mill.—perennial herb; throughout the Middle East; widely used substitute for dill in flavoring and conserving pickles; young, green stems eaten fresh (Zohary, 1972)

B. Forage Plants

The foliage of some firewood trees and shrubs is often used as forage, for example, *Balanites aegyptiaca* (Balanitaceae), *Haloxylon persicum* (Chenopodiaceae), *Acacia nilotica, A. seyal, A. tortilis* (Leguminosae), *Ziziphus spina-christi* (Rhamnaceae), and *Populus euphratica* (Salicaceae) (see Ayensu, 1980, 1983).

Chenopodiaceae

Anabasis articulata (Forsk.) Moq.—small shrub; most desert regions of the Middle East; heavily browsed by camels and goats (Zohary, 1966); pasturage for goats and camels (Schulz and Whitney, 1985)

Atriplex halimus Linn.—shrub; Mediterranean to Arabia; rather palatable browse shrub [The salt content of the leaves increases with the aridity of the habitat, which makes the plant less palatable (Zohary, 1966).]

Compositae

Carthamus oxyacantha M. Bieb.—annual thorny herb; Arabia, Iraq; plant used as fodder after spines rounded by beating; grazed by animals in early stages when spines are soft (Chakravarty, 1976)

Gundelia tournefortii Linn.—in central Turkey plants collected in large quantities and dried for winter fodder (Feinbrun-Dothan, 1978)

Ifloga spicata (Forsk.) Sch. Bip.—small annual; coastal and inland arid regions of the Middle East; heavily browsed by grazing animals (Schulz and Whitney, 1985)

Rhanterium epapposum Oliv.—hot desert of Arabia nand adjacent countries (Zohary, 1973); heavily browsed by goats (Schulz and Whitney, 1985)

Convolvulaceae

Convolvulus lanatus Vahl—richly branches spinescent shrub; Egypt, Israel, Jordan; heavily browsed by camels and sheep (Schulz and Whitney, 1985)

Cruciferae

Eremobium aegyptiacum (Spreng.) Aschers. and Schweinf. *ex* Boiss. [= *E. diffusum* (Decne.) Botsch.]—sandy deserts of Egypt, Israel, Jordan, and Arabia; good pasturage (Schulz and Whitney, 1985)

Farsetia aegyptia Turra—small shrub; deserts of Egypt, Israel, Jordan, and Arabia; heavily browsed by camels (Schulz and Whitney, 1985)

Dipsacaceae

Cephalaria syriaca (Linn.) Schrad.—annual herb; throughout the Middle East; weed in wheat fields; harvested by Arab farmers in Palestine and used as a fodder plant (Zohary, 1973)

Geraniaceae

Erodium circutarium (Linn.) L'Hérit.—a valuable pasture plant (Zohary, 1972)

Gramineae

Aegilops crassa Boiss.—annual grass; Jordan, Israel, Syria, Lebanon, Turkey, Iraq, Iran, Afghanistan; a useful forage plant (Bor and Guest, 1968)

Aegilops kotschyi Boiss.—annual grass; Mediterranean to Iran; useful forage grass, eaten in Kuwait by most animals as dry hay during summer (Bor and Guest, 1968)

Aegilops speltoides Tausch—annual grass; Jordan, Israel, Syria, Lebanon, Turkey, Iran; useful forage plant for sheep (Bor and Guest, 1968)

Aeluropus lagopoides (Linn.) Thwaites—perennial grass; common in salt marshes of the Middle East; a good fodder plant (Chakravarty, 1976)

Aeluropus litoralis (Gouan) Parl.—perennial grass; salt marshes and sand dunes of the Middle East; grazed by animals (Chakravarty, 1976)

Alopecurus myosuroides Huds.—annual herb; Mediterranean to Afghanistan; reported to be useful fodder grass in Iraq (Bor and Guest, 1968)

Bromus rubens Linn.—annual grass; Mediterranean to Iran; seems to be a good fodder grass (Chakravarty, 1976)

Bromus tectorum Linn.—annual grass; throughout the Middle East; grazed by sheep, goats, and mules; eaten when dry by donkeys, horses, and other animals (Bor and Guest, 1968)

Cenchrus ciliaris Linn.—perennial tufted grass; throughout the Middle East, except Lebanon and Turkey; a fine fodder rich in carbohydrates and proteins; stands cutting well, drought resistant (Chakravarty, 1976), and easily grown from seed (Bor and Guest, 1968)

Cymbopogon olivieri (Boiss.) Bor—useful fodder plant in the arid steppe after annual grasses have withered away (Bor and Guest, 1968)

Dactylis glomerata Linn.—perennial grass; Mediterranean to Afghanistan; a valuable grass for pasture and hay, can withstand long spells of drought in deep soil where root system is fully developed; yield less affected than that of most other grasses on poorer types of soil; gives good yield of feeding matter and quickly recovers after grazing (Bor and Guest, 1968)

Dichanthium annulatum (Forsk.) Stapf—perennial with thick, woody rhizome; throughout the Middle East, except Turkey; a useful fodder grass, eaten by horses and other animals (Bor and Guest, 1968)

Echinochloa colonum (Linn.) Link—much valued as a quick-growing fodder grass; greedily grazed by all kinds of livestock, relished by cattle at all stages; nutritive value enhanced when in grain; hay can be stacked for 5–6 years; said to fatten cattle and help their healthy growth (Chakravarty, 1976)

Echinochloa crus-galli (Linn.) Beauv.—good grass for grazing and cattle fodder, usually fed green but not used for hay (Chakravarty, 1976)

Enneapogon persicus Boiss.—perennial grass forming dense tufts; Egypt, Israel, Jordan, Arabia, Iran, Afghanistan; a useful pasture and fodder plant (Bor and Guest, 1968)

Hordeum bulbosum Linn.—tall perennial grass with a bulbous swelling at the base; Mediterranean to Afghanistan; grazed by most animals when young; useful constituent of pasture when food supply short; sometimes dried and preserved as winter hay in the mountains of Iraq (Bor and Guest, 1968)

Lasiurus hirsutus (Forsk.) Boiss.—perennial grass; Egypt, Arabia, Iraq, Afghanistan; an excellent desert fodder grass, grazed by sheep, camels, and other animals (Bor and Guest, 1968)

Lolium perenne Linn.—perennial grass; throughout the Middle East; a very good fodder grass for pasture or hay, superb as grazing grass [Many strains have been raised from this grass, which gives heavy yields of nutritious food for cattle (Chakravarty, 1976).]

Panicum repens Linn.—perennial grass; Mediterranean to Iraq; a good forage grass (Bor and Guest, 1968)

Panicum turgidum Forsk.—good fodder for camels, horses, donkeys, and other animals (Bor and Guest, 1968)

Pennisetum divisum (Gmel.) Henrard—perennial grass; deserts of the Middle East; good fodder for camels, goats, and sheep

Phalaris minor Retz.—tufted annual grass; throughout the Middle East; a useful forage grass, generally grazed by sheep, cattle, and other livestock, though reported to be injurious to horses and young animals (Bor and Guest, 1968); grains used as feed for poultry because of high protein content (Chakravarty, 1976)

Poa bulbosa Linn.—perennial grass, bulbous at base; Mediterranean to Afghanistan; though rather tough and wiry, provides spring grazing for livestock in upper plains, foothills, and lower mountain pastures of Iraq (Bor and Guest, 1968)

Poa pratensis Linn.—perennial grass; Mediterranean to Afghanistan; an important fodder grass, numerous strong-growing rhizomes enable it to withstand extremes of drought, heat, and cold better than many other useful grasses; thrives best on loose soils and less suitable for heavy compact soils (Bor and Guest, 1968)

Poa sinaica Steud.—perennial grass; throughout the Middle East; provides invaluable pasture for sheep; in desert between Damascus and Baghdad some travellers in March (after winter rains) noted that it was the commonest grass and an excellent fodder; provides the main winter grazing in the desert (Bor and Guest, 1968)

Sorghum halepense (Linn.) Pers.—perennial grass; Mediterranean to Afghanistan; normally considered a good fodder for all livestock, also suitable for hay; if grazed in times of drought may cause death to young animals, while larger animals usually appear less seriously affected (Bor and Guest, 1968)

Stipa barbata Desf.—tough perennial grass; Mediterranean to Afghanistan; culm and foliage provide good grazing and nutritive value and make satisfactory hay (Bor and Guest, 1968)

Stipa capensis Thunb.—annual grass, branches at base; throughout the Middle East; important constituent of the herbage, particularly for sheep, which are grazed over immense tracts of land in dry steppes and subdesert pastures (Bor and Guest, 1968)

Stipagrostis ciliata (Desf.) de Winter—perennial grass; Egypt, Israel, Jordan, Arabia, Iraq, Iran, Afghanistan; good fodder for camels, goats, etc. (Bor and Guest, 1968)

Stipagrostis plumosa (Linn.) Anders.—perennial grass; Egypt to Afghanistan; a favorite and very fattening food for horses in Kuwait, and a good fodder for sheep and other animals (Chakravarty, 1976); good pasturage (Schulz and Whitney, 1985)

Stipagrostis raddiana (Savi) de Winter—perennial grass; Egypt, Jordan, Arabia, Iraq, southern

Iran, Afghanistan; [Täckholm and Drar (1941) noted that it was an excellent fodder grass in Sinai, and used to be plaited and exported to northern Arabia.]

Themeda triandra Forsk.—perennial grass; Egypt, Syria, Lebanon, Turkey, Arabia; highly valued as fodder plant, very nutritious and well liked by all kinds of livestock, and if cut before ripe makes excellent hay (Bor and Guest, 1968)

Leguminosae

Alhagi graecorum Boiss.—whole plant eaten by camels; considered a good fodder

Medicago ciliaris (Linn.) All.—annual weed; Mediterranean; good fodder for cows and buffaloes

Medicago lupulina Linn.—perennial; biennial or annual herb; throughout the Middle East; according to Zohary (1972), plant sometimes grown as green fodder or hay crop

Medicago orbicularis (Linn.) Bart.—annual herb; Mediterranean to Afghanistan; excellent pasture plant, also cultivated for hay and green manure (Zohary, 1972)

Medicago scutellata (Linn.) Mill.—annual herb; Mediterranean; sometimes grown as a hay or pasture plant (Zohary, 1972)

Melilotus alba Medic.—annual or biennial herb; throughout the Middle East; cattle come to like plant as a graze, but sometimes get colic after grazing it (Townsend and Guest, 1974); and may also be cut for hay

Trifolium fragiferum Linn.—creeping perennial; throughout the Middle East; useful pasture plant (Townsend and Guest, 1974)

Trifolium pratense Linn.—perennial decumbent herb; Turkey, Iran, Afghanistan; a nutritous fodder just before blooming, soon becoming tough and hairy; cultivated for pasture, hay, and silage (Townsend and Guest, 1974)

Trifolium resupinatum Linn.—annual herb; throughout the Middle East; widely recognized as a useful pasture, and tolerates a certain degree of salinity (Townsend and Guest, 1974)

Trifolim subterraneum Linn.—prostrate annual; Mediterranean to Iran; a valuable pasture (Townsend and Guest, 1974)

Several other *Trifolium* species, which are known from different regions of the Middle East, are valued for their grazing or pasture use, for example *Trifolium ambiguum* M. Bieb., *T. arvense* Linn., *T. campestre* Schreb., *T. echinatum* M. Bieb., *T. hirtum* All., *T. hybridum* Linn., *T. purpureum* Lois., *T. repens* Linn., *T. scabrum* Linn., *T. spumosum* Linn., *T. stellatum* Linn., and *T. tomentosum* Linn. (cf. Townsend and Guest, 1974).

Malvaceae

Althaea ludwigii Linn.—perennial herb; Mediterranean to Afghanistan; plant palatable to sheep (Boulos, 1977)

Malva parviflora Linn.—palatable to sheep (Boulos, 1977)

Polygonaceae

Calligonum comosum L'Hérit.—leafless, richly branched shrub; most deserts of the Middle East; preferred pasturage of goats and camels (Schulz and Whitney, 1985)

Resedaceae

Ochradenus baccatus Del.—glabrous shrub, old branches spinescent, leafless; southeastern Mediterranean and Arabia; browsed by camels (Zohary, 1966) and other animals (Schulz and Whitney, 1985)

Salvadoraceae

Salvadora persica Linn.—shoots and leaves used as fodder (Zohary 1972)

Zygophyllaceae

Fagonia olivieri DC.—small glabrous shrub; Egypt (Sinai), Arabia; heavily browsed by animals (Schulz and Whitney, 1985)

C. Medicinal Plants

Acanthaceae

Blepharis ciliaris (Linn.) B. L. Burtt—spiny, herbaceous, short-lived perennial; southeastern Mediterranean to Iran and Arabia; seeds antiinflammatory, for wound healing (Boulos, 1983)

Anacardiaceae

Rhus tripartita (Ucria) Grande—dioecious shrub with spiny branches; Mediterranean, with extensions eastward; infusion of fruits and leaves recommended for gastric and intestinal ailments (Boulos, 1983)

Apocynaceae

Nerium oleander Linn.—evergreen shrub or small tree in water course rocky areas; Mediterranean to Arabia and Iraq; poisonous, not used orally; macerated leaves used for treating itch and loss of hair; decoction of leaves and bark used as antisyphilitic; leaves toxic and cause paralysis by contact or ingestion, hence used as insecticide; extracts of plant show cardiotonic activity; pounded dry leaves used for ulcers in animals (Boulos, 1983)

Asclepiadaceae

Calotropis procera (Ait.) Ait. f.—shrub or small tree with succulent leaves, rich in latex; Egypt, Israel, Jordan, Arabia to Afghanistan; dry leaves smoked as cigarettes for asthma; cataplasm of fresh leaves used for sunstroke; leaf extracts cardiotonic; root used as emetic and expectorant; root bark used for dysentery, elephantiasis, syphilitic ulcers, and as a stomachic and diaphoretic; latex used for scabies of camels and goats, and applied on teeth to loosen them, also for toothache; decoction of bark and latex used for scabies and leprosy (Boulos, 1983)

Leptadenia pyrotechnica (Forsk.) Decne.—Egypt, South Israel, South Jordan, and Arabia; branches diuretic; infusion of branches used to alleviate retention of urine and to help expel uroliths (Boulos, 1983)

Solenostema arghel (Del.) Hayne—small shrub; Egypt, Israel, Jordan, Syria, Arabia; effective remedy for cough; infusion of leaves used for gastrointestinal cramps, colds, urinary tract problems, and as a stomachic and anticolic medication; antisyphilitic if used for prolonged periods of 40 to 80 days (Boulos, 1983)

Balanitaceae

Balanites aegyptiaca (Linn.) Del.—spiny tree; Egypt, Israel, Jordan, Arabia; anthelmintic, purgative, malaria emetic, vermifuge, febrifuge; used to treat boils, leucoderma, herpes, wounds, syphilis, colds, liver and spleen problems, and aches (Ayensu, 1979); kernel extracts have low activity against a snail that harbors schistosomal worms; fruit kernels and fruits used as mild laxative; fruit used as antidote to arrow poison; leaves used to clean malignant wounds; bark fumigant used to heal circumcision wound; root extract used to treat malaria (Boulos, 1983)

Berberidaceae

Leontice leontopetalum Linn.—perennial herb with large subterranean tuber; Mediterranean to Iran; tubers used as remedy for epilepsy (Zohary, 1966)

Boraginaceae

Heliotropium bacciferum Forsk.—richly branched perennial with woody base; Egypt, southern Israel, southern Jordan, Arabia; plasters from dried, powdered leaves used to treat abscesses, boils, sprains, contusions, edema, and swellings of all kinds; plasters also used for antiscabies preparations in veterinary medicine (Boulos, 1983)

Capparaceae

Capparis decidua (Forsk.) Edgew.—spiny shrub or small tree; southern Egypt, southern Israel, Southern Arabia; ash of bark used as hemostatic, astringent, and disinfectant for wounds and sores; green branches used as analgesic for rheumatism; also used to treat scabies in camels, cardiac problems, boils, swellings, toothache, cough, asthma, inflammation, fever, rheumatism; astringent, laxative, diaphoretic, anthelmintic (Ayensu, 1979; Boulos, 1983)

Maerua crassifolia Forsk.—small tree or shrub; Egypt, Israel, and Jordan in the wadis and oases of the Dead Sea area (Zohary 1966), and Arabia; infusion of leaves used for intestinal diseases; decoction of leaves and bark used as a febrifuge, and for cephalalgia, toothache, infected hairy skins; mixture of powdered leaves with henna leaves (*Lawsonia inermis*) and fat used for rapid healing of wounds and sores, cataplasm of this mixture reduces pain of bone fractures and aching parts of the body (Boulos, 1983)

Stellaria media (Linn.) Vill.—annual delicate herb; moist and shaded parts of gardens and orchards in most of the Middle East; essence of fresh plant taken to relieve rheumatic pains and psoriasis; externally used as a rub to ease rheumatic pains; used as a diuretic, emollient, and mild astringent; decoction used externally for wounds and ulcers (Boulos, 1983)

Vaccaria pyramidata Medic.—annual glabrous herb; Mediterranean extending westwards, and Arabia; root used as vulnerary for abscesses, furuncles, ulcers, scabies, mastitis, lymphangitis, and as an emmenagogue and galactogogue; contraindicated in pregnancy; decoction of root used to take care of wounds, sores, scabies, and different dermal infections, with consideration given to its toxicity; leaves and roots contain saponins provoking, in strong doses, a general paralysis of muscles (Boulos, 1983)

Chenopodiaceae

Atriplex halimus Linn.—roots cut into long, narrow pieces used as toothbrushes; remedy for hydrops; ash of plant rich in alkaline salts, taken with water for gastric acidity; seeds in small doses used as emetic, large doses poisonous (Boulos, 1983)

Cornulaca monacantha Del.—shrub with sharp, spiny leaves; Egypt, Israel, Jordan, Arabia; docoction of leaves used for treatment of jaundice (Boulos, 1983)

Haloxylon articulatum (Cav.) Bunge—plant used to treat internal ulcers (Chakravarty, 1976)

Compositae

Achillea santolina (Linn.—perennial herb; Egypt to Afghanistan; leaves used to treat dysentery, intenstinal colic, and for expulsion of gases; insect repellent, kills flies on which it acts as a poison (Chakravarty, 1976); pain of toothache reduced by rubbing young flowering branches against teeth, or by chewing them; anthelmintic, stomachic (Boulos, 1983)

Ambrosia maritima Linn.—annual or perennating aromatic herb; Mediterranean; contains ambrosin and damsin, both cytotoxic principles (Ayensu, 1979; decoction of plant used for rheumatic pains, asthma, bilharziasis, diabetes, and to expel kidney stones; flowering branches used as stimulant, stomachic, emolient, vulnerary, diuretic, for renal troubles, and are slightly astringent (Boulos, 1983)

Artemisia herba-alba Asso—fragrant shrub; Mediterranean to Iran; yields a volatile oil that is anthelmintic and promotes digestion; dry powder used for healing wounds and burns (Boulos, 1970); and to treat toothache (Feinbrun-Dothan, 1978); leaves and flowers used as a febrifuge, as a calmative for the stomach, and to treat cephalalgia; also cures nervous troubles, calms the emotions, and used for ophthalmic diseases; a component of mixtures used for treating hemorrhagic wounds; infusion of flowering branches used as vermifuge, emmenagogue, tonic, stomachic; dry powdered plant used for healing wounds and burns, and as a diuretic; infusion used for rheumatism and bronchitis; cataplasm of boiled flowers used to ripen and cure abscesses; antidiarrhetic; essential oil distilled from plant used as good antiseptic and insecticide, also used as parasiticide in veterinary medicine (Boulos, 1983)

Artemisia judaica Linn.—richly branched perennial with woody base, strongly aromatic, tomentose; Egypt, Israel, Jordan, Arabia; vapors from leaves inhaled to relieve congestion of colds; smoke of burnt branches keeps snakes away; believed to prevent skin diseases in camels when eaten by same; infusion of capitula relieves gastrointestinal cramps; infusion of capitulescences used as stomachic (Boulos, 1983)

Centaurea calcitrapa Linn.—annual or biennial herb; Mediterranean; whole plant used as bitter astringent, appetizer, vulnerary, antiophthalmic, antifebrile, stomachic, and for intermittent fever and eye diseases; leaves used for cephalalgia; capitula used as febrifuge; root and fruits diuretic; seeds used as vulnerary, febrifuge, and for renal stones and pains (Boulos, 1983)

Cotula cinerea Del.—annual glabrous herb; sandy desert regions of Egypt, Israel, Jordan, and Arabia; infusion of flower heads aromatic and very agreeable, used as stomachic and to flavor tea, replacing peppermint; useful for bronchopulmonary conditions, and for treating scorpion bites, rheumatism, vomiting, nausea, and stomach pains (Boulos, 1983)

Gundelia tournefortii Linn.—white latex causes nausea (Feinbrun-Dothan, 1978)

Inula viscosa (Linn.) Ait.—perennial herb; Mediterranean; infusion used for treatment of bronchitis, tuberculosis, and anemia; also used as astringent (Nauroy, 1954)

Lactuca serriola Linn.—annual or biennial herb; Mediterranean, Arabia, extending eastwards; plant used as calmative; decoction used as antivenin for scorpion and snake bites (acting by hypotonic effect of the sap); emollient, antispasmodic, diuretic (Boulos, 1983)

Silybum marianum (Linn.) Gaertn.—tincture produced from seeds used for liver disorders, jaundice, gallstones, peritonitis, coughs, bronchitis, congestion of the uterus, and varicose veins (Schauenberg and Paris, 1977)

Convolvulaceae

Convolvulus arvensis Linn.—twining herb; throughout the Middle East; cathartic, purgative; root extracts possess antihemorrhagic and purgative effects (Ayensu, 1979)

Convolvulus glomeratus Choisy—perennial herb; southern Israel, southern Jordan, southern Iran; purgative (Ayensu, 1979)

Convolvulus hystrix Vahl—spiny shrub; Egypt, Arabia to Afghanistan; purgative (Ayensu, 1979)

Convolvulus scammonia Linn.—perennial twining or trailing herb; Israel, Jordan, Syria, Iraq, Turkey; hydrogogue, cathartic, and used to treat dropsy and anasarca (Ayensu, 1979)

Cruciferae

Anastatica hierochuntica Linn.—annual herb; Egypt, Israel, Jordan, Syria, Arabia, Iraq, Iran; believed to facilitate childbirth and reduces pain during delivery; infusion taken against colds, and used to treat epilepsy; emmenagogue (Boulos, 1983)

Capsella bursa-pastoris Linn.—annual herb; throughout the Middle East; astringent (Fourment and Roques, 1941); effective against uterine hemorrhages and nasal and internal bleeding (Schauenberg and Paris, 1977); infusion of dried plants effective remedy for menorrhagia, dysmenorrhea, and uterine hemorrhages; homeopathic tincture used for nasal and internal bleeding, also used in cases of cystitis and for stones in the urinary tract (Boulos, 1983)

Nasturtium officinale R. Br.—antiscorbutic, aphrodisiac (Nauroy, 1954)

Sisymbrium irio Linn.—annual herb; Egypt, Arabia, Israel, Jordan, Syria, Lebanon, Turkey, Iraq, Iran (Zohary *et al.*, 1980); extracts of plant lower body temperature (Ayensu, 1979)

Zilla spinosa (Linn.) Prantl—perennial spiny shrub; Egypt, Arabia, Israel, Jordan, Syria, Lebanon, Iraq (Zohary *et al.*, 1980); useful remedy in the treatment of ailments such as kidney stones (El-Menshawi *et al.*, 1980)

Cucurbitaceae

Bryonia cretica Linn.—dioecious perennial, rarely monoecious; Mediterranean; contains cucurbitacins, which have cytotoxic and cathartic effects (Ayensu, 1979)

Citrullus colocynthis (Linn.) Schrad.—perennial herb; widely distributed in the arid Middle East; veterinary preparations used to treat itch very often contain colocynth; cataplasm of green or dried plant used as a remedy for leukoderma, and as a resolvent and astringent; hot sap of plant used to cure certain skin diseases of camels; decoction of roots mixed with garlic externally used to treat snake bites; sap of unripe green fruits used against scorpion stings; dry fruits used as effective insecticide, especially for mites and weevils; fruit pulp violent purgative, diuretic, and antiepileptic, and used to treat gonorrhoea; garlic added to extract of boiled seeds used to treat bites of poisonous snakes; seeds used as anthelmintic, abortive, and gonorrhoea; one seed daily, swallowed without chewing, for 21 days, useful remedy for diabetes (Boulos, 1983)

Ecballium elaterium (Linn.) A. Rich.—perennial succulent herb; Mediterranean, extending to the east, on neglected land and roadsides; whole plant toxic and fruit juice violent purgative, diuretic, and emetic; fruit dipped in oil used as suppository against piles; fruit juice used as nose drops, and for jaundice through nasal instillation, once a day for one week, during which period the patient eats one raw egg with an equal volume of oil (Boulos, 1983)

Cupressaceae

Juniperus phoenicea Linn.—monoecious shrub or tree, leaves evergreen; Egypt (northern Sinai), southern Jordan, northwestern Arabia (Zohary, 1966); hot infusion of leaves used to treat children's diarrhea, and as sedative for abdominal pains; dry powdered leaves used to cure mild dermal inflammations, especially on babies, and as a dilator for the urinary tract, laxative, and intestinal disinfectant; leaves used as emmenagogue, antidiarrhetic, and help during childbirth by increasing contractions of the uterus (Boulos, 1983)

Cynomoriaceae

Cynomorium coccineum Linn.—succulent unbranched perennial; most deserts of the Middle East; entire plant aphrodisiac, spermatopoietic, tonic, astringent; dried plant powdered and mixed with butter used for biliary obstruction; powder also added to meat dishes as a condiment (Boulos, 1983)

Cyperaceae

Cyperus rotundus Linn.—perennial herb; widely distributed weed in damp ground of gardens, orchards, and fields, especially in warmer regions of the Middle East; tubercles contain 0.5% essential oil used as aromatic stomachic to treat nervous gastralgia, dyspepsia, diarrhea, and as an emmenagogue, sedative, analgesic (for dysmenorrhea, amenorrhea, and chronic metritis), diuretic, carminative, and stimulant; also used as a colic remedy, and to remove renal calculi; tonic, aphrodisiac, anthelmintic, analeptic, condiment, stomachic, and to increase body weight; in fresh state used as diaphoretic, astringent, and for scorpion stings; decoction of tubercles diaphoretic; infusion of tubers used for intestinal pains (Boulos, 1983)

Datiscaceae

Datisca cannabina Linn.—bush dioecious herb with aspect of hemp plant (*Cannabis*); Mediterranean to Afghanistan; plant is bitter; used as a diuretic, expectorant, purgative, and to treat fevers, and gastric and scrofulous troubles (Chakravarty, 1976)

Ephedraceae

Ephedra alata Decne.—shrub; Egypt, Israel, Jordan, Arabia, Iraq; bronchodilator, antiasthmatic activity due to ephedrine and related alkaloids (Ayensu, 1979); whole plant used as a depurative, and for treatment of hypertension; also used as antiasthmatic, sympathomimetic, and astringent; branches chewed for cephalalgia, and cooked in butter and eaten by women for miscarriage (Boulos, 1983)

Euphorbiaceae

Euphorbia falcata Linn.—annual herb; Mediterranean to Iran; infusion mixed with milk recommended for colds and rheumatic pains (Boulos, 1983)

Euphorbia hirta Linn.—annual herb; weed in irrigated fields of Eygpt, Israel, Jordan, Lebanon, Syria, and hot areas of the Middle East; extracts have antitumor, antihistamine, and immunostimulant activities, and are free of irritant effects; extracts used to treat sores, eye trouble, dysentery, cough, warts, colic, wounds, tumors, boils, asthma, deficiency of breast milk, snake bite, worms, gonorrhea, and depressed heartbeat and respiration; antibacterial, antitubercular, tonic, cathartic, diuretic, purgative (Ayensu, 1979)

Euphorbia peplis Linn.—annual herb; Mediterranean, in coastal sand; latex used as diuretic, depurative. purgative, emetic, expectorant, and for asthma attacks, liver disorders, hydrops, gout, and chest diseases (Boulos, 1983)

Mercurialis annua Linn.—dioecious annual herb; Mediterranean to Afghanistan; tincture of plant recommended for rheumatism and gastric disorders; fresh plant sometimes used as laxative, purgative, and diuretic (Boulos, 1983)

Ricinus communis Linn.—glabrous, richly branched shrub; native to southeastern Egypt, but cultivated and naturalized in most of the Middle East; root bark used as purgative; root decoction used for rheumatism, inflammatory affections, chronic enlargements, skin disease, abdominal pains, diarrhea, nervous disorders, toothache, lumbago, sciatica, jaundice, and kidney and bladder trouble; leaves rubbed on joints to relieve pain, and used as an emmenagogue; decoction used as purgative, has useful action on liver and kidneys, and is heated and placed on head for feverish headache, on breast to stop milk secretion, on abdomen to promote menstrual flow, and also used as emetic for poisoning; cataplasm of leaves used for carbuncles, wounds, lacteal tumors, indurations of the mammary gland, swellings, stomachache, rheumatism, lumbago, sciatica; cataplasm of fresh leaves cures boils; infusion used to treat itch; powdered seeds toxic, applied externally to abscesses and carbuncles, prescribed in small doses for persistent constipation, and used to treat buccal and pharyngeal inflammations, gastrointestinal hemorrhages, headaches, dizziness, stupor, and hypothermy; pollen is strong allergen; dried flowers mixed with honey used as antidiarrhetic; seed oil used as purgative, emetic, emollient, and used externally as a rub to treat rheumatic pains and paralysis of arms and legs; Seed oil also used for tumors, boils, worms, headache, ophthalmia, wounds, rash, epilepsy, stomachache, colic, itch, lumbago, jaundice, rheumatism, fever, diarrhea, toothache, kidney and bladder trouble, asthma, cancer, leprosy, and venereal and skin diseases; seeds used to treat scorpion sting, fever, inflammatory affections, parasitic skin diseases, leprosy, cancer, and as a cathartic and emollient; seeds destroy mosquitos (Boulos, 1983)

Fumariaceae

Fumaria judaica Boiss—delicate glabrous annual herb; in damp places, southeastern Mediterranean; entire plant used for treatment of jaundice, hepatic fever, and spleen obstructions; used externally for scabies and dermatitis (Boulos, 1983)

Fumaria parviflora Lam.—glabrous annual herb; weed in fields, gardens, and orchards throughout most of the Middle East; entire plant astringent, antipruriginous, and sedative; infusion recommended for children to maintain their vitality, and as a depurative, laxative, and diuretic (Boulos, 1983)

Gentianaceae

Centaurium erythraea Rafn. (= *C. umbellatum* Gibil.)—annual or biennial herb; Mediterranean to Afghanistan; flowering and fruiting plants used to make a bitter tonic; infusion used as a febrifuge and depurative, and as a treatment for anemia, gastralgia, biliary stones, gastric troubles, and for intermittent fevers; decoction useful for diabetes; infusion used for headache and fever (Boulos, 1983)

Centaurium spicatum (Linn.) Fritsch—annual herb; marshes and moist ground of many regions of the Middle East; flowering and fruiting plants used for hypertension, elimination of ureter and kidney stones, and as a healing agent for wounds; a component of ointments used for sciatica; infusion useful for diabetes (Boulos, 1983)

Geraniaceae

Erodium cicutarium (Linn.) L'Hérit.—Flowering branches used as hemostatic, astringent, antidiarrhetic, as a healing agent for wounds, and helps during childbirth by increasing contractions of the uterus (Boulos, 1983)

Gramineae

Cymbopogon proximus (Hochst.) Stapf—tufted, perennial, aromatic grass; Egypt, Arabia; dried tufts of plant burnt and fumes inhaled to treat influenza and some neurotic diseases, or tufts soaked in water for one week and cold infusion drunk; decoction of lower part of plant (without the flowers) used for colic and fever; decoction of a mixture of the plant and *Ambrosia maritima* used for diabetes; infusion of leaves used as diuretic, stomachic, and lowers blood pressure (Boulos, 1983)

Cymbopogon schoenanthus (Linn.) Spreng.—perennial, densely tufted grass; Egypt and Arabia; infusion of plant used as diuretic, emmenagogue, astringent, carminative, sudorific, and antirheumatic; cataplasm used for wounds of camels; infusion of flowers used as febrifuge (Boulos, 1983)

Cynodon dactylon (Linn.) Pers.—perennial rhizomatous herb; throughout the Middle East; decoction of rhizomes used for renal and urinary troubles, and as a depurative, emmenagogue, diuretic, refreshing agent, sudorific, emollient, and disinfectant vulnerary; decoction also used for cough, supression of urine and vesical calculus, and for purifying the blood (Boulos, 1983)

Echinochloa crus-galli (Linn.) Beauv.—used for treating spleen diseases and for checking hemorrhage (Chakravarty, 1976)

Imperata cylindrica (Linn.) Beauv.—perennial rhizomatous herb; hot regions of the Middle East; rhizome used as antipyretic, diuretic, and hemostatic (Keys, 1976)

Lolium perenne Linn.—plant contains several alkaloids, mainly perlolidine and perloline; perloline sometimes suggested as treatment for rheumatism; infusion of plant drunk with an astringent wine for stopping diarrhea and hemorrhage

Lolium temulentum Linn.—tall annual grass; throughout the Middle East; decoction of entire plant prescribed for hemorrhage and incontinence or urine; mature grains used to prepare a tincture in homeopathy to treat neuralgia, rheumatism and arthritis, nausea, nose bleeds, intestinal cramps, and trembling limbs

Phragmites australis (Cav.) Trin. *ex* Steud.—perennial grass; in wet places throughout the Middle East; rhizome used as stomachic, antiemetic, antipyretic, diaphoretic, diuretic, and for acute arthritis, jaundice, pulmonary abscesses, and food poisoning (Boulos, 1983)

Juncaceae

Juncus acutus Linn.—perennial tufted herb; in marshy places, mainly Mediterranean; infusion of fruits mixed with barley grains useful for colds (Boulos, 1983)

Juncus rigidus Desf.—perennial rhizomatous herb; throughout the Middle East; seeds used as diuretic and antidiarrhetic, and also used for dysuria and hydrops (Täckholm and Drar, 1950)

Labiatae

Ajuga iva (Linn.) Schreb.—small villous perennial herb; Mediterranean; cold infusion of herb used as anthelmintic and depurative, especially in springtime; dried green plants (powdered) taken to treat sinusitis and cephalalgia, powder made into small balls then digested for its useful action against stomach and intestinal pains, digestive tract upsets, and enteritis; considered a general preventive for all diseases; powdered dried plant or its hot infusion taken after meals to treat diabetes; infusion of flowering branches used as antidiarrhetic, vulnerary, depurative, and very effective vermifuge, and for feminine sterility, colds, and troubles of the digestive tract (Boulos, 1983)

Lavandula dentata Linn.—densely branched shrub; Arabia; infusion of inflorescences taken first thing in the morning and before going to bed for cases of urine retention, and to remove kidney or ureter stones (Boulos, 1983)

Marrubium vulgare Linn.—perennial, richly branched, white, densely tomentose herb; throughout the Middle East; infusion of leaves recommended for cases of jaundice, diabetes, fever, typhoid, and typhus, and also takes part in the treatment of otitis and eczema, reputed diuretic and emmenagogue; infusions or decoctions, mainly of the fresh plant (easy to get throughout the year), mixed with dry raisins or honey used as febrifuge in the treatment of malaria, typhus, and liver infections (particularly jaundice); used to treat pulmonary troubles and headaches, often by inhaling fresh juice of plant; tonic and stimulant in cases of anemia and convalescence; antidiabetic; has regulatory action in cases of cardiac troubles; infusion taken after meals as antidiabetic, antirheumatic, and for colds; strong doses of infusion purgative and emetic; infusion of inflorescences used as expectorant for various respiratory disorders; tincture used for gastrointestinal, hepatic, and biliary disorders; hot infusion used as febrifuge, antidiabetic, pectoral, tonic, cholagogue, and for stomach troubles (Boulos, 1983)

Mentha pulegium Linn.—perennial herb; in moist ground usually near canals, Mediterranean, extending eastwards; infusion of leaves and flowering branches used as antispasmodic, antiseptic, cholagogue, bechic, and insecticide; infusion, cataplasm, or inhaling vapors of fresh plant used for colds and catarrh, and for infections of the throat, bronchials, and lungs; flowering branches used as carminative, expectorant, anticatarrhal, and disinfectant; infusion used for abdominal ailments (Boulos, 1983)

Ocimum kilimandscharicum Guerke—aromatic herb; Turkey; used to treat colds, coughs, sore eyes, measles, abdominal pains, and diarrhea (Ayensu, 1979)

Otostegia fruticosa (Forsk.) Schweinf. *ex* Penzig—shrub; Egypt, Arabia; infusion or cataplasm of flowering branches good remedy for sunstroke (Boulos, 1983)

Salvia fruticosa Mill. (= *S. triloba* Linn. f.)—aromatic shrub; Israel, Jordan, Syria; infusion of leaves and young shoots used for influenza, cough, and rheumatic pains (Boulos, 1983)

Teucrium polium Linn.—richly branched aromatic shrub; Mediterranean to Iran; hot infusion of tender parts of plant taken for stomach and intestinal troubles; plant used in steam bath to treat colds and fevers, useful against smallpox and itch; infusion used as vermifuge, stimulant, tonic, astringent, vulnerary, depurative, and for feminine sterility and colds (Boulos, 1983)

Thymus capitatus (Linn.) Hoffmanns. and Link—dwarf aromatic shrub; Mediterranean; cold infusion of flowering branches used for coughs and as a tonic; drinking hot infusion helps in cure of some skin diseases (Boulos, 1970)

Leguminosae

Acacia modesta Wall.—deciduous small tree; Afghanistan; seeds in diet of rats produced a 25% drop in blood-sugar levels (Ayensu, 1979)

Acacia nilotica (Linn.) Willd. *ex* Del., including *A. arabica* (Lam.) Willd.—large evergreen tree; Egypt to Iran; gum exudates used as antidiarrhetic, stem-bark extract used as antiamebic, antispasmodic, and hypotensive; infusion of fruits used for diarrhea, especially in children; powdered fruits used to treat fevers, as a cure for sore gums and loose teeth, and for diabetes by taking a teaspoonful first thing in the morning; nubians in southern Eygpt believe that diabetic patients may eat as much carbohydrate-rich food as they please without risk if they regularly take powdered pods; mixture of powdered pods and henna (*Lawsonia inermis*) leaves used in a plaster effective against some skin diseases and irritated parts between toes; applied plaster must be left to dry for a few days, then washed with cold water (Boulos, 1983)

Alhagi graecorum Boiss.—infusion of branches used as laxative, vermifuge, purgative, and used to cure rheumatic pains and bilharziasis (Boulos, 1983)

Anagyris foetida Linn., shrub or small tree; Mediterranean, extending eastwards, Arabia; leaves, branches, and root used as purgative, vermifuge (in moderate doses), and emetic, and as an emmenagogue in higher doses; powdered leaves used externally to treat migraine, tumors, edema, scrofula, and ulcers; cataplasm of leaves used as resolvent; fruit toxic; seeds used as emmenagogue, emetic, purgative, and for renal affections (Boulos, 1983)

Astragalus gummifer Labill.—small shrub; Turkey, Iran; major source of gum tragacanth, a suspending agent and emulsifier (Ayensu, 1979)

Cassia italica (Mill.) Lam. *ex* F. W. Andrews—shrub; Egypt, Israel, Jordan, Arabia, Iraq, Iran, Arabia; leaflets and pods well known for purgative effect; crushed seeds used for ophthalmic diseases (Boulos, 1983)

Cassia senna Linn. (= *C. acutifolia* Del.)—small shrub; Egypt, Israel, Jordan, Arabia, Iraq; infusion of powdered leaflets and pods used as popular laxative and purgative; infusion of leaves purgative, often mixed with leaves of roses (Boulos, 1983)

Ceratonia siliqua Linn.—entire pod or seeds alone used principally against diarrhea in children and babies, also adults; fruit astringent, diuretic, laxative, and bechic (Boulos, 1983)

Glycyrrhiza glabra Linn.—perennial herb, woody at base; Mediterranean; fields licorice extract, and flavoring; estrogenic, relaxes uterine muscles (Ayensu, 1979)

Glycyrrhiza glandulifera Waldst. and Kit.—Iran; medicinal uses probably similar to those of *G. glabra*

Lupinus albus Linn.—annual villous herb; Mediterranean; seeds used as vermifuge; liquid left after soaking bitter seeds in water used as parasiticide and emollient of the skin for scurf, tinea, and itch; cataplasm of seeds used as emollient and resolvent; powdered dry seeds taken to treat diabetes (Boulos, 1983)

Melilotus indica (Linn.) All.—annual herb; weed in fields, gardens, orchards, etc., widely distributed in the Middle East; infusion of flowering branches emollient and antispasmodic; seeds used for diseases of the genital organs of both sexes (Boulos, 1983)

Ononis spinosa Linn.—perennial spinescent shrub; Mediterranean to Iran; root used as diuretic; infusion of flowers depurative; decoction of flowering branches externally used as antiseptic for eczema; infusion used for kidney stones and disorders of the urinary tract (Boulos, 1983)

Retama raetam (Forsk.) Webb.—leafless shrub; widespread and characteristic element of most deserts of the Middle East; used for making eyewash for eye troubles; root used against diarrhea;

branches used as febrifuge, and for treatment of wounds; powdered branches mixed with honey are emetic, also given as purgative and vermifuge, abortive in large doses; fruits toxic (Boulos, 1983)

Liliaceae

Aloe perryi Baker—succulent perennial; dried sap of leaves (the drug) detersive, desiccative, emmenagogue, powerful purgative, cathartic, insecticide, antiseptic, vermifuge, cholagogue, bitter tonic, and improves digestion; drug rarely used by itself (toxic in moderate doses), usually mixed with gum Arabic, honey, or sugar; also used as tonic, stomachic, and for amenorrhea, jaundice, and dyspepsia (Boulos, 1983)

Androcymbium gramineum (Cav.) McBride—herb with short stem and small corm; Egypt, Israel, Jordan (Täckholm and Drar, 1954); plant toxic to man and animals; corms antimitotic and contain colchicine, which is known to cause pain and congestion if applied to skin; when inhaled causes violent sneezing; internally, increases amount of bile poured into the intestine

Asparagus stipularis Forsk.—perennial spiny shrublet; Mediterranean and Arabia; roots used for syphilis; infusion of tuberous roots used to remove renal stones; infusion of pounded roots for headache; decoction of whole plant diaphoretic; young, tender shoots used as appetizer, stomachic, diuretic, and for jaundice, liver ailments, and rheumatism; berries, shoots, and roots used as stomachic and appetizer; decoction of shoots and riots used for treatment of syphilis; whole plant fried with eggs and camel's fat spermatogenic and aphrodisiac, and facilitates secretions and releases obstructions; decoction of seeds used to treat hemorrhoids (Boulos, 1983)

Asphodelus aestivus Brot. (= *A. microcarpus* Salzm. and Viv.)—perennial herb, root with spindle-shaped fleshy tubers; Mediterranean; cataplasm of leaves antirheumatic; rubbing body with roasted tubers and drinking decoction of leaves and scapes good remedy for withering and paralysis; for toothache, dried tubers cooked in oil, then oil poured into ear on opposite side of head from aching tooth; same oil used as ear drops to treat ears that are discharging liquids; powdered dried tubercles mixed with barley flour used as dressing against ulcers and abscesses of the breast; ash of tubercles diuretic and often used as eye powder against leucoma of the cornea; powdered, dry, tuberous roots locally applied to abscesses; exposing body to fumigated, dry, tuberous roots said to help against jaundice; fruits used for treating earache and acute toothache (Boulos, 1983)

Urginea maritima (Linn.) Baker—perennial herb with large bulbs, sometimes several kilograms each; Egypt, Israel, Jordan, Syria, Lebanon; handling all parts of plant causes irritation to skin; bulb toxic to man and rats; two varieties of bulbs are known: red and white, red squill mainly dried and powdered to furnish rat poison, white squill used in treatment of heart disease; and to prepare cough mixtures, also diuretic; fresh slimy bulbs applied to wounds and tumors for healing, and used as expectorant for bronchitis, chronic catarrh, and pneumonia; in strong doses emetic, cathartic, and upsets the nerves; fresh bulb vesicant, rubefacient, anthelmintic, and useful for rheumatism, edema, and gout; cardiac effect similar to *Digitalis,* slowing down the pulse and increasing its strength; emmenagogue, abortive, aphrodisiac; dried powdered bulbs made into tablets, sucked slowly in the mouth to treat internal tumors; infusion of dried bulb strong purgative (Boulos, 1983)

Malvaceae

Malva parviflora Linn.—annual herb; throughout the Middle East; root chewed or rubbed into gums for pyorrhea; infusion of flowering and fruiting branches used as gargle for astringent properties, prescribed for people on diets, and for gastrointestinal ailments, used as bechic and emollient; seeds and leaves used as cataplasm, rectal injection or gargle according to the case (Boulos, 1983)

Malva sylvestris Linn.—annual herb; throughout the Middle East; leaves and flowers soothing, pectoral, mildly astringent, intestinal stimulant, and laxative in large doses; infusion used for digestive and urinary diseases; decoction used externally for bathing, as a gargle, and for inflammations of the skin and mucous membranes; emollient; used to treat all occular ailments; infusion used for cough and diarrhea (Boulos, 1983)

Myrtaceae

Myrtus communis Linn.—evergreen aromatic shrub; Jordan, Israel, Syria, Lebanon, Turkey; infusion of whole plant used as stimulant and antidiarrhetic; roots astringent; infusion of leaves and young branches excellent remedy for asthma; leaves used as stomachic, astringent, and antiseptic for the respiratory system; infusion of leaves used for respiratory ailments and as cataplasm for painful organs, also antidiarrhetic; decoction of leaves used to blacken hair; dry leaves and flowers mixed with other aromatic herbs used for preparing hair lotions; leaves used as tonic, aromatic, and for preparation of "Water of the Angel" by distillation, for extraction of an antiseptic essence, and in perfumery; leaves used as fumigant or cataplasm to soothe pains of abscesses and boils; infusion of leaves used as eye lotion; leaves mixed with shampoo serve as hair tonic; essential oil (obtained by distillation) used as pulmonary antiseptic; dry flower buds used for smallpox; decoction of flowers has beneficial regulatory action on blood circulation (Boulos, 1983)

Nyctaginaceae

Boerhavia repens Linn.—perennial puberulent herb, woody at base; Egypt, Israel, Jordan, Arabia; purgative, emetic, antisyphilitic (Bellakhdar, 1978)

Oleaceae

Olea europaea Linn.—tree; cultivated in the Mediterranean region and elsewhere in the Middle East; decoction of olive wood used as remedy for mouth ailments, such as aphtha, stomatitis, etc.; same decoction applied on abdomen has curious effect of stopping diarrhea; decoction of leaves used for cough; leaves astringent, hypotensive, antidiabetic, and used as a cholagogue; extracts of leaves hypoglycemic, diuretic, and antibacterial; olive oil used as cholagogue, and for hepatic ailments, chronic constipation, asthenia, and bad appetite; oil from wild fruits used as antidote against all poisons, and as hair tonic; olive oil used as excellent purgative with no side effects; oil mixed with flour used as plaster for boils and abscesses (Boulos, 1983)

Cistanche phelypaea (Linn.) Coutinho—perennial succulent herb, root parasite on woody Chenopodiaceae; deserts of the Middle East; diuretic; thick lower part of plant dried and powdered for use against diarrhea; aphrodisiac and tonic in spermatorrhea impotence; dried, powdered plant mixed with camel's milk employed as cataplasm against contusions (Boulos, 1983)

Palmae

Hyphaene thebaica (Linn.) Mart.—doum palm, stem branched; Egypt and Israel; thick roots used for treatment of bilharziasis; resin from tree diuretic, diaphoretic, and recommended for tapeworm; resin used to treat bites of poisonous animals; fruit astringent (Boulos, 1983)

Phoenix dactylifera Linn.—date palm; throughout the Middle East; cultivated and subspontaneous; date palm wood used as toothbrush; terminal bud "djourmmar" used for intestinal hemorrhage, diarrhea, and jaundice; dates were used internally in medications designed to purge, to clear the enigmatic, or to regulate the urine; in vaginal pessaries, with other ingredients, they enhance fertility; used to relieve cough and are fleshforming; juice of boiled dates given to invalids to restore strength and assuage thirst; green dates reputed aphrodisiac and tonic; kernels of dates made into cataplasm used to treat ulcers of the genital organs; ash of kernels used to prepare eye lotion to treat blepharitis (Boulos, 1983)

Papaveraceae

Papaver dubium Linn.—annual herb; Egypt, Arabia, Syria, Lebanon, Turkey, Iraq, Iran (Zohary *et al.*, 1980); infusion of capsules given to children in cases of measles (Boulos, 1983)

Papaver rhoeas Linn.—annual herb; Egypt, Arabia, Syria, Lebanon, Iraq, Iran (Zohary *et al.*,

1980); capsules (often mixed with those of *Papaver somniferum*) used as soporific for babies and act against cardiac affections; flowers and grains used as pectoral, calmative, emollient, sudorific, expectorant, and are slightly narcotic; syrup produced from flowers used as sedative to soothe coughs (Boulos, 1983)

Pistaciaceae (Anacardiaceae)

Pistacia lentiscus Linn.—shrub or small tree; Mediterranean; resinous exudation of tree known as mastic employed as analgesic and sedative in gastralgia, cardiodynia, mastitis, peptic ulcer, boils, carbuncles, and also as antitussive and expectorant; resin prescribed as nervous calmative (infusion in tea) and emmenagogue, chewed to purify breath, and constituent of cosmetic products and depilatory creams; resin boiled with milk used for throat troubles; to treat ulcers, a teaspoonful of mastic pounded and mixed with equal amount of honey taken first thing in the morning for three weeks; decoction of root used for cough; dried and fumigated bark of trunk and branches locally applied to facilitate childbrith; infusion of leaves diuretic, astringent, and an emmenagogue; peeled nuts yield oil by hot extraction, which is effective against itch and used as a rub for treating rheumatism (Boulos, 1983)

Plantaginaceae

Plantago afra Linn.—annual herb; Mediterranean extending eastwards, Arabia; leaves and roots have same uses as those of *Plantago coronopus;* seeds, previously mixed with milk overnight, used against all types of dysentery, and gastroduodenal ulcers, diarrhea, and chronic constipation; seeds emollient, mechanical purgative; decoction of seeds or seeds soaked in water used as emollient and cooling agent for renal and urinary affections and inflammations, as well as for cases of internal hemorrhoids (Boulos, 1983)

Plantago coronopus Linn.—annual herb; deserts of the Middle East; roots used for hemorrhoids, malaria, and fevers; powdered leaves locally applied to wounds, burns, abscesses, bites, and inflammations; leaves used as astringent, hemostatic, vulnerary, and as an immediate analgesic (Boulos, 1983)

Plantago ovata Forsk.—annual herb; Mediterranean to Iran, Arabia; used to treat constipation and dysentery, and as a demulcent and laxative (Ayensu, 1979)

Polygonaceae

Polygonum aviculare Linn.—annual herb; weed in cultivated ground in most of the Middle East; plant astringent, vulnerary, and hemostatic; decoction of entire plant recommended for enteritis, diarrhea, dysentery, and bronchitis; leaves and stems diuretic, anthelmintic, and antidiarrhetic; externally employed as an emollient, and as an astringent for hemorrhoids, pruritus, and chancroid (Boulos, 1983)

Polygonum maritimum Linn.—annual herb; Mediterranean; infusion of leaves and roots astringent and antidiarrhetic; decoction or cataplasm used to reduce swellings and soothe burns (Boulos, 1983)

Rumex vesicarius Linn.—leaves used as laxative and tonic (Zohary, 1966); plant eaten fresh to treat jaundice, hepatic conditions, constipation, calculi, and bad digestion (Bellakhdar, 1978)

Portulacaceae

Portulaca oleracea Linn.—whole plant used as antiphlogistic and bactericide for bacillary dysentery, diarrhea, hemorrhoids, and enterorrhagia; used in prescriptions as antidiabetic; fresh leaves externally used as cataplasm for maturing of abscesses; whole plant used as aphrodisiac, emollient, calmative, diuretic, refreshing agent, antiscorbutic, and vermifuge; seeds used as calmative and slake thirst (Boulos, 1983)

Ranunculaceae

Ranunculus sceleratus Linn.—annual herb; in moist ground, Mediterranean and eastwards; tincture from fresh plant used in treatment of skin conditions such as herpes, eczema, erysipelas, and pruritus, and in rheumatic conditions such as sciatica, arthritis, and rhinitis; also used as plant appetizer and tonic, and useful for asthma, gonorrhea, and ulcers (Boulos, 1983)

Resedaceae

Reseda luteola Linn.—annual or biennial herb; mainly Mediterranean, extending eastwards; infusion of entire plant used for stomachache and diarrhea; infusion of leaves used as antidiarrhetic (Boulos, 1983)

Rhamnaceae

Frangula alnus Miller (= *Rhamnus frangula* Linn.)—shrub; Turkey to Afghanistan; bark widely used in extract form as a cathartic, purgative, and laxative (Ayensu, 1979)

Paliurus spina-christi Miller (= *Rhamnus paliurus* Linn., *Paliurus aculeatus* Lam.)—spiny shrub; Syria, Turkey, Iraq, Iran (Davis *et al.*, 1967); used as an astringent, diuretic, tonic, and anticathartic (Zohary, 1972)

Rhamnus alaternus Linn.—dioecious, evergreen, unarmed shrub or small tree; Mediterranean; bark used as laxative; leaves astringent (Zohary, 1972); bark emetocathartic; leaves and root rich in tannin; berries mild purgative and laxative (Boulos, 1983)

Rhamnus catharticus Linn.—spinescent shrub or small tree; Turkey, Iran; extracts of fruit used as cathartic, purgative, and laxative (Ayensu, 1977)

Ziziphus lotus (Linn.) Lam.—northwestern Egypt, Israel, western Syria, southern Turkey; fruits used for making pectoral cough drops and for preparing a tonic, and as a febrifuge; tonic prescribed for convalescents; active against furuncles, measles, and smallpox; emollient (Boulos, 1983)

Ziziphus spina-christi (Linn.) Willd.—ash of wood mixed with vinegar, prescribed for local application in treatment of serpent bites; infusion of leaves astringent, anthelmintic, and antidiarrhetic; Cataplasm of leaves used to treat abscesses and furuncles; cataplasm of fresh green leaves put on swollen eyes (probably due to certain inflammations) before going to bed; fruits component of infusions used as febrifuges, emollients, and laxatives; infusion of fruits reputed remedy for measles (Boulos, 1983)

Ziziphus zizyphus (Linn.) Meikle (= *Z. jujuba* Miller.)—tree; Turkey to Afghanistan; emollient, tonic; a 50% ethanol–water extract of shoots lowered blood pressure in dogs (Ayensu, 1979); leaves used as anthelmintic and antidiarrhetic; fruits used to treat cough for pectoral ailments, and used as emollient, calmative, anticatarrhal, and are slightly diuretic (Boulos, 1983)

Rosaceae

Rosa canina Linn.—shrub; Mediterranean; fresh fruit used as diuretic, antipyretic, and mild astringent (Zohary, 1972); fruit also antidiarrhetic, diuretic, and antiscorbutic (Boulos, 1983)

Rutaceae

Haplophyllum tuberculatum (Forsk.) A. Juss.—perennial aromatic herb, woody at the base; Mediterranean, extending eastwards, Arabia; flowering and fruiting branches used as febrifuge and aphrodisiac, and as a local antipoison, also for vomiting, nausea, constipation, malaria, difficult childbirth, anemia, rheumatism, gastric pains, intestinal worms, and eye and ear troubles; decoction used for rheumatic pains (Boulos, 1983)

Ruta chalepensis Linn.—shrub with pungent odor; Mediterranean; fresh plant used as scorpion repellant due to strong smell; leaves and seeds boiled in oil, oil then used as rub for rheumatic pains and swellings; infusion of entire plant used for colds, intestinal troubles, and earaches; hot infusion of plant cooled and used as ear drops for earaches; entire plant boiled in milk recommended for treating cases of nervousness; dried plant used as snuff for nasal diseases; its infusion used as nose drops to treat vomiting and fevers of children and babies (Boulos, 1983)

Salvadoraceae

Salvadora persica Linn.—plant used to treat gonorrhea, spleen ailments, boils, sores, gum diseases, and stomachache; wood boiled in oil constitutes a liniment against contusions; bark used as tonic; powdered bark used for bites of poisonous animals; leaves, roots, bark, and flowers contain oil, diuretic; powdered leaves mixed with millet flour and honey made into small balls and taken every morning for 40 days as antisyphilitic; fruits edible, used as stomachic, carminative, and febrifuge, and also to fortify the stomach and bring good appetite (Boulos, 1983)

Sapindaceae

Dodonaea viscosa (Linn.) Jacq.—water and ethanol extracts lethal to earthworms (Ayensu, 1979)

Scrophulariaceae

Verbascum sinuatum Linn.—biennial herb; Mediterranean, extending eastwards; roots and leaves popular drug for hygiene of eyes, having reputation of ameliorating sight; also used for inflammations, and as an antipoison; dried powdered plant used to treat all eye diseases by putting powder under cyclids; plant burnt, powdered, and mixed with alcohol, then applied to eyes in small doses for treating weakness of sight (probably due to cataracts) (Boulos, 1983)

Solanaceae

Hyoscyamus desertorum (Boiss.) Täckh.—perennial herb; Egypt, Israel, Jordan; reaction similar to belladona; contains alkaloids, which are classical parasympatholytic agents (Ayensu, 1979)

Hyoscyamus muticus Linn.—large herbaceous perennial; Egypt, Arabia to Iran; plant provides relief from painful spasmodic conditions of the nonstriated muscles, characteristic of lead colic and irritation of the bladder; also used to allay nervous irritation, hysteria, and irritable cough; cataplasm of fresh leaves used to allay pain; dried leaves smoked as cigarettes to treat asthma (Boulos, 1983)

Solanum nigrum Linn.—annual polymorphic herb; weed on cultivated land throughout the Middle East; plant toxic; cataplasm of entire plant used as calmative, emollient for burns, and for dermal affections; decoction of plant used as wash for burns, and for vaginal injection; diluted infusion of berries used as mydriatic eye lotion, ear drops, and as an emollient for external use; Berries narcotic, analgesic if used externally, and sedative; seeds (mixed with food) used as aphrodisiac (Boulos, 1983)

Withania obtusifolia Täckh.—shrub; Egypt; narcotic, antiepileptic (Ayensu, 1979)

Withania somnifera (Linn.) DUnal—shrub; throughout the Middle East; plant narcotic; employed against female sterility; antiepileptic, hypotonic; used to treat stomachache, ulcers, colds, rashes, and gonorrhea; roots used for treatment of rheumatic pains, and as a hypotonic and calmative; dried, powdered roots taken in samll doses by women for sterility; leaves and fruits used as febrifuge, diuretic, and antirheumatic; berries mildly laxative; seeds toxic, emetic, diuretic, and anesthetic (Boulos, 1983)

Sterculiaceae

Glossostemon bruguieri Desf.—perennial herb with thick cylindrical roots; Iraq, Iran; plant said to yield aphrodisiac principle; root decoction used against cough (Chakravarty, 1976); roots have high nutritive value; decoction of roots and other herbs traditionally offered to women after delivery in Egypt and some other countries in the Middle East

Tamaricaceae

Tamarix aphylla (Linn.) Karst.—tree or large shrub; Egypt, Israel, Jordan, Arabia, Iraq, Afghanistan (Baum, 1978); decoction of leaves and young branches used for edema of the spleen; same decoction mixed with ginger used for uterus affections; bark of large branches, boiled in water with vinegar, used as lotion against lice; infusion of galls astringent, and used for enteritis and gastralgia (Boulos, 1983)

Tamarix dioica Roth.—dioecious tree; Iran and Afghanistan; a 50% ethanol–water extract of roots showed antiviral and antispasmodic activity *in vitro*, and hypotensive activity *in vivo* (Ayensu, 1979)

Umbelliferae

Ammi majus Linn.—annual herb; Mediterranean; first used in Egypt for the treatment of leucoderma; seeds used as diuretic, carminative, tonic, digestive, stomachic, used for angina pectoris and asthma (Boulos, 1983)

Ammi visnaga (Linn.) Lam.—annual herb; Mediterranean; seeds used as diuretic, appetizer, carminative, stimulant, emmenagogue, lithotriptic, vasodilator, antispasmodic, emetic, purgative, and relieve congestion of the prostate gland; also used for urinary disorders, angina pectoris, asthma, and gastric ulcers; infusion of seeds releases renal stones (come out with urine), and used as gargle for toothache (Boulos, 1983)

Crithmum maritimum Linn.—perennial herb, woody at the base; Mediterranean; leaves aromatic, depurative, diuretic, and possibly anthelmintic; fruits and root stimulant, depurative, appetizer, antiscorbutic, and diuretic; essential oil from seeds and juice from leaves used as excellent vermifuge (Boulos, 1983)

Dorema ammoniacum D. Don—perennial; Iran; principal source of ammoniac, a gum resin; used as an antispasmodic, expectorant, and stimulant, and for bronchial, liver, and spleen ailments (French, 1971)

Dorema aucheri Boiss.—perennial; Iran; source of gum ammoniac, a gum resin; antispasmodic, expectorant (Ayensu, 1979)

Eryngium campestre Linn.—perennial herb; Mediterranean to Afghanistan; diaphoretic, expectorant, emetic, and used for urinary and uterine ailments (French, 1971); plant used as diuretic, appetizer, and laxative; root used as diaphoretic, expectorant, and as an emetic for urinary and uterine ailments; decoction of the root and flowering parts of plant used to treat kidney stones, to remove excess chlorine from the blood, and for some skin diseases (Boulos, 1983)

Eryngium creticum Lam.—annual prickly herb; Mediterranean to Iraq; leaves and roots used to treat anemia, dropsy, and colic; prolong perturbation period (Chakravarty, 1976)

Ferula assa-foetida Linn.—perennial herb; Arabia to Iran; expectorant, carminative, laxative, antispasmodic, stimulant, emmenagogue, and anthelmintic; used to treat hypochondriasis; used in veterinary medicine (French, 1971)

Ferula sumbul (Kauffmann) Hook. f.—perennial herb; Iran, Afghanistan; used as antispasmodic, stimulant, carminative, and to treat nervous-system disorders and cholera (French, 1971)

Foeniculum vulgare Miller—infusion of entire plant used for lumbago and abdominal pains; juice of fresh plant largely used in preparation of eye lotions for cataracts; green plant edible; plant cooked with food or infusion of plant used to treat liver ailments; fresh plant or infusion of dried plant used as vermifuge; root diuretic; fruit used as stimulant, carminative, stomachic, galactagogue, expectorant, antispasmodic, digestive, aphrodisiac, appetizer, tonic, calmative, and emmenagogue (Boulos, 1983)

Peucedanum aucheri Boiss.—perennial herb; Iran, Afghanistan; seeds and leaves used for indigestion (French, 1971)

Pituranthos tortuosus (Desf.) Benth. and Hook. f.—aromatic desert shrub; Mediterranean; mannitol, to which its diuretic action may be attributed, isolated from plant in appreciable quantities; plant also used as flavoring agent; tender parts eaten fresh to stimulate appetite (Boulos, 1983)

Scandix pecten-veneris Linn.—annual herb; Mediterranean to Afghanistan; used to treat bleeding amenorrhea and bladder affections (French, 1971)

Tordylium officinale Linn.—annual herb; Mediterranean; fruits used as emmenagogue, and component of antidote for animal (e.g. snake) poisons (French, 1971)

Urticaceae

Urtica pilulifera Linn.—annual monecious herb; Mediterranean, extending eastwards; mixture of plant juice with oil prescribed to cure sore joints; contents of stinging hairs provide cure for rheumatism and hemorrhage; decoction of summits used as diuretic and depurative; seeds used to treat renal stones and inflammation of the bladder, and as a diuretic and aphrodisiac (Boulos, 1983)

Urtica urens Linn.—annual monoecious herb; Mediterranean, extending to the east; fresh plants used as effective but painful rub to treat rheumatism, and as an antidiarrhetic and aphrodisiac; used for hemorrhage, kidney ailments, and as a revulsive; infusion of entire plant used as an antihemorrhagic and galactagogue; infusion and decoction of leaves diuretic; extract of fresh leaves used in homeopathy for eczema, dysmenorrhea, metrorrhagia, and nosebleeds, and in lotions to promote hair growth; used externally, it soothes wounds and ulcers (Boulos, 1983)

Verbenaceae

Verbena officinalis Linn.—perennial herb; damp regions of the Middle East; ash of plant used to cure burns and boils; leaves used as antispasmodic, febrifuge, emmenagogue, diuretic, galactagogue, stimulant, and antidiarrhetic; infusion or decoction used for insomnia, and as a stomachic; cataplasm used as vulnerary (Boulos, 1983)

Zygophyllaceae

Fagonia cretica Linn.—spiny perennial herb; Egypt; a 50% ethanol–water extract of aerial parts showed antiviral and antispasmodic activity *in vitro* as well as hypotensive and central nervous system (CNS) stimulant activity *in vivo* (Ayensu, 1979)

Peganum harmala Linn.—perennial, woody at base; throughout the Middle East; plant used as emetic and diuretic; fresh plants digested in sheep's fat used as rub to treat rheumatism; vapors of burnt plant inhaled for headache and neurotic pains; dried powdered plant used for treatment of purulent conjunctivitis; its decoction in oil taken first thing in the morning effective against hemorrhoids, and as a depurative; fresh branches used as revulsive; seeds anthelmintic; powdered seeds mixed with honey and ginger rubbed on skin for articular pains and rheumatism; cataplasm analgesic; powdered seeds boiled in olive oil used to ameliorate the quality of hair by making it thicker and stronger, and used as a rub for alopecia; seeds CNS stimulant, cause paralysis, and poisonous in strong doses; oil extract from seeds used for some infectious eye diseases, rheumatic pains, and some skin diseases; seeds eaten in small doses for asthma; seeds sudorific, emmenagogue; powdered roasted seeds taken after meals to treat bad digestion and diabetes; infusion of seeds useful for cardiac diseases, and prolonged use helps in cases of sciatica; pounded roots and seeds mixed with tobacco smoked in pipes for toothache (Boulos, 1983)

Tribulus terrestris Linn.—prostrate annual with spiny fruits; throughout the Middle East; fruits detersive, astringent, purgative, and diuretic (Zohary, 1972); extract of plant antispasmodic; fruit used as tonic for spermatorrhea, neuroasthenia, vertigo, and as an astringent for oral inflammations, also for dysentery, and pains of the bladder; detersive (Boulos, 1983)

Zygophyllum coccineum Linn.—perennial shrub with succulent leaves; Egypt, Israel, Jordan, western Arabia; ethanol extracts show antipyretic activity in rats, antihistaminic activity in guinea pigs, hypotensive activity in rats, and diuretic activity in rabbits (Ayensu, 1979); fruits used in treatment of rheumatism, gout, asthma, and hypertension, and as an anthelmintic and antidiabetic (Boulos, 1983)

D. Industrial Plants

The following list of industrial plants from the Middle East is given by Zohary (1973):

Oils, Gums, and Resins

Amygdalus spp.	*Ferula galbanifera*
Artemisia judaica	*Ferula szowitsiana*
Astragalus spp. (scores of tragacanthic species)	*Lallemantia iberica*
	Lavandula stoechas
	Liquidambar orientalis
Balanites aegyptiaca	*Mentha* spp.
Boswellia spp.	*Moringa peregrina*
Brassica spp.	*Origanum* spp.
Cistus creticus (ladanum-gum)	*Opapanax chironium*
	Pinus spp.
Commiphora spp.	*Pistacia lentiscus* (chias)
Cymbopogon schoenantus	*Ruta chalepensis*
Eruca sativa	*Styrax officinalis*
Ferula asa-foetida	*Thymus* spp., etc.

Especially noteworthy is the "tapping" of valuable gums from tragacanthi astragali, widely used in Iran and elsewhere.

Dyes

Alkanna tinctoria	*Euphorbia helioscopia*
Ammi visnaga	*Glycyrrhiza glabra*
Anchusa italica	*Indigofera* spp.
Anthemis tinctoria	*Lawsonia alba*
Arnebia hispidissima	*Punica granatum*
Asperugo procumbens	*Reseda luteola*
Chrozophora plicata	*Tephrosia apollinea*
Chrozophora tinctoria	*Teucrium polium*
Echium italicum	*Verbena supina*

Tannins and Detergents

Acacia spp.	*Quercus macrolepis*
Anabasis spp.	*Quercus* spp.
Hammada spp.	*Pistacia palaestina*
Limonium spp.	*Pistacia terebinthus*
Nuphar luteum	*Punica granatum*
Nymphaea alba	*Salix* spp.
Glycyrrhiza echinata	*Salsola* spp.
Glycyrrhiza glabra	*Tamarix* spp.
Quercus boissieri	

Some of the above-listed species will be treated in more detail under their appropriate groups.

1. FIBER AND TIMBER PLANTS

In some particularly isolated arid and hyperarid regions of the Middle East, where transport is a major problem, many inferior timbers, not listed here, are used for diverse purposes such as building, carpentry, etc.; the following list represents some examples of good and inferior quality fibers and timbers.

Balanitaceae

Balanites aegyptiaca (Linn.) Del.—attractive wood is easily worked, fine grained, durable, resistant to insects, saws and planes well, and used for bowls, mortars and pestles, tool handles, gunstocks, and cabinetwork (Ayensu, 1983)

Ericaceae

Arbutus andrachne Linn.—shrub or tree; Jordan, Israel, Syria, Lebanon, northern Iraq, Turkey; wood compact and suitable for turning (Feinbrun-Dothan, 1978)

Fagaceae

Quercus ithaburensis Decne.—deciduous tree with thick trunk; Israel, Jordan, Syria, Lebanon, Turkey; timber used in turnery, and for agricultural tools (Zohary, 1966)

Gramineae (Poaceae)

Desmostachya bipinnata (Linn.) Stapf—plant used to make ropes, baskets, and mats (Täckholm and Drar, 1941); possible raw material for paper industry if mixed with *Imperata cylindrica* (Bor and Guest, 1968)

Imperata cylindrica (Linn.) Beauv.—tests revealed that paper of satisfactory quality could be manufactured from this grass mixed with *Desmostachya bipinnata* (Bor and Guest, 1968); Täckholm and Drar (1941) stated that the white cotton-like floss (inflorescence) has been used for stuffing cushions in tropical Africa

Juncaceae

Juncus acutus Linn.—pulp successfully tested for paper manufacture (Zahran *et al.*, 1979)

Juncus rigidus Desf.—pulp successfully tested for paper manufacture (Zahran *et al.*, 1979)

Leguminosae

Acacia nilotica (Linn.) Willd. *ex* Del.—wood used for making waterwheels (*sakias*), wood gets harder the more it is exposed to water (Boulos, 1966); hard, tough wood also resistant to termites, impervious to water, and popular for railroad ties (sleepers), tool handles, carts, and oars; an attractive wood, good for carving and turnery, and still used for boatbuilding, as it was in ancient Egypt (Ayensu, 1980)

Sesbania sesban (Linn.) Merrill—shrub or small tree; Egypt, and cultivated elsewhere; bark fiber used for making ropes; stems used as roofing for huts (Ayensu, 1983)

Moraceae

Ficus sycomorus Linn.—large tree, Egypt, Israel, Jordan, Arabia; famed for durable timber, which was used for sarcophagi in Ancient Egypt (Zohary, 1966); wood used in carpentry and for agricultural tools in Egypt

Palmae

Hyphaene thebaica (Linn.) Mart.—fibers used for making ropes, baskets, mats, bags, etc.; timber strong, compact, and heavy, and largely used in upper Egypt for posts, beams, doors, water pipes, and furniture (Täckholm and Drar, 1950)

Phoenix dactylifera Linn.—leaf petioles used for making cages, fences, and roofs; thick basal parts of petioles used for fuel; leaflets used for making bags, ropes, mats, and baskets; brown fibers at the leaf base constitute the raw material for making ropes and substitute sponges for bathing; wood used for different building purposes, especially for roofs; although light and soft, timber lasts for centuries (El-Hadidi and Boulos, 1979)

Salicaceae

Populus euphratica Oliv.—fast-growing deciduous tree; Israel, Jordan, Syria, Iraq, Iran; timber soft and suitable for some kinds of carpentry (Zohary, 1966); wood easy to saw and works to a good finish; a good turnery wood, can be peeled off on a rotary cutter; used for planking, lacquer work, artificial limbs, matchboxes, and splints; also suitable for plywood, cricket bats, shoe heels, and can be used for pulp (Ayensu, 1983)

Salix alba Linn.—tree; Mediterranean to Iran; wood used for inferior carpentry (Zohary, 1966)

Tamaricaceae

Tamarix aphylla (Linn.) Karst.—leafless tree; Egypt, Israel, Jordan, Arabia, Iraq, Iran, Afghanistan (Baum, 1978); tree cultivated by cuttings to obtain straight stems of better wood quality for carpentry (chairs, cupboards, boats, etc.); wild plants give inferior quality wood and are usually cut for fuel (Boulos, 1966)

Tiliaceae

Corchorus olitorius Linn.—source of commercially valuable jute fiber (Zohary, 1972)

Thymelaeaceae

Thymelaea hirsuta (Linn.) Endl.—evergreen shrub; Mediterranean; branches yield strong fibers, used by bedouins for making ropes

2. FIREWOOD AND FUEL PLANTS

In many parts of the Middle East, especially in arid or hyperarid regions, all dry plants or parts of plants are practically used as firewood or for fuel, the following are a few examples:

Balanitaceae

Balanites aegyptiaca (Linn.) Del.—tree; wood is hard, heavy, and tough; makes excellent firewood and good-quality charcoal; regenerates vigorously after it is cut (Ayensu, 1983)

Berberidaceae

Leontice leontopetalum Linn.—tubers sometimes attain large sizes; used as a fuel (Boulos, 1959)

Chenopodiaceae

Anabasis articulata (Forsk.) Moq.—collected and used as fuel (Zohary, 1966; Schulz and Whitney, 1985)

Cornulaca monacantha Del.—desert shrub; Egypt, Arabia, Iraq, Iran, Afghanistan; often the only firewood in certain hyperarid regions, especially in Egypt and Arabia

Haloxylon persicum Bunge—shrub or small tree; Egypt, Jordan, Israel, Arabia, Iraq, Iran, Afghanistan; in desert areas of central Asia it is the major supplier of fuel wood; an excellent fuel, thermal efficiency compares favorably with bituminous coal, used extensively for making charcoal (Ayensu, 1980)

Salsola baryosma (Roem. and Schult.) Dandy—desert shrub; Egypt, Israel, Jordan, Arabia; often a characteristic element in desert wadis of extreme aridity; constitutes a useful local firewood

Leguminosae

Acacia nilotica (Linn.) Willd. *ex* Del.—wood used for firewood and for making good quality charcoal (Boulos, 1966); a very popular fuel, has been used extensively to fuel locomotives and river steamers in India; powers the boilers of some small industries as well (Ayensu, 1980)

Acacia tortilis (Forsk.) Hayne—thorny tree in desert wadis of the southern hot regions of the Middle East; dense, red heartwood of this species has high calorific value and makes superior firewood and charcoal; plant coppices well, so no need to replant trees after every harvest (Ayensu, 1980); in Egypt, wood used by bedouins as firewood and for making charcoal

Alhagi graecorum Boiss.—plant uprooted and dried in piles to be used as fuel by nomadic tribes and peasants (Chakravarty, 1976)

Pinaceae

Pinus halepeniss Mill.—evergreen tree; Israel, Jordan, Syria, Lebanon, Turkey; rather light wood with large and apparent resiniferous canals; dries quickly, and used as a fuel (Nahal, 1963)

Salicaceae

Populus euphratica Oliv.—formerly, tree constituted dense forests along watercourses throughout a vast range, but forests have been almost completely destroyed to supply firewood; a few remnant forests still exploited for firewood, but now mainly by coppice on short rotations of 1 or 2 years (Ayensu, 1983)

Rhamnaceae

Ziziphus spina-christi (Linn.) Desf.—wood much used for fuel; hard and dense, burns with intense heat (Ayensu, 1980)

Rosaceae

Sarcopoterium spinosum (Linn.) Sp.—spiny dwarf shrub, Mediterranean; widely used for fuel (Zohary, 1972)

Sapindaceae

Dodonaea viscosa (Linn.) Jacq.—dark brown wood, close grained, hard, heavy, and resists termite attack satisfactorily; a good firewood, easily ignited (Chakravarty, 1976)

Tamaricaceae

Tamarix spp.—among the many species known from the Middle East, the following are used for firewood and, often, for making charcoal [distribution records after Baum (1978)]:

T. aphylla (Linn.) Karst.

T. arborea (Ehrenb.) Bunge—Egypt, Israel, Jordan

T. aucherana (Decne.) Baum—Arabia, Iraq, Iran, Afghanistan
T. dubia Bunge—Iran, Afghanistan
T. macrocarpa (Ehrenb.) Bunge—Egypt, Israel, Jordan, Syria, Arabia, Iran
T. nilotica (Ehrenb.) Bunge—Egypt, Israel, Jordan, Lebanon
T. ramosissima Ledeb.—Iraq, Iran, Afghanistan

3. OIL PLANTS

Compositae

Carthamus lanatus Linn.—glandular spiny annual; Egypt, Turkey, Iraq, Iran; pale yellow oil (16%) obtained from seeds compares favorably with that of safflower (*Carthamus tinctorius* Linn.) (Chakravarty, 1976)

Carthamus oxyacantha M. Bieb.—annual thorny herb; Arabia, Iraq; oil obtained from seeds resembles safflower oil in properties; oil edible, and used for illumination (Chakravarty, 1976)

Carthamus tinctorius Linn.—prickly annual herb, probably native of Afghanistan; widely cultivated in the Middle East; seeds of cultivated plants contain 30–40% soil, wild plants yield ~20%; oil used in paints, varnishes, linoleum, etc., possesses a quick-drying property (Chakravarty, 1976)

Cruciferae

Brassica tournefortii Gouan—annual herb; Mediterranean to Iraq and Arabia; seed contains 30.7% fixed oil, similar to mustard oil (Chakravarty, 1976)

Dipsacaceae

Cephalaria syriaca (Linn.) Schrad.—seeds contain 24% oil; plant cultivated in Turkey and Soviet Union as an oil crop (Zohary, 1973; Feinbron-Dothan, 1978)

Euphorbiaceae

Ricinus communis Linn.—all-important industrial oil plant cultivated since ancient times in tropical and subtropical countries (castor beans); oil constitutes up to 70% of the seed, and is used in the chemical industry, and as an aviation lubricant (Zohary, 1972)

Juncaceae

Juncus acutus Linn.—seeds rich in oils; not tested for human consumption, might have toxic effects; possibility of use as a raw material in various chemical industries (Zahran *et al.*, 1979)

Juncus rigidus Desf.—seeds rich in oils, as for *J. acutus* Linn. (Zahran *et al.*, 1979)

Salvadoraceae

Salvadora persica Linn.—leaves, roots, bark, and flowers contain oil (Zohary, 1972)

4. GUM AND RESIN PLANTS

Burseraceae

Commiphora opobalsamum Engl.—aromatic shrub; Arabia; "Balm of Gilead" produced from plant (Willis, 1966), often called "Balsam of Mecca" (Bedevian, 1936); several *Commiphora* species yield myrrh, incense, bdellium, and other resins, which exude and collect in lumps (Willis, 1966)

Leguminosae

Astragalus adscendens Boiss. and Hausskn.—subshrub forming low cushions; Iraq, Iran; gum tragacanth collected from this species (Chakravarty, 1976)

Astragalus brachycalyx Fisch. *ex* Hoh.—subshrub forming low cushions; Turkey, Iran; one of the Iranian tragacanth gums (Chakrvarty, 1976).

Astragalus gummifer Labill.—low shrub with woody branches; Syria, Lebanon, Iraq; the official gum tragacanth-yielding species (Chakravarty, 1976)

Pinaceae

Pinus halepensis Mill.—each tree gives 1.5–3.8 kg per year of good quality resin, widely appreciated in the international market (Nahal, 1963)

E. Other Plants

1. DUNE STABILIZATION, EROSION CONTROL, AND AFFORESTATION

Chenopodiaceae

Haloxylon persicum Bunge—lower trunk establishes new roots when covered by sand; excellent plant for stabilizing sandy areas (Ayensu, 1980)

Fagaceae

Quercus ithaburensis Decne.—a comparatively fast-growing oak, and quite promising broad-leaved tree for afforestation

Gramineae

Desmostachya bipinnata (Linn.) Stapf—a useful soil and sand binder (Tackholm and Drar, 1941)

Lasiurus hirsutus (Forsk.) Boiss.—a good sand binder (Bor and Guest, 1968)

Panicum repens Linn.—a useful sand binder along the banks of creeks and marshes (Bor and Guest, 1968)

Panicum turgidum Forsk.—used in sand dune fixation in some desert regions in Egypt

Poa bulbosa Linn.—fibrous rootmat of plant helps to bind soil and impede erosion in the hills (Bor and Guest, 1968)

Poa sinaica Steud.—a most useful grass in dry steppe and subdesert, forming a soil-binding surface mat of fibrous roots and bulbils over large tracts (Bor and Guest, 1968)

Pennisetum orientale Rich.—perennial grass; Mediterranean to Afghanistan; a useful soil binder (Bor and Guest, 1968)

Stipagrostis ciliata (Desf.) de Winter—a good sand binder (Bor and Guest, 1968)

Stipagrostis plumosa (Linn.) Anders.—has great ability as a sand binder (Chakravarty, 1976)

Pinaceae

Pinus halepensis Mill.—a most successful timber tree, suitable for afforestation in dry, calcareous regions

Polygonaceae

Calligonum comosum L'Hérit.—an excellent desert sand binder (Zohary, 1966); used as sand binder because of extensive root system (Schulz and Whitney, 1985)

Rhamnaceae

Ziziphus spina-christi (Linn.) Desf.—develops an extremely deep taproot and spreading laterals, useful for stabilizing sand dunes and other unstable soils; also makes useful windbreaks and shelter-belts (Ayensu, 1980)

Tamaricaceae

Tamarix aphylla (Linn.) Karst.—common forest tree in arid zone of Middle East; thrives well on sand dunes under extreme conditions of low rainfall, but only when widely spaced, otherwise soon impoverishes water sources and stops growing (Zohary, 1962); used for reforestation (Schulz and Whitney, 1983)

2. ORNAMENTS

Alliaceae

Allium ampeloprasum Linn.—bulb with elegant reddish flowers; Mediterranean; scape 1–1.5 m; may provide a good ornamental bulb

Amaryllidaceae

Stenbergia clusiana Ker-Gawl.—bulbous herb, golden yellow flowers, showy; Israel, Jordan, Syria, Lebanon, Turkey; good ornamental bulb suitable for rock gardens

Boraginaceae

Alkanna orientalis (Linn.) Boiss.—perennial herb, corolla golden yellow, inflorescence showy; Sinai to Turkey, Iran; used in flower beds and rock gardens

Echium judaeum Lacaita—annual herb; Israel, Jordan, Lebanon, Syria; suitable plant for winter flower beds in gardens

Compositae

Belis perennis Linn.—small perennial herb; Mediterranean; Cultivars grown for ornament (Feinbrun-Dothan, 1978)

Chrysanthemum coronarium Linn.—annual herb, up to 1 m high; Mediterranean; forms excellent flower beds in gardens, also a cut flower

Iridaceae

Crocus spp.—several *Crocus* species are known from the Mediterranean part of the Middle East, of which some are endemic to Syria and Lebanon (Mouterde, 1966); the following are candidates for garden ornmanetals: *Crocus ochroleucus* Boiss. and Gaill., *C. kotschyanus* C. Koch, *C. theibauti* Mout., *C. haussknechtii* Boiss. and Reut., *C. macrobolbos* Jovet and Gombault, *C. cancellatus* Herbert, *C. aleppicus* Baker, *C. hyemalis* Boiss. and Bl., and *C. graveolens* Boiss. and Reut.

Iris spp.—the section *Oncyclus* (Siems.) Baker, of the genus *Iris* Linn., comprises ~65 species with extraordinary, large, and colorful flowers, the majority of which are strict endemics growing in hard-to-reach localities distributed over open herbaceous formations in southwestern Asia (Avishai and Zohary, 1977); the following is a selection of some species with high ornamental value that are known from Jordan, Israel, Syria, Turkey, and Iran: *Iris lycotis* Woron, *I. calcarea* Dinsm., *I. basaltica* Dinsm., *I. elegantissima* Sosn., *I. damascena* Mout., *I. hermona* Dinsm., *I. nazarena* Foster, *I. maribilis* Gaw., *I. samaria* Dinsm., *I. sari* (Schott) Bak., *I. maculata* Bak., *I. barnumae*

Bak., and Foster, *I. polakii* Stapf, *I. antilibanotica* Dinsm., *I. zonobieae* Mout., *I. jordana* Dinsm., *I. gileadensis* Dinsm., *I. nigricans* Dinsm., *I. petrana* Dinsm., *I. atropurpurea* Baker, *I. mariae* Babey, etc.

Leguminosae

Calycotome villosa (Poir.) Link—perennial shrub with thorny branches, flowers yellow, showy; Mediterranean; flourishes on calcareous soils, a good ornament in rock gardens

Lupinus varius Linn.—annual herb up to 80 cm high, inflorescence showy, flowers blue; Mediterranean; good candidate as cut flower, may be used for flower beds

Retama raetam (Forsk.) Webb—leafless shrub, flowers white, forms long conspicuous inflorescences; throughout the deserts of the Middle East; good cut winter flower

Spartium junceum Linn.—leafless shrub, flowers yellow, showy; Mediterranean; beautiful ornamental plant

Liliaceae

Asphodeline damascena Boiss.—biennial herb with white flowers; endemic to Syria and Lebanon (Mouterde, 1966); suitable for rock gardens

Asphodeline lutea (Linn.) Reichenb.—perennial herb with tuberous roots; Jordan, Israel, Syria, Lebanon; showy inflorescence with yellow flowers makes the plant a good ornament in dry, rocky regions

Colchicum hierosolymitanum Feinbr.—perennial herb; Israel, Lebanon, Syria; numerous pink flowers make the species a suitable ornamental rock-garden plant

Fritillaria spp.—glabrous, bulbous perennials; some species restricted to the Mediterranean part of the Middle East (Mouterde, 1966); good potential ornamentals, e.g., *F. alfredae* Post, endemic to Syria and Lebanon; *F. elwesii* Boiss., Syria, Lebanon, Turkey; *F. libanotica* (Boiss.) Baker, Jordan, Israel, Syria, Lebanon; *F. persica* Linn., Turkey, Iraq, Iran; and *F. crassifolia* Boiss. and Huet., Syria, Lebanon, Turkey

Scilla hanburyi Baker—bulbous herb; Jordan, Israel, Syria, Lebanon; possibly a good ornament for rock gardens in arid and semiarid regions

Tulipa spp.—bulbous perennials; the following are good potential ornamentals for their showy, attractive flowers: *T. lownei* Baker, endemic to Syria and Lebanon; *T. aleppensis* Regel, endemic to Syria and Lebanon; *T. polychroma* Stapf, Egypt, Jordan, Iran; and *T. stylosa* Stapf, Egypt, Israel, Syria, Iraq, Iran

Papaveraceae

Glaucium aleppicum Boiss.—perennial herb, flowers showy, red; Syria, Lebanon, Turkey, Iraq; suitable for flower beds

Glaucium grandiflorum Boiss. and Heut.—perennial shrub, flowers orange-yellow, showy; good cut flower

Primulaceae

Androsace maxima Linn.—small annual herb; Mediterranean to Iran; a handsome plant suitable as ornamental rock or potted plant (Chakravarty, 1976)

Cyclamen coum Mill.—perennial herb; Mediterranean to Iran; grown as ornamental in pots and rock gardens; protected by law in Israel (Feinbrun-Dothan, 1978)

Cyclamen persicum Mill.—perennial herb; Israel, Jordan, Lebanon, Syria, Turkey (not known in the wild from Iran); grown as ornamental (numerous cultivars); protected by law in Israel (Feinbrun-Dothan, 1978)

Ranunculaceae

Adonis aleppica Boiss.—annual herb with conspicuous purple-red flowers; Israel, Jordan, Syria, Lebanon, Turkey, Iraq; good candidate for winter flower beds in semiarid regions

Adonis palaestina Boiss.—annual herb; Israel, Jordan, Syria, Lebanon; conspicuous flowers class this species among those suitable for winter flower beds in gardens

Anemone coronaria Linn.—perennial tuberous herb; Mediterranean; flowers showy, in various bright colors that make the plant a good spring cut flower

Ranunculus asiaticus Linn.—perennial tuberous herb; Mediterranean to Iran; bright multcolored flowers, may be introduced as a garden plant and cut flower

Thalictrum isopyroides C. A. Mey.—perennial herb; Jordan, Syria, Turkey, Iraq, Iran, Afghanistan; a handsome plant with showy foliage and inflorescence, may be introduced as a rock-garden ornamental in arid and semiarid regions

Umbelliferae

Eryngium maritimum Linn.—rigid perennial; Israel, Syria, Lebanon, Turkey; grown in gardens for its colored stem, prickly coriaceous leaves, and showy inflorescence; considered excellent for borders and rock gardens; *E. creticum* Lam. and *E. glomeratum* Lam., mainly Mediterranean species, used for same purpose

3. MISCELLANEOUS USES

Amaranthaceae

Achyranthes aspera Linn.—perennial herb; Egypt, Israel, Jordan, Arabia; plant ashes employed as source of alkali in dyeing (Zohary, 1966)

Aerva javanica (Burm. f.) Schultes—perennial shrub; hot deserts of the Middle East; woolly spikes used for stuffing saddles and pillows (Zohary, 1966)

Anacardiaceae

Rhus tripartita (Ucria) Grande—bark used for tanning and dyeing (Schulz and Whitney, 1985)

Balanitaceae

Balanites aegyptiaca (Linn.) Del.—because of saponin content of roots, bark, fruits, and wood chips, all these parts used as "soap" for washing clothes (Ayensu, 1983)

Berberidaceae

Leontice leontopetalum Linn.—tubers contain saponin and used in the Middle East for cleansing purposes (Zohary, 1966)

Boraginaceae

Alkanna tinctoria (Linn.) Tausch.—perennial herb; Mediterranean; roots contain *alkanna,* used for coloring toilet preparations of an oily or volatile nature; red coloring matter in alcohol solution used as microscopic stain for detection of oils and fats (Chakravarty, 1976)

Chenopodiaceae

Anabasis articulata (Forsk.) Moq.—contains potassium and therefore used as a detergent (Zohary, 1966)

Atriplex halimus Linn.—ash of plant used for manufacture of soap (Zohary, 1966)

Haloxylon articulatum (Cav.) Bunge—perennial shrub; throughout the Middle East; plant dried and used as crude soap for washing clothes; Chakravarty (1976) reported that the plant yields sufficient crude carbonate for making glass and soap

Compositae

Carthamus tinctorius Linn.—dried flowers supply safflower carmine, a red dye used like saffron (Feinbrun-Dothan, 1978)

Datiscaceae

Datisca cannabina Linn.—roots yield a yellow color used for dyeing cotton, wool, and silk; coloring matter also present in leaves and twigs (Chakravarty, 1976)

Gramineae (Poaceae)

Phragmites australis (Cav.) Steud.—stout rhizomatous perennial grass with rigid culms, sometimes attining a height of 3.5 m; throughout the Middle East; mature reeds put to many uses by marsh people, especially in light construction and the provision of mats and screens; tall stems bound together in bundles and bent into arched pillars provide the framework of houses, stores, and cattle byres in the marshes, with the walls and roof then filled in with matting; the canes were tried as raw material for the manufacture of compressed "reed-board" slabs for wall linings and partitions (Bor and Guet, 1968)

Leguminosae

Acacia nilotica (Linn.) Willd.—bark and pods widely used in the leather industry; tannin content varies from 12 to 20% (Ayensu, 1980)

Rosaceae

Sarcopoterium spinosum (Linn.) Sp.—used for broom manufacture and hedging (Zohary, 1966)

Salicaceae

Salix alba Linn.—bark used for tanning (Zohary, 1966)

Salvadoraceae

Salvadora persica Linn.—woody twigs used as toothpicks (Zohary, 1972); bedouins use thick fibrous twigs as toothbrushes

VI. SUMMARY

Although the arid lands of the Middle East are not richly productive in terms of plant biomass per unit area, and need to be utilized with care to maintain a balance between productivity and harvest, i.e., collection, cutting, grazing, etc., these lands have floras that contain a notable variety of species with products that are often of special value. These species represent a treasure of genetic resources that needs to be conserved.

There is need for a scientific program that would (1) systematically survey the genetic resources of biotypes and ecotypes, and the material resources of special uses, special chemicals, and compounds; (2) study means for propagation and husbandry of these species and their biotypes with special attributes; (3) establish a regional center for collecting and disseminating information on these plants; and (4) plan and implement the establishment of a network of nature reserves that would provide protection to endangered plants.

ACKNOWLEDGMENTS

I wish to thank Dr. Kamel Hanna Soliman, formerly of the Meteorological Department, Cairo, for the data and information on climate; Dr. Bahay Issawi, of the Geological Survey of Egypt, for the geological information; and Professor M. Kassas, of Cairo University, for reading some parts of the manuscript. Mrs. Vivienne Armer, of Kuwait University Herbarium, kindly read the typescript.

REFERENCES

Abd-El-Gawad, M. (1969). New evidence of transcurrent movements in Red Sea area and petroleum implications. *Bull. Am. Assoc. Pet. Geol.* **53,** 1466–1479.

Avishai, M., and Zohary, D. (1977). Chromosomes in the *Onocyclus* Irises. *Bot. Gaz.* **138,** 502–511.

Ayensu, E. S. (1979). Plants for medicinal uses with special reference to arid zones. *In* ''Arid Land Plant Resources'' (J. R. Goodin and D. K. Northington, eds.), pp. 117–178. Texas Tech Univ., Lubbock.

Ayensu, E. S., ed. (1980). ''Firewood Crops, Shrub and Tree Species for Energy Production,'' Vol. 1. Natl. Acad. Sci., Washington, D.C.

Ayensu, E. S., ed. (1983). ''Firewood Crops, Shrub and Tree Species for Energy Production,'' Vol. 2. Natl. Acad. Sci., Washington, D.C.

Baum, B. R. (1978). ''The genus Tamarix.'' Isr. Acad. Sci. Humanit., Jerusalem.

Bellakhdar, J. (1978). ''Médecine Traditionelle et Toxicologie Ouest-Sahariennes, Contributions a l'Étude de la Pharmacopée Marocaine.'' Editions Techniques Nord-Africaines, Rabat, Morocco.

Blackburn, N., ed. (1970). ''The Middle East and North Africa 1970–71.'' Europa Publ. Ltd., London.

Bor, N. L., and Guest, E. (1968). Gramineae, In ''Flora of Iraq,'' Vol. 9. Ministry Agric., Baghdad.

Boulos, L. (1959). ''A Contribution to the Flora of Gaza Zone.'' Agric. Extension Dep., Minist. Agric., Cairo.

Boulos, L. (1966). Flora of the Nile region in Egyptian Nubia. *Feddes Repert.* **73,** 184–215.

Boulos, L. (1970). Medicinal herbs in Libya, *In ''Al Hasad,''* Vol. **16.** Esso Standard Libya Publ., Beirut.

Boulos, L. (1977). Studies on the flora of Jordan. 5. On the flora of El-Jafr-Bayir Desert. *Candollea* **32,** 99–110.

Boulos, L. (1978). Materials for a flora of Qatar. *Webbia* **32,** 369–396.

Boulos, L. (1983). ''Medicinal Plants of North Africa.'' Reference Publ., Algonac, Michigan.

Boulos, L., and Lahham, J. (1977). Studies on the flora of Jordan 3. On the flora of the vicinity of the Aqaba Gulf. *Candollea* **32,** 73–80.

Chakravarty, H. L. (1976). "Plant Wealth of Iraq," 1. Minist. Agric. and Agrarian Reform, Baghdad.

Clawson, M., Landsberg, H. H., and Alexander, L. T. (1971). "The Agricultural Potential of the Middle East." Am. Elsevier, New York.

Davis, P. H. *et al.*, eds. "Flora of Turkey and the East Aegean Islands," 1 (1965), 2 (1967), 3 (1970), 4 (1972), 5 (1975), 6 (1978), 7 (1982) (In progress). Univ. Press, Edinburgh.

El-Hadidi, M. N., and Boulos, L. (1979). "Street Trees in Egypt," ed. 2. Cairo.

El-Menshawi, B., Karawya, M., Wassel, G., Reisch, J., and Kjaer, A. (1980). Glucosinolates in the genus *Zilla* (Brassicaeae). *J. Nat. Prod.* **43,** 534–536.

El-Shazly, E. M., Saleeb, G. S., and Zaki, N. (1974). Quaternary basalts in Saint John's Island, Red Sea, Egypt. *Egypt. J. Geol.* **18,** 137–147.

Feinbrun-Dothan, N. (1978). "Flora Palaestina," Vol. 3. Isr. Acad. Sci. Humanit., Jerusalem.

Fisher, W. B. (1963). "The Middle East," ed. 5. Methuen, London.

Fourment, P., and Roques, H. (1941). "Répertoire des Plantes Médicinales et Aromatiques d'Algérie." Alger.

French, D. H. (1971). Ethnobotany of the Umbelliferae. *In* "The Biology and Chemistry of Umbelliferae" (V. H. Heywood, ed.), pp. 385–412. Academic Press, New York, for the Linnean Soc. of London.

Gischler, C. E. (1979). "Water resources in the Arab Middle East and North Africa." Middle East and N. Afr. Studies Press, Cambridge.

Kassas, M., and Imam, M. (1957). Climate and microclimate in the Cairo Desert. *Bull. Soc. Géogr. Egypte* **30,** 25–52.

Keys, J. D. (1976). "Chinese Herbs, Their Botany, Chemistry and Pharmacodynamics." Rutland, Vermont, and Tokyo.

Koller, D. (1959). Germination. *Sci. Am.* **132,** 75–83.

Kunkel, G. (1984). "Plants for Human Consumption, An Annotated Checklist of the Edible Phanerogams and Ferns." Koenigstein.

Lexique Stratigraphique International (1968). Arabie Saoudite, *In* "Asie," vol. *3,* fasc. 1061.

Mouterde, P. "Nouvelle Flore du Liban et de la Syrie," Vol. 1 (1966), 2 (1970). Dar El-Machreq, Beyrouth.

Nahal, L. (1963). "Le Pin d'Alep (*Pinus halepensis* Mill.) Étude Taxonomique, Phytogéographique, Ecologique et Sylvicole." Nancy.

Nauroy, J. (1954). "Contributions a l'Étude de la Pharmacopée Marocaine Traditionnelle (Drogues Végétales)." Thèse. Paris.

Paylore, P., and Greenwell, J. R. (1979). Fools rush in: pioneering the arid zones. *Arid Lands Newsl.* **10,** 17–18.

Paylore, P., and Greenwell, J. R. (1980). Fools rush in 2: selected arid lands population data. *Arid Lands Newsl.* **12,** 14–18.

Picard, L. (1970). On Afro–Arabian graben tectonics. *Geol. Rundsch.* **59,** 337–381.

Said, R. (1962). "Geology of Egypt." Elsevier, New York and Amsterdam.

Schauenberg, P., and Paris, P. S. F. (1977). "Guide to Medicinal Plants." Lutterworth, Guildford and London.

Schulz, E., and Whitney, J. W. (1985). "Vegetation on the Arabian Shield and Adjacent Sand Seas." Report 85–116, U.S. Geological Survey.

Soliman, K. H. (1978). "Climate of Arab Republic of Egypt." Cairo. [In Arabic].

Täckholm, V., and Drar, M. Flora of Egypt. *Bull Fac. Sci., Cairo Univ.* **17** (1941), **28** (1950, **30** (1954). (Vol. 1 with G. Täckholm).

Townsend, C. C., and Guest, E. (1974). Leguminosae, *In* "Flora of Iraq," Vol. 3. Minist. Agric. and Agrarian Reform, Baghdad.

UNESCO (1977). *Tech. Notes* **7.**
Willis, J. C. (1966). ''A Dictionary of the Flowering Plants and Ferns,'' ed. 7. Cambridge Univ. Press, Cambridge.
Zahran, M. A., Abdel-Wahed, A. A., and El-Demerdash, M. A. (1979). Economic potentialities of Juncus plants. *In* ''Arid Land Plant Resources'' (J. R. Goodin and D. K. Northington, eds.), pp. 244–260. Texas Tech Univ., Lubbock.
Zohary, M. (1966, 1972). ''Flora Palaestina,'' Vols. 1 and 2. Isr. Acad. Sci. Humanit., Jerusalem.
Zohary, M. (1973). ''Geobotanical Foundations of the Middle East,'' 2 vols. Gustav Fischer Verlag, Stuttgart; Swets and Zeitlinger, Amsterdam.
Zohary, M., Heyn, C. C., and Heller, D. (1980). ''Conspectus Florae Orientalis, An Annotated Catalogue of the Flora of the Middle East.'' Isr. Acad. Sci. Humanit., Jerusalem.

5

NORTH AMERICA

C. M. McKell
NPI
Salt Lake City, Utah

I. INTRODUCTION

Many agricultural ecologists have noted that most of the main crop species used in North America have been introduced from other continents (Wilsie, 1962). A major reason may well be that although the indigenous people made moderate use of crop plants in their cultures, they were overwhelmed in numbers by the migration of European settlers who brought their crop plants with them and largely ignored the potential of native plants used by the indigenous people.

Further, the intensive cultivation practices used by the immigrants were not congruent with the indigenous people's relatively extensive practice of gathering fruits, seeds, and other parts of wild plants. Intensive development of North American arid and semiarid lands suitable for cultivation generally involved land clearing, leveling, tilling, irrigating, and mechanical harvesting. Thus the land was made suitable for the introduced crops rather than taking advantage of crops already suited to the land.

The purpose of this chapter is to examine the geography of the arid lands of North America and to discuss selected native species that have a potential for development. Some of the more important species or groups of plants are viewed in terms of their favorable characteristics, areas of adaptation, and constraints to development. In cases where some development is underway, an indication is given of the magnitude and kinds of actions being taken.

II. PHYSIOGRAPHY

The North American continent covers ~24,287,528 km^2 and includes Greenland, Canada, the United States of America, Mexico, Central America, and the islands of the Caribbean (Table I). Approximately 3,550,500 km^2 may be considered arid or semiarid (<50 cm precipitation per annum), most of which is located west of the hundredth meridian in Mexico and the United States. Probably the most distinctive feature of the North American drylands is the great variety in topography, including high plains and plateaus, closed basins, and deep canyons (Petrov, 1976); mountains and hills are common features. Rainfall is irregular and evaporation generally is very high; the effective precipitation is extremely low throughout most of the year. Historically, the arid and semiarid lands have had sparse human populations, which generally concentrate near mountains and rivers. Mineral resources are abundant and play an increasingly important role in the economic development of the countries. As these resources are developed, the population increases steadily. Water is probably the greatest limiting factor to the development of the arid and semiarid lands. Most streams are ephemeral and problems with salinity and dissolved solids are common. Good water is usually found near mountains, in streams, and in underground supplies. All major rivers have been dammed, usually in several locations, providing water and hydroelectric power. Most of the land is used for grazing, alhtough sizable tracts are farmed where water is available. Overgrazing and poor farming practices have deteriorated many grasslands and desert transitional areas (MacMahon, 1979).

The North American arid and semiarid lands are here divided into 10 regions;

Table I

Demographic Data for the North American Continent

Country	Population	Density (per km^2)	Area (km^2)	Annual % population increase (1970–1977)
Bahamas	220,000	19.29	11,406	3.6
Belize	150,000	6.53	22,963	3.1
Canada	23,320,000	2.34	9,976,139	1.3
Costa Rica	2,070,000	40.67	50,901	2.6
Cuba	9,460,000[a]	82.60	114,524	1.7
Dominican Republic	5,000,000	103.21	48,443	2.9
El Salvador	4,120,000[a]	192.59	21,393	—
Greenland	49,666	0.02	2,175,590	2.7
Guatemala	6,440,000	59.14	108,888	2.9
Haiti	4,750,000	171.18	27,749	1.6
Honduras	2,830,000	25.25	112,087	—
Jamaica	2,090,000	182.95	11,424	1.6
Mexico	64,590,000	32.74	1,972,537	3.5
Nicaragua	2,310,000	15.61	148,000	3.4
Panama	1,770,000	23.77	74,470	3.1
Puerto Rico	3,337,000	375.07	8,897	2.8
Trinidad and Tobago	1,100,000[a]	214.59	5,126	—
USA	216,820,000	23.07	9,396,647	0.8
Virgin Islands	100,000[b]	290.70	344	—

[a] Estimate for 1976.
[b] 1978 census.

each will be described separately. Generalized information about each is found in Table II. It must be recognized that the boundaries are arbitrary, and, in reality, indefinite, since climatic influences on the land are seldom abrupt and boundaries can move from year to year, depending on climatic conditions.

A. The Great Basin Desert

The Great Basin is the largest of the North American deserts, occupying ~500,000 km^2. It is entirely within the United States and occupies most of the state of Nevada, the western half of Utah, southern Idaho, southeastern Oregon, and parts of eastern California (Fig. 1). It is an arid upland region with enclosed valleys and broad basins lying between high, steep mountains (Jaeger, 1957).

The Great Basin is considered a ''cold'' desert because of its high elevation, northern position, and predominantly winter precipitation (>60%) as snow (MacMahon, 1979). In the northern part, winters are cold and summers hot; the

Table II

Precipitation Parameters for Major Arid Regions of North America

Desert or semiarid region	Approximate area (km^2)	Average annual precipitation (cm)	Principal season of precipitation
Great Basin Desert	500,000	13–25	Winter–early spring
Mojave Desert	65,000	3.5–13	Winter–early spring
Sonoran Desert	300,000	Trace–32	Summer and winter
Chihuahuan Desert	350,000	6–25	Summer
Columbia–Snake River Plateau	256,000	25–50	Fall, winter, and spring
California Valley	51,000	<25	Winter
Wyoming Basin	109,557	<25	Winter and spring
Colorado Plateau	330,000	25–50	Late summer and winter
Great Plains	1,472,000	35–50	Summer
Southern temperate grassland	250,000	40–50	Late summer

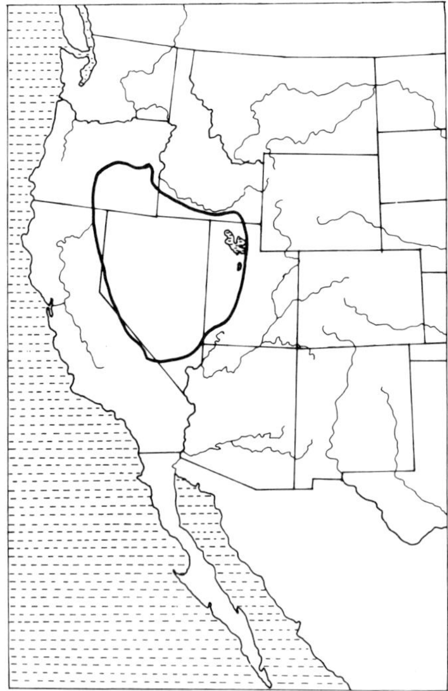

Fig. 1. The Great Basin.

southern region has milder winters with hot summers. Winds blow mainly from the northwest throughout the year. Average annual precipitation varies from less than 13 cm in the west to less than 25 cm in the east. Although, on the average, most precipitation falls during the winter, the melted snow is very important in recharging soil moisture in the deeper soil horizons for use during the summer by the shrubby vegetation. The aridity of the Great Basin is largely due to the rainshadow effect from the Sierra Nevada.

Hunt (1974) divided the region we describe as the Great Basin into four areas: the central area, the Bonneville Basin, the Lahontan Basin, and the lava and lake area. The central area comprises elevated basins, usually more than 1525 m high, and mountains. Some of the valleys are closed, but there are no perennial lakes in this region. Much of the area drains into the Lahontan Basin via the Humboldt River. The mountains in the eastern and northern parts of the central area are mostly limestone, while to the west they are sandstone, siltstone, and shale.

The Bonneville Basin, east of the central area, is lower, with most basins less than 1525 m in elevation. There is no exterior drainage; all water drains either into the smaller enclosed basins or the three major lakes: the Great Salt Lake, Utah Lake, or Sevier Lake. The Great Salt Lake has no outlet and is about four times as salty as the ocean (Fenneman, 1931). Utah Lake is a freshwater lake only because it has an outlet and drains into the Great Salt Lake. Sevier Lake is dry much of the time, although it would be ephemeral were its tributaries not used for irrigation. These three lakes are what remain of the ancient Lake Bonneville, which, at its peak during the Pleistocene, covered $\sim$51,800 km^2 (Hunt, 1974). The mountains in the Bonneville Basin are largely limestone, though volcanic rocks form some mountain ranges. The mountains cover $\sim$25% of the basin, playas and alluvial flats cover $\sim$40%, and gravel fans make up the rest (Hunt, 1974).

The Lahontan Basin, located to the west of the central area and east of the Sierra Nevada, is similar to the Bonneville Basin. It, too, is lower than the central area, with the elevation of the basin generally less than 1525 m. Alluvial flats and playas make up the greatest part of the basin; the mountain ranges are made up of volcanic rock. Water from this area drains into several large lakes. During the Pleistocene, Lake Lahontan covered most of this region. It was centered at Carson Sink and extended almost to the Sierra Nevada.

The lava and lake area is located in the northwestern part of the Great Basin and is topographically higher than the Lahontan Basin. This area is lava plateau with many volcanic cones. There are several lakes near the base of the mountains at the edge of the Great Basin, as well as some exterior drainage by the Pit and Klamath Rivers.

Because of the lack of drainage outlets, saline soil is a common problem in most basins. The lowest depressions are salt- and alkali-encrusted playas or dry lakes. In general, the plants in these basins are distributed according to their tolerance to salts and alkali. The vegetation is mostly shrubby with sagebrush (*Artemisia* spp.), shad scale (*Atriplex confertifolia*), greasewood (*Sarcobatus vermiculatus*), and winter fat (*Ceratoides lanata*) dominating much of the landscape. Piñon–juniper woodlands are common on the hills and mountains. The Great Basin is primarily grazing land, although irrigated farming occurs in limited areas where ample water is available. There are very few perennial rivers; these usually originate in the mountains surrounding the Great Basin. These rivers and aquifers are important in providing good water for drinking and irrigation.

B. The Mojave Desert

The Mojave Desert is located to the south of the Great Basin Desert and occupies ~65,000 km^2 in extreme southern Nevada, southeastern California, and a very small part of northwestern Arizona (Fig. 2). It is an upland desert with elevations mostly between 610 and 1525 m. Precipitation is very scanty, averaging 3.5–13 cm per year, and comes principally during the winter and early spring. Winter in the Mojave is quite mild but summers are very hot. The arid

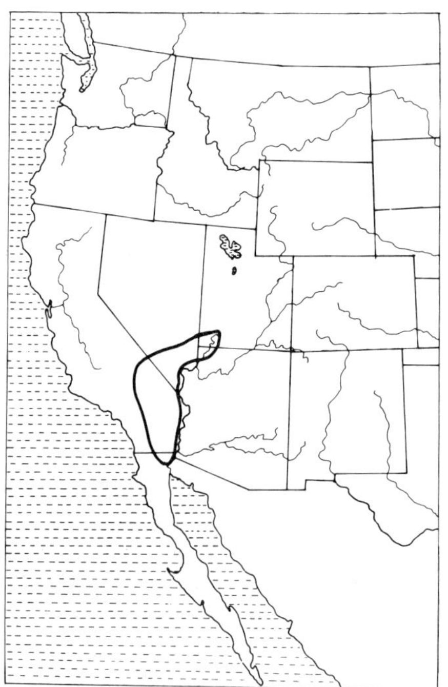

Fig. 2. The Mojave Desert.

climate here is primarily a result of the rain shadow effect of the southern California mountains.

The middle and eastern sections of the Mojave Desert are dominated by mountain ranges with extensive, gently sloping alluvial fans. The western area has a comparatively smooth surface (Jaeger, 1957). Salt flats and salt marshes are often found in the valleys and basins between the mountains. A small part of the Mojave drains to the sea through the Colorado River, the only major river flowing through the desert. Most of the remainder drains to enclosed basins; the lowest of these is Death Valley, which lies 150 m below sea level.

The soils of the Mojave are aridisols and entisols originating on such materials as volcanic lava flows, limestone, and sandstone. Extensive areas of sand dunes, composed of granitic particles, are found there. The soils are impregnated with lime salts, with great concentrations occurring in the dry lakes and salt flats of the many basins. There are very few perennial streams in the Mojave, and other surface water sources such as springs and tanks are scarce. The Colorado River is dammed for irrigation and hydroelectric power in several locations. Common vegetation of the Mojave includes creosote bush (*Larrea tridentata*), Joshua tree (*Yucca brevifolia*), and white bursage (*Franseria dumosa*).

C. The Sonoran Desert

The lowest, hottest, and most diverse of the North American deserts is the Sonoran Desert (Fig. 3). It covers more than 300,000 km^2, including southeastern California and southern Arizona, in the United States, and western Sonora and most of Baja California, in Mexico. The Sonoran Desert principally has biseasonal rainfall in summer and winter, although it tends toward summer precipitation to the southeast and winter precipitation to the north and west (MacMahon, 1979). Average annual precipitation also varies from a trace in the western parts to ~32 cm on the eastern range (Jaeger, 1957).

Because the Sonoran Desert is so varied, it is probably best described by region rather than as a whole. Jaeger (1957) classified the desert in six major subdivisions: the desert plains and foothills of Sonora, the Arizona upland or saguaro desert, the Yuman Desert, the Colorado Desert, the Vizcaino–Magdalena Desert, and the gulf coast desert.

1. THE DESERT PLAINS AND FOOTHILLS OF SONORA

The desert plains and foothills of Sonora occupy the southernmost region of Sonora east of the Gulf of California. The desert plains is an undulating area, most of which receives an average of 30 cm of precipitation, primarily in summer. The vegetation is principally a mixture of shrubs and low-growing trees, including palo verde (*Cercidium sonorae*), ironwood (*Olneya tesota*), honey bean mesquite (*Prosopis juliflora*), creosote bush, and tree ocotillo (*Fou-*

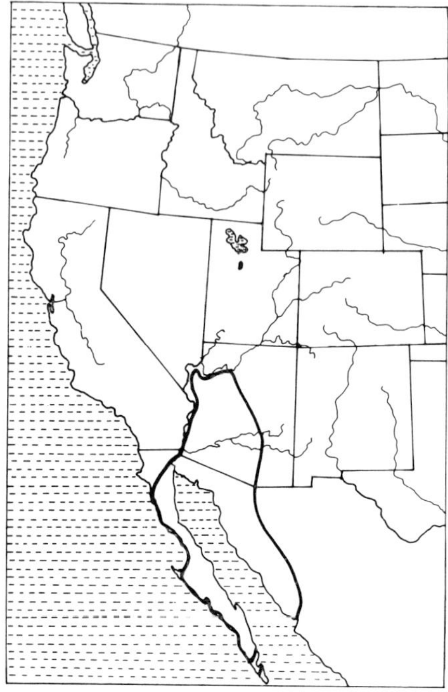

Fig. 3. The Sonoran Desert.

quieria macdougalii); a variety of cactus species also are evident (Jaeger, 1957). In the foothills to the east are commonly found many shrubby cacti, *Acacia* spp., mesquite, paloverde, ocotillo, and brittlebush (*Encelia farinosa*). Four principal rivers drain the state of Sonora, Mexico; none of these reaches the sea except during an occasional heavy rain. The main population is concentrated along these waterways.

2. THE ARIZONA UPLAND OR SAGUARO DESERT

The Arizona upland desert occupies the northeastern region of the Sonoran Desert and is located mostly in Arizona with a small part in Sonora. It has high plateaus, large, flat, arid valleys, small granitic hills, and rugged volcanic mountains. Precipitation is biseasonal in winter and summer and averages 30 cm. Vegetation includes saguaro cactus (*Carnegiea gigantea*), honey bean mesquite (*Prosopis juliflora*), paloverde, creosote bush, and several grass genera, including *Bouteloua, Hilaria,* and *Sporobolus* (Jaeger, 1957). The Gila, Bill Williams, Santa Cruz, San Pedro, and Sonoyta Rivers drain this desert area.

3. THE YUMAN DESERT

The Yuman Desert is found from southwestern Arizona to the Colorado River and south into the northwestern part of Sonora. This is a region of extensive, low, sandy plains; large dunes; scattered, volcanic hills; and low, barren mountains. Precipitation averages ~10 cm per year. Typical vegetation includes creosote bush, brittlebush, white bursage, ocotillo, saguaro, and other cacti. The Gila and Colorado Rivers drain this region; other important sources of water are natural tanks, though these are usually few and far between.

4. THE COLORADO DESERT

The Colorado Desert is found west of the Colorado River and occupies southeastern California and northern Baja California. This is a very low, basin-like desert with much of it lying near or below sea level. It drains into either the Colorado River or the Salton Sea, which is 83 m below sea level. To the north and south of the Salton Sea lie two important agricultural valleys, the Coachella Valley and Imperial Valley, which have deep, rich, alluvial soils. Much of the rest of the Colorado Desert is sandy and rocky. The vegetation is dominated by creosote bush with white bursage and brittlebush. The San Andreas fault runs through the desert, giving rise to springs and seeps, both hot and cold, every few miles. A large amount of water is diverted from the Colorado River for irrigation of these agricultural lands.

5. THE VIZCAINO–MAGDALENA DESERT

The Vizcaino–Magdalena Desert occupies the middle and lower portions of Baja California, excluding the gulf coast of California. A region of varied topography, this area includes broad sandy valleys and outwash plains, scattered granitic hills, low mountainous ridges, broad volcanic tablelands, lava flows, cinder cones, and elevated, lava-capped mesas dissected by canyons (Jaeger, 1957). Soils are often alkaline. Precipitation is sparse but is considered biseasonal. It is not unusual for several years to pass without effective rainfall. Fog is common from June to August and the prevailing winds are sea breezes from the west. The aridity of much of this desert is due, in part, to the effects of coastal currents (MacMahon, 1979). There are few springs there, and streams cease flowing before ever reaching the sea.

6. THE GULF COAST DESERT

The gulf coast desert occupies a narrow coastal strip along both sides of the Gulf of California. It consists of low mountains rising from the shore, sandy plains, salt flats, gravelly plains, and alluvial fans (Jaeger, 1957). All of the gulf coast desert is in the rain shadow of the peninsular upland (Jaeger, 1957) and is probably the driest part of the Sonoran Desert. Summer rainfall is important here; winter storms provide little precipitation. Fresh water comes from a few permanent springs. The population is very sparse. Vegetation is also sparse and includes cardón (*Pachycereus pringlei*), smoke tree (*Dalea spinosa*), ironwood, fourwing saltbush (*Atriplex canescens*), and jojoba (*Simmondsia chinensis*).

D. The Chihuahuan Desert

The second largest desert in North America is the Chihuahuan Desert, covering ~350,000 km^2. It occupies a major portion of the Mexican states of Chihuahua, Coahuila, San Luis Potosí, and northwestern Nuevo León, and, in the United States, portions of western Texas and southern New Mexico (Fig. 4).

The average annual precipitation over the region is between 6 and 25 cm, although it often deviates widely from the average during any one year. July, August, and September are the months of highest precipitation; winters are dry. Rainfall comes in small but intense storms, resulting in poor distribution of

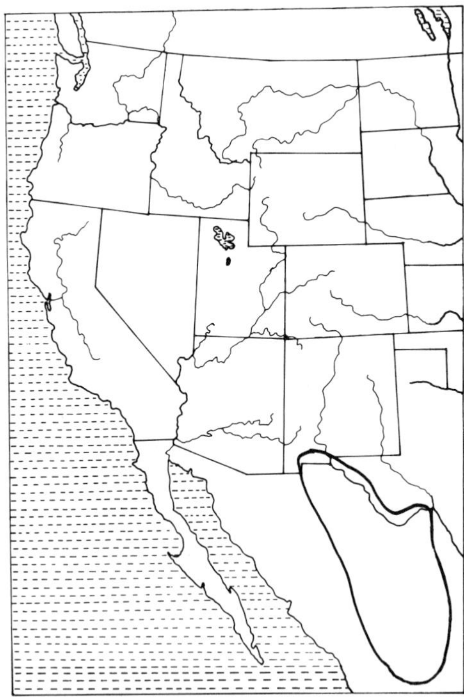

Fig. 4. The Chihuahuan Desert.

moisture over the region (MacMahon, 1979). Winters are mild, although temperatures may occasionally go below freezing; summers are very hot.

The Chihuahuan Desert is highest in elevation in the south and lowest northward along the Rio Grande and U.S. border. It is mostly a vast intermountain plateau; elevations are generally between 915 and 1675 m (Jaeger, 1957). The landscape consists of many steep-walled mountains separated by broad and often undrained basins. The desert is flanked on the east by the Sierra Madre Oriental and on the west by the Sierra Madre Occidental, effectively blocking moisture from both directions. Most of the soils originate from limestone; volcanic rock also is very common, especially in central and western Chihuahua. Varying concentrations of gypsum may be found in some limestone soils. Extensive dunes of silica sand are found in northern Chihuahua; gypsum sand dunes, such as those of the White Sands National Monument near Alamagordo, New Mexico, are also found in a few localities.

The Chihuahuan Desert has some exterior drainage in the northeast by the Rio Grande. There are many undrained basins, many of which have dry lake beds with layers of limestone clay and alkali or salt encrustations. The few perennial rivers are important as water sources both for agricultural and domestic use; groundwater supplies are also important.

The vegetation is varied and often determined by the soil type, salt and gypsum concentration, and alkalinity. In the northern part, creosote bush, agave (*Agave lecheguilla*), sotól (*Dasylirion wheeleri*), and barrel cactus (*Echinocactus wislizenii*) are common; in the south are found candelilla (*Euphorbia antisyphilitica*) and guayule (*Parthenium argentatum*) (Jaeger, 1957).

E. The Columbia–Snake River Plateau

The Columbia–Snake River Plateau shares its southern boundary with the Great Basin Desert and extends northward into Canada. It covers more than 256,000 km^2 and includes within its boundaries southern and western Idaho, eastern Oregon, eastern Washington, a small part of northeastern California, northwest Nevada, northern Utah, and southern British Columbia (Fig. 5).

The surface of the plateau averages ~915 m in elevation. It includes most of the lava fields of the northwest, irregular mountains, scablands, some closed basins, and an area known as Palouse prairie. The climate is semiarid and cool. Precipitation averages less than 25 cm in the west and increases eastward to more than 50 cm in the mountains. The Columbia–Snake River Plateau is in the rain shadow of the Cascade Mountains. Precipitation occurs throughout the year, except in summer. Thunderstorms and hail are common in the east but rare in the west. Generally, winds are from the north throughout most of the region.

Most of the plateau is drained by the Snake and Columbia Rivers. Each river is large and has several large tributaries. Most water problems there have been

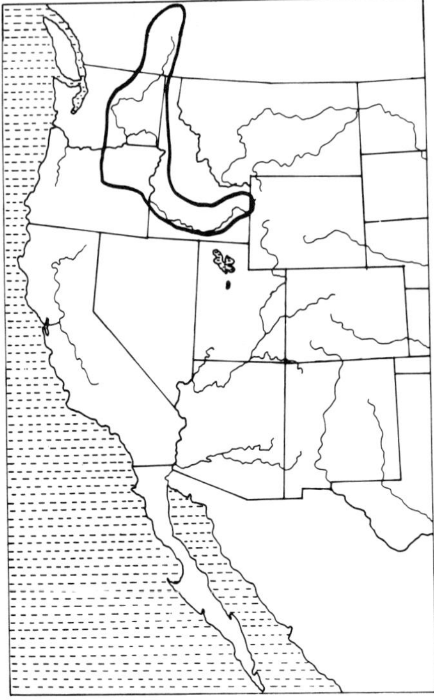

Fig. 5. The Columbia–Snake River Plateau.

related to the distribution of floods (Hunt, 1974); the dams and reservoirs built along both rivers have eliminated many problems and provided hydroelectric power. Hard, alkaline water is common in the southern tributaries of the Snake River. The southwestern region of the plateau has no exterior drainage except at the edge. The playas are alkaline or saline, with intermittent streams draining into them. Groundwater is generally plentiful throughout most of the plateau, and the quality of water is good, for the most part, but some is unsuitable for use. Much of the good groundwater is under heavy use. Human population centers are concentrated along the rivers that flow through the plateau.

Soils have developed on volcanic ash, lava flows, extensive loessial deposits, and some alluvium. The loess deposits in the northern region of the plateau may be up to 45 m thick (Hunt, 1974). These are the soils of the Palouse prairie. In the western part of the plateau the soils are only weakly developed, but become deeper and more fully developed eastward, where conditions are less arid. Extensive areas of bare lava are found in the southeastern region, known as the Snake River Plain. Limited areas of dune sand are scattered over the plateau.

The natural vegetation of the plateau is mostly sagebrush–grassland with regions of grassland and woodland. Palouse prairie is found in areas receiving more than 25 cm of precipitation annually. The natural vegetation here is dominated by bluebunch wheatgrass (*Agropyron spicatum*). Since the aeolian soils provide excellent cropland, most of the Palouse prairie is now used to grow wheat.

F. The California Valley

The California Valley includes more than 51,000 km^2 in California between the Sierra Nevada and the coastal ranges (Fenneman, 1931) (Fig. 6). Elevations are mostly less than 120 m, with the highest, near 500 m, at the southern end. In general, the valley slopes westward. It includes three large basins, the Sacramento, San Joaquin, and Tulare. Most of the region drains to the Pacific Ocean principally by means of the Sacramento and San Joaquin Rivers, except for the majority of the Tulare Basin. The Tulare and Bueno Vista Lakes are the playas found in this southern closed basin. Surrounding the valley on all sides are chaparral covered hills and mountains supporting a dense cover of sclerophyll shrubs.

Most of the soils of the California Valley are alluvial deposits of sand, gravel, and clay; many are limey or saline. Precipitation generally comes during the mild

Fig. 6. The California Valley.

winters; summers are hot and dry. Average annual precipitation is generally less than 25 cm. Warm winds from the southwest predominate in the winter and cool winds from the northwest in summer (Landsberg, 1974). Dominant vegetation includes bunchgrass (*Stipa* and *Elymus*), greasewood, and saltbush. Large tracts of land are under irrigation and contribute a sizable share of the world's food production. Groundwater supplies are extensive throughout the California Valley and are particularly important for irrigated croplands in the San Joaquin Valley. Huge demands, however, are being made on the groundwater supply in the California Valley, resulting in a lowered water table and settling of the land.

G. The Wyoming Basin

The Wyoming Basin (Hunt, 1974) covers 109,557 km^2 and is found almost totally within the state of Wyoming. It is nearly enclosed by mountains and dissected by low hills; elevations range from 1800 to 2400 m. It connects with both the Great Plains to the east and the Colorado Plateau to the south (Fig. 7). Average annual precipitation is less than 25 cm, most of which comes in spring. The winters are cold and summers hot. Summer precipitation usually comes as severe thunderstorms. Winds are mostly from the northwest.

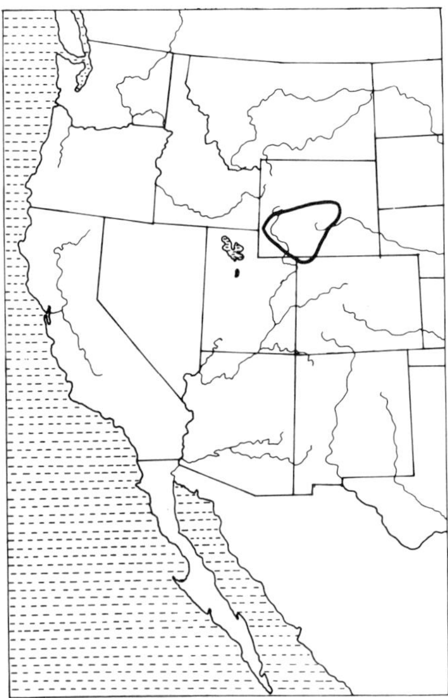

Fig. 7. The Wyoming Basin.

The Wyoming Basin has extensive alluvial deposits, as well as sandstone and shale formations. Soils are alkaline and often salty; the surface layer usually has very little organic matter and is calcareous. Lime or gypsum is usually present in the deeper layers and may develop into a caliche hardpan.

The Wyoming Basin straddles the Continental Divide; though hardly noticeable, the drainage of the basin is divided to the east and west. Several large rivers drain the region; none of them passes through Wyoming without artificial controls (Hunt, 1974). Lakes are abundant, especially at the base of mountain ranges. Many lakes are depressions in the interior of the basin; most are small and dry, although some are perennial lakes and marshes. The population concentrates near mountains and along river valleys. Many towns have grown up alongside the railroads and highways that cross the southern part of Wyoming. The vegetation is dominated by sagebrush, shad scale, greasewood, Gardner saltbush (*Atriplex gardneri*), and western wheatgrass (*Agropyron smithii*). Most lands are grazed, though some areas are irrigated where water is available.

H. The Colorado Plateau

Probably the most colorful of North America's dry lands is the Colorado Plateau. Located in western Colorado, southwestern Wyoming, eastern Utah, northeastern Arizona, and northwestern New Mexico, it covers more than 330,000 km^2 (Fig. 8). Within it lies the Painted Desert and nearly two dozen national parks and monuments, including the Grand Canyon, Mesa Verde, and Canyonlands. High altitudes are characteristic; the plateau surface is generally higher than 1500 m, with some peaks as high as 3350 m. The plateau slants to the northeast and the topography is quite varied; Hunt (1974) divided the region into six sections. The Grand Canyon is located in the high southwest region (Grand Canyon section), and the south rim (Datil section) is covered by thick lava. East of these two areas is a large structural depression (Navajo section), the deepest part of which is called the San Juan Basin. Broad flats on shale formations separated by low cuestas (depressions) are characteristic of this region; numerous volcanic rocks indicate past volcanic activity. North of the Navajo section is the Canyonlands section, where, as the name would imply, deep canyons are the dominant feature. The canyons have been carved out of sandstone by rivers. Also included in this area are shale badlands and sandy deserts. Northward lies the Uinta Basin section, which is a large structural bowl. The High Plateaus section lies at the western edge of the Colorado Plateau. This is a region of high (2750–3350 m), often lava-topped, plateaus separated by wide, flat-bottomed valleys.

Average annual precipitation is less than 25 cm in the interior and greater than 50 cm along the high southwest rim. A major portion comes in the summer as thunderstorms. The aridity of the interior is partially due to the rain shadow effect of the rim, and also to the high rate of evaporation. Summers are hot, with winds from the northwest, and winters cold, with winds mainly from the south.

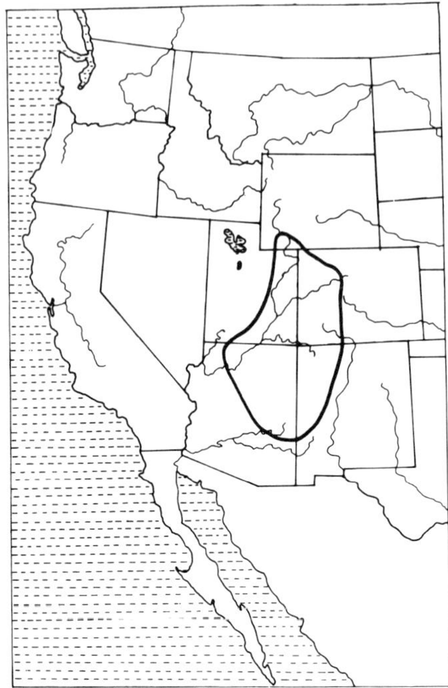

Fig. 8. The Colorado Plateau.

Most of the Colorado Plateau is drained by the Colorado River. Rivers and streams crossing the plateau lose water to evaporation and seepage, and those arising on the plateau are usually intermittent. Other water sources are springs, natural tanks, and wells, most of which provide good water, though dissolved solids may be a problem.

Soils are usually classed as entisols and are usually alkaline; caliche is common. There are large areas of bare rock and extensive sand dunes. Vegetation varies wtih elevation and soil type and includes sagebrush, saltbush, blackbrush (*Coleogyne rammissima*), and grama grasses (*Bouteloua* spp.). Piñon–juniper woodlands are common at higher elevations. Pine and spruce–fir forests are found on high mountains and the south and southwest rims. Very little of the Colorado Plateau is suitable for agriculture. The lands are mostly grazed, and overgrazing has become a serious problem leading to erosion and arroyo cutting (Hunt, 1974).

I. The Great Plains

The semiarid interior plains region of North America is known as the Great Plains (Fig. 9). It covers ~1,472,000 km^2 and extends from Canada to the

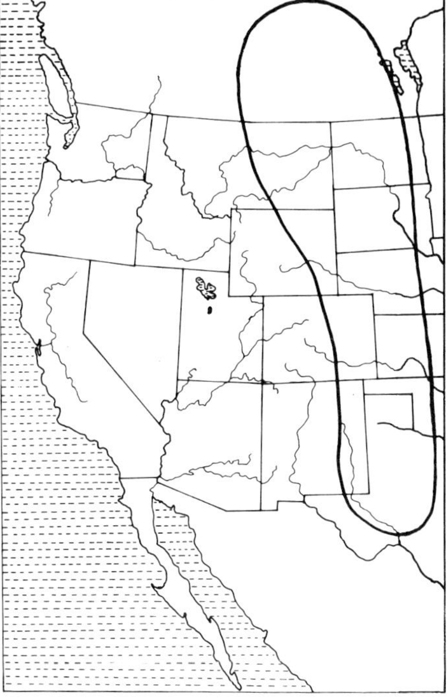

Fig. 9. The Great Plains.

United States–Mexico border. The only distinct boundary is to the west at the Rocky Mountains. The others are somewhat arbitrary because of the gradual merging of one region with another. Other boundaries are the Athabasca River region in Alberta, Canada, to the north, the United States–Mexico border to the south, and the hundredth meridian to the east. The Great Plains slopes eastward from ~1675 m at the foot of the Rocky Mountains to ~600 m in the east (Hunt, 1974). Average annual precipitation varies from ~35 cm in the west to 50 cm in the east; most comes during summer. Year-to-year precipitation is highly variable, and periodic extended drought is inevitable. Temperature differences between seasons are extreme; winters are cold and summers hot. Tornadoes, hailstorms, hot winds, and frost, in addition to years of drought, are hazards that make agriculture in the Great Plains risky. Winds are generally from the north (Landsberg, 1974); chinooks are important in the northern part of the Great Plains in moderating temperatures in the winter and early spring.

The topography of the Great Plains is varied. The Canadian portion is hilly with elevation averaging 760–1050 m, and hills as high as 1430 m. The northern portion of the Great Plains in the United States is a large upland plateau region that has been incised by small streams and rivers and has several dome-shaped mountains rising 450–600 m above it. Shales from badlands and red clinker, caused by the spontaneous ignition of lignite beds that baked the overlying shale, are distinctive features of this region. Further south is an area of high plains, large parts of which are sandy with extensive dunes. In northeastern New Mexico is found a region Hunt (1974) calls the Raton section. It is unique in having lava-capped mesas or plateaus, with some 1830–2130 m high. In southeastern New Mexico and in the Texas Panhandle lies one of the most nearly level parts of the United States. These plains are nearly featureless; they are highest at the western edge (~1675 m) and lowest to the east (~760 m). The southern part of the Great Plains, in Texas, is a large plateau formed on limestone. Surface streams are few; drainage is mostly below ground.

The Great Plains, in the United States, drains from the Rocky Mountains eastward to the Missouri and Mississippi Rivers. The Canadian portion drains toward Hudson Bay via the Saskatchewan and Nelson Rivers. Water is a definite limiting factor to the growth of population, agriculture, and industry in the Great Plains, with supplies inadequate to meet the anticipated demands of the future (Hunt, 1974). Cities near the western boundary get their water supplies from the mountains. Further south and east, rivers and streams lose a great deal to evaporation and seepage, with a resulting increase in the concentration of dissolved solids. Most large rivers are dammed in one or more places for flood control and irrigation. Groundwater from the Ogallala Aquifer is being depleted in many areas with virtually no recharge (Walsh, 1980).

Soils vary widely and include extensive loess deposits, sand, sand dunes, and shale; they are very generally classed as mollisols, entisols, alfisols, and aridisols. They are often poor in organic matter, though some regions are suitable for agriculture if given sufficient water. The horizon of lime accumulation occurs at progressively shallower depths, and the pH of the soil tends to increase with increasing aridity (Eyre, 1963). Short, warm-season, drought-resistant grasses, especially species of *Bouteloua* and *Buchloe,* predominate over the Great Plains. *Agropyron* and *Stipa* species are also important. Most lands are used for grazing, though irrigated agriculture focusing on wheat, hay, and corn is also important. Because of overuse of the underground aquifer, some irrigated crops will have to be replaced with nonirrigated species having low energy requirements.

J. The Southern Temperate Grassland

Between the Chihuahuan Desert on the east and the Sierra Madre Occidental on the west lies a transitional region known as the Southern Temperate Grasslands (Fig. 10). It extends northward into southwestern Arizona and southeastern New Mexico, covering ~250,000 km^2. The grassland occurs at fairly high elevations (1200–2100 m); very cold temperatures can occur in winter and summers are hot. Precipitation averages 40–50 cm annually and falls mostly in

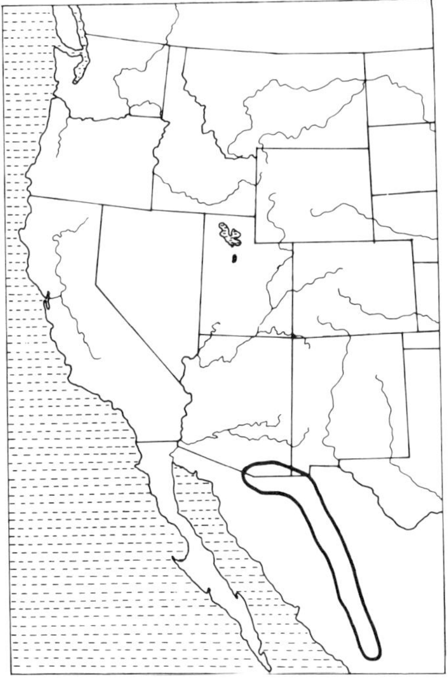

Fig. 10. The Southern Temperate Grassland.

the late summer. Soils vary considerably and include deep, fine-textured soils, sandy soils, clay, and soils mixed with rock and gravel (Shelford, 1963). Limestone soils also occur throughout the region. Water is scarce, but is available from the few rivers or streams and underground supplies. Vegetation is mostly of the shortgrass type (species of *Bouteloua* predominate) but also includes some taller grasses (Dice, 1943) as well as mesquite, *Yucca,* and cacti (*Opuntia* spp.).

III. PLANT RESOURCES

Many plants of North America's arid and semiarid lands have a potential for use. Why they have not been exploited is largely a result of the circumstances of settlement and need. Recent changes in energy costs, availability of water for agriculture, and availability of raw materials from conventional crops have stimulated a reappraisal of arid land plant resources and the extensive areas in which they grow. The emphasis of this discussion will be on relatively unexploited native North American plants with a high potential for food, forage, industrial products, and medicines. The discussion will describe principal plant characteristics, distribution, the nature of their useful products, and some notion as to the problems inherent in their development at the present time.

Information concerning potentially useful plant resources of North America has been developed in various reports, such as those prepared by Theisen *et al.* (1978), Revelle (1978), McKell *et al.* (1968), the National Academy of Sciences (1975a), and Ritchie (1979). A broad perspective of useful plant resources of the world may be found in Usher's *A Dictionary of Plants Used by Man* (Usher, 1974) and from Bailey's *Manual of Cultivated Plants* (Bailey, 1975).

Selection of plants for inclusion in this chapter was based on a moderate-to-high potential for use and sufficient information to encourage further development or experimentation with the plant resource. Groups of plants useful for food, forage crops, industrial products, and medicinal values are compared in tables at the beginning of each section.

A. Food Plants

In an emergency, some part of almost any plant can be prepared as a food. Obviously, some species would be less desirable or nutritious than others, and a very few may contain toxic constituents. Information on toxicities has accumulated during a long period of trial and error. Numerous species found in arid and semiarid lands have been used as food at various times and appear to have some potential for development (Table III).

Buffalo berry [*Shepherdia argentea* (Pursh) Nutt.] is a small tree or thorny shrub growing up to 5.5 m tall; it produces clusters of drupelike red to yellow fruits ~0.6 cm in diameter. The main area of distribution is in the semiarid western margin of the Great Plains from Colorado to Manitoba, Canada. Wright (1943) reported that buffalo berry plants can tolerate salinity and dry soils. Temperature extremes also are tolerated. Seeds germinate quickly in response to a combination of high and low temperatures. The high vitamin C content of the berries makes them useful for preparation of jams and jellies, although cooking destroys some of the vitamin (Knowles and Wilk, 1943). Development potential is moderate because of competition from more desirable fruit species. Ecologically equivalent areas of the world may find this species to be an attractive alternative crop.

Buffalo gourd (*Cucurbita foetidissima*) is a perennial vine that is well adapted to growth on sandy to droughty desert soils of the southwestern United States and Mexico. The extensive root system and fleshy tubers are effective in water uptake and storage, thus making the plant suited to warm long-day areas where rainfall is less than 30 cm. Frost causes the plants to go into dormancy (Bemis *et al.*, 1979).

Seeds of the buffalo gourd were favored by the Indians of the southwestern United States for food. Extrapolating the productivity of single plants on an area

Table III

Native Food Plants with a Potential for Development and Use

Name							
Common	Scientific	Part used	Use or product	Land type adaptation	Cultural practices possible	Yield/range	Development potential
Buffalo berry	*Shepherdia argentea*	Fruit	Used to make jellies and jams	Rangelands of the Great Plains	Hand labor	Collected from scattered natural stands	Small
Buffalo gourd	*Cucurbita foetidissima*	Seed, starch, root	Seeds contain 35% protein and 35% edible oil	Marginal lands, desert areas	Mechanization	1000 kg/ha protein, 1000 kg/ha oil	Large
Bush cherry	*Prunus* spp.	Fruit	Fruit	Pasture	Mechanized	3–4.5 kg/plant, calculated at 1000 kg/ha	Small
Foxtail millet	*Setaria italica*	Seed	Grain	Marginal cropland	Mechanized	1000–1500 kg/ha seeds, 3000–4000 kg/ha forage	Small
Metcalf's bean	*Phaseolus metcalfei*	Seed	Grain	Range	Hand labor	Collected from native stands	Small
Olneya bean	*Olneya tesota*	Fruit, seed	Nut, bean	Desert	Hand labor	Collected from scattered plants	Small
Piñon Pine	*Pinus edulis*, *P. monophylla*	Seed	Nut	Dry scrub forest, range, SW United States	Hand labor	300 kg/ha, 180 kg clean seed/ha	Medium
Spear scale	*Atriplex hortensis*	Seed, leaves	Grain or pot vegetable	Range	Hand labor, mechanization	5000–6000 kg/ha in one trial	Small
Tepary bean	*Phaseolus acutifolius* var. *latifolius*	Seed	Edible bean food	Range	Mechanized	400 kg/ha dry, 1200 kg/ha with irrigation	Medium
Western sand cherry	*Prunus besseyi*	Fruit	Fruit	Idle pasture	Hand collection	Scattered stand collection	Medium
Prickly pear cactus, Nopal	*Opuntia* spp.	Fruit	Sweet, fleshy inner tissue of fruit	SW United States and the Mexican desert	Hand collection	Yields not defined	Medium

basis suggests yields of 1000 kg/ha of protein and a similar amount of oil from the seeds. The fleshy starch-containing tuber is the mainstay for sustaining above-ground production of seed-containing grouds, but harvesting the fleshy roots kills the plant.

A National Academy of Science (1975a) report suggested that further research on buffalo gourd is needed to clarify its potential for producing seeds of high oil and protein content. In studies conducted in Lebanon, Curtis (1972) attempted to grow and genetically develop buffalo gourd as a desert crop. The potential of this crop appears to be high for marginal croplands with sandy soils. Development of appropriate cultural practices and better characterization of the range of variability would help to bring this plant into greater use.

Another species with some similarity in its potential for development is the wild watermelon (*Citrullus vulgaris*). This plant produces a large quantity of seeds that are high in protein and oil and suitable for use as a human or animal feed. In addition, the pulp is rich in sugars that are suitable for ethanol production.

Bush cherry (*Prunus japonica, P. glandulosa, P. pumila, P. tomentosa*) is found throughout a wide geographic area, featuring cold winters, and summers with dry periods; this area includes the northern Great Plains, corn belt, and Great Lakes states (Theisen *et al.*, 1978). The red fleshy fruits mature early and thus escape the effects of summer dry periods. Plants grow in a shrubby form to ~1 m in height, but can grow as high as 3 m under good conditions in hedgerows. Plants are long-lived and have deep root systems, a probable reason for their apparent drought hardiness. Cultural practices have received little attention except for the development of higher-yielding types. Plants respond to fertilizer, but specific requirements are not known because they are grown on marginal soils, in small plantings (North Dakota Agricultural Experimental Station, 1935). The potential for expanded use of bush cherry is moderately high, but requires research on higher yielding varieties with improved resistance to brown rot and moisture stress. Bush cherries might prove to be a more competitive crop than bush berries and strawberries because of a slightly lower labor requirement for harvesting.

Foxtail millet (*Setaria italica*) has a high yield because a large proportion of the plant, especially the stem and ear, remains photosynthetic into maturity. Plants are tolerant to high temperatures and have a high water-use efficiency, which is even greater than that of sorghum (Sandhu *et al.*, 1974).

Foxtail millet is already grown in the United States as a forage or seed crop, but finds greater favor as a food crop in Asia, Africa, and Canada (Malm and Rachie, 1971). In the Great Plains it is grown as a catch crop when other crops fail. As a shift to more water- and energy-efficient crops becomes necessary with the depletion of groundwater reserves, such as the Ogallala Aquifer (Walsh, 1980), foxtail millet may replace crops that are cultivated more intensively.

Drought is tolerated by this plant, but recovery from drought is poor and seedling vigor is low. High seeding rates of 10 to 30 kg/ha are recommended for optimum stands and weed control by competition. Early harvesting provides grain with a higher protein content, as opposed to waiting for the carbohydrate maximum in seeds. Foxtail millet is also desired for its forage value, which is best if plants are not allowed to mature. Grain yields range from 1000 to 1500 kg/ha, and if cut for forage, yields average 3000–4000 kg/ha.

Metcalf's bean (*Phaseolus metcalfei*) is a trailing perennial herb native to the arid southwestern United States and adjacent parts of Mexico in areas where annual rainfall averages between 250 and 375 mm. The deep, fleshy roots assist in sustaining plants through summer drought periods.

Pods are flat, 3.8–5.4 cm long, and contain flat seeds more than 1.3 cm long. The seeds are harvested green or dry and are valued for their high nutritional values, which are similar to those of lima and pinto beans. Early use was limited to livestock feeding, but the bean has potential value for human use. Research is needed on cultural practices and utilization (Theisen *et al.*, 1978).

Olneya bean (*Olneya tesota*) is sometimes referred to as desert ironwood and grows as a small tree in the arid southwestern United States in Arizona, New Mexico, and southeastern California, and in adjacent Mexico. High water-use efficiency is one of its important adaptations. It can drop or retain its leaves as an adaptation to periods of moisture stress and thus exist through periods of drought.

Seeds from the *Olneya* bean were used as food by the American Indians. They also provide livestock feed because of high palatability. Productivity of individual plants is not known, and only limited research has been done on the physiology and ecology of the species (Szarek and Woodhouse, 1977). Research is needed on cultural practices to encourage the use of this species. Spines present a deterrent to development but could be overcome by selective breeding (Theisen *et al.*, 1978).

Piñon pine (*Pinus edulis* and *P. monophylla*) is widely distributed over semi-arid mountain slopes and mesas of the Great Basin and Colorado Plateau; piñon trees grow in association with *Juniper* spp. in a low density pygmy forest, on shallow, rocky soils, in a rainfall zone of 250 to 375 mm (West *et al.*, 1973).

Seeds (''nuts'') are harvested in the late fall by collecting the unopened cones and letting them dry out on plastic or canvas sheets. As the cones open, the seeds are collected, packaged, and sold in local markets. Yields of 0 to 300 kg/ha vary from year to year, and are influenced by a favorable environment during pollination and seed development during the 2-year maturity period. The thin-walled seeds are valued for their flavor and are rich in protein, fat, vitamins, and phosphate (Botkin and Shires, 1948).

Intensive development of piñon nuts does not appear to be feasible becausc of the extensive and scattered nature of distribution. Piñon–juniper stands that have encroached on valley grazing lands or lost their understory shrub and grass species have been cleared by heavy equipment and seeded to forage grasses, thus eliminating pine nut production from many hectares of the intermountain West.

Spear scale (*Atriplex hortensis,* also *A. patula*) is a little-used, arid-land annual that may be a substitute for spinach. Drought, salinity, and high temperature resistance makes this an attractive subsistence potherb and seed crop for arid regions. The range of distribution in the United States includes the western Great Plains and the intermountain deserts. Growth is rapid during the spring months and only 30 days is required for vegetative growth. Seed production requires ~200 days. The protein content (~20%) makes it attractive as a potherb. Height may range from 12 to 60 cm (Theisen *et al.*, 1978).

Cultural practices such as seeding in rows spaced 50–100 cm apart and spraying to control leaf-eating insects can increase yields. Potential for development in the United States is not attractive because of more suitable alternatives, but in areas of similar ecological conditions this plant may be very useful.

Tepary bean (*Phaseolus acutifolius* var. *latifolius*) can be grown without irrigation in an arid location of 200 to 300 mm annual rainfall, at temperatures up to 37°C (Theisen *et al.*, 1978). Beans are produced on bush vines in small pods that reach maturity in 60 to 90 days. The plant has a deep taproot and grows well on deep alluvial soils, thus making it suitable for runoff farming practices.

Tepary beans were cultivated in Mexico more than 5000 years ago and wild varieties were harvested by American Indians. The beans contain high protein levels (23–25%), equal to most economic legumes, but are deficient in sulfur-containing amino acids. In addition to regular use in soups, they may be baked or parched and made into meal.

Potential for development is conditioned by other alternatives for use of dryland areas. Adaptation to aridity, short growing seasons, and high temperatures make them an ideal crop for dryland farming. Interbreeding—with species that are less subject to causing flatulence—in order to obtain drought resistance may be of value.

Western sand cherry (*Prunus besseyi*) has a wide range of distribution westward from the Great Plains to the Great Basin. Plants grow as stoloniferous shrubs from 30 to 150 cm tall and produce small fruits similar to cherry and plum. They are resistant to drought and cold (Theisen *et al.*, 1978). Fruit may be gathered at the end of the 120-day growth period. Quality varies from sweet to bitter, depending on the plant. Fruits may be eaten fresh or cooked. Sand cherry may be used as a drought-resistant rootstalk in addition to its fruit yield. Competition from conventional fruit crops reduces the potential for developing sand cherry production in the United States, but ecologically equivalent areas in developing countries may be considered.

Prickly pear cactus (*Opuntia* spp.) is well adapted to growth on sandy soils of the arid southwest deserts, including northern Mexico and the bordering states of the United States where rainfall is less than 300 mm annually. Drought resistance is accomplished by binding water in succulent pads and crassulacean acid metabolism (CAM), which reduces stomatal loss of water vapor.

In arid regions of Mexico, young prickly pear pads are used as a vegetable, either boiled or as salad greens. The fleshy fruits that develop after flowering are peeled and eaten as a delicacy. Cactus plants also find considerable use as a dairy supplement that is high in sugar but low in crude protein (Rojas-Mendoza and Malo, 1965). In the southwestern United States, prickly pear cactus has been used as an emergency range feed (Reed, 1960). In texas, ranchers have burned off the objectionable spines with a torch to enable cattle to eat the pads during drought years.

The potential for increased production will depend on alternatives. Spineless varieties of prickly pear cactus exist and are under development. A combination of low-protein cactus with high-protein forage, such as *Atriplex,* has been recommended as a way to produce a balanced animal diet (Bonsma and Mare, 1942)

B. Forage Plants

The list of forage plants with potential for development could include many species that are both palatable and nutritious (Committee on Feed Composition, 1958). However, the reader should refer to regular sources of information on these conventional forage species. Instead, a few highly productive shrubs and innovative methods for their use will be discussed here (Table IV).

Various species of saltbush (*Atriplex*) have a wide distribution in arid western rangelands of the United States where annual precipitation is between 150 and 350 mm. Many research results have shown that some shrubs are highly competitive with understory grasses and cause a reduction in grass yields (Frischknecht, 1963). However, Rumbaugh *et al.* (1981) have shown that fourwing saltbush has a favorable effect on the growth of seeded grass species. Saltbush species are highly tolerant of salinity and drought. Plants retain their leaves through the winter, are multistemmed, dioecious, and range in height from 15 to 150 cm. Some of the major species of this far-ranging genus, which is valued for its forage production capability, include *A. canescens, A. lentiformis, A. polycarpa, A. confertifolia, A. cuneata, A. gardneri,* and *A. corrugata.* The order of species in this list also indicates their reduction in size and increase in salinity tolerance.

Salinity and drought tolerance are important attributes of *Atriplex* species. Osmotic adjustment (Richardson and McKell, 1980) and accumulation of salts in giant trichomes on leaves (Mozafar and Goodin, 1970) are two major adaptations permitting species to thrive in harsh environments.

Goodin (1979) gave several reasons for the high forage value of *Atriplex.* These include a high capacity for production during summer feed shortages; a

Table IV

Native Forage Plants of Unusual Potential for Expanded Use

Name							
Common	Scientific	Part used	Nutritional value	Land type adaptation	Season of year best used	Yield (kg/ha)	Development potential
Saltbush	*Atriplex canescens, A. lentiformis, A. confertifolia, A. gardnerii*	Leaves and twigs	Leaves, twigs 12–24% crude protein	Range	Opposite season from grass use	600–900	High
Bitterbrush and other shrubs	Various species	Foliage and twigs	Foliage 11–15% crude protein	Range	Grazing season	—	High
Prosopis	*Prosopis glandulosa*	Stems, leaves, pods	Leaves 21%, pods 13% crude protein	Range, marginal land	Dry months in grass dormancy	5000	Medium
Ramon	*Brosimum alicastrum*	Leaves, pods		Range, dry tropics	Dry season	—	Low

low water requirement, which indicates high-production efficiency; a deeply penetrating root system capable of extracting moisture from depths not explored by shallow rooting grasses and forbs; and a high content of crude protein and phosphorus. *Atriplex* forage compares well with alfalfa in yield (Goodin and McKell, 1970) and crude protein content, which ranges from 15 to 24% (Committee on Feed Composition, 1958), as well as energy value, thus justifying the use of *Atriplex* for rangeland grazing. Although the oxalic acid content of some members of the family Chenopodiaceae occurs in sufficient concentration in species such as *Halogeton glomeratus* to be toxic to grazing animals, no toxic levels have been reported for *Atriplex* species (Goodin, 1979).

Where range improvement programs or intensive management practices are possible, fourwing saltbush or other *Atriplex* species can be used to increase both animal and rangeland productivity. Van Epps and McKell (1977) proposed interseeding *Atriplex* and other high-protein fodder shrubs in existing stands of crested wheatgrass (*Agropyron cristatum*). A major benefit of this species combination can be obtained in a postgrazing season period of use (fall and winter in the Great Basin) when the perennial forage grasses are dormant and have lost their protein value but still contain adequate levels of energy value. At this time, *Atriplex* has adequate protein and carotene to fill the gestation requirements of pregnant ewes (Cook, 1972). In a preliminary study Otsyina *et al.* (1980) compared October sheep grazing on an *Agropyron cristatum–Atriplex canescens* pasture with a pasture of only *Agropyron cristatum* and found that the sheep lost weight on the straight grass pasture, whereas they nearly maintained their weight on the shrub–grass pasture. More testing of shrub interplanting is needed to answer the questions concerning species combination, palatability, and animal management.

Fourwing saltbush and other shrubby species of the Chenopodiaceae have been recognized for many years as good sources of crude protein in the nutrition of range livestock (Cook *et al.*, 1956). However, drastic and nondiscriminatory range improvement practices and a preponderance of agronomic influence based on a grass–forage philosophy have fostered programs of shrub control that have ignored the use of palatable shrubs to supplement the off-season grazing of livestock.

Interplantings of *Atriplex* in North American rangelands could be used to increase range carrying capacity during periods of low-feed quantity and quality. Such plantings were made in Syria (Draz, 1974) and resulted in an added benefit of improved range condition.

Bitterbrush (*Purshia tridentata* and *P. glandulosa*) is among several fodder shrubs native to the desert lands of the western United States (Guinta *et al.*, 1978). Bitterbrush is valued for its nutritional qualities, especially during winter or periods of drought. Medin and Ferguson (1980) reported a high yield (696 kg/ha) of oven-dry browse from a deer range that had been seeded to bitterbrush in southwestern Idaho. They also cited an increase in the use of shrubs to restore disturbed rangelands for multiple-use management. According to Cook (1972), shrubs provide a high amount of digestible protein, carotene, and phosphorus, from vegetative stages through maturity.

Some of the more well-known palatable shrubs (Plummer *et al.*, 1968) of the Great Basin are curlyleaf mountain mahogany (*Cercocarpus ledifolius*), valued for wildlife browse; winter fat (*Ceratoides lanata*), valued for winter grazing of livestock in the desert; cliff rose (*Cowania mexicana* subsp. *stansburiana*), useful for both wildlife habitat and animal browse; big sagebrush (*Artemisia tridentata*), with some ecotypes having better than average palatability; and rubber rabbitbrush (*Chrysothamnus nauseosus*), an aggressive shrub with a wide range in palatability and adaptability.

Recent interest in biomass for fuel energy has prompted a study of rangeland shrubs for their maximum yield under extensive production conditions. Whereas Goodin and McKell (1970) showed that yields of 7000 to 10,000 kg/ha could be obtained from 3-year-old *Atriplex* plants under marginal agricultural conditions, Van Epps *et al.* (1982) measured individual plant weights of large *Atriplex canescens* on favorable range sites ranging from 12 to 38 kg per plant. Based on plant densities observed in favorable sites with deep soil, the individual plant weight could be extrapolated to 9800 kg/ha.

The potential for greater utilization of rangeland shrubs is high. A multiple-use management policy for publicly owned desert rangelands should provide an incentive for including selected shrubs in seeding mixtures for improvement of livestock forage, wildlife habitat, biomass, and reclamation of disturbed areas.

New methods of rangeland and livestock management are needed to make optimum grazing use of shrubs. From their research in central Utah, Otsyina *et al.* (1980) suggested interplanting shrubs in existing stands of crested wheatgrass in adequate proportions to create a balance of energy value and protein content. Other techniques that should be studied include rangeland interplanting, selective control of shrubs and forbs of low palatability and nutritional value to allow the natural increase of desirable shrubs, and seeding and planting an ecologically balanced but highly useful forage community (Gasto and Contreras, 1972). At the Browse in Africa Symposium sponsored by the International Livestock Center for Africa (ILCA) (LeHouerou, 1980), the most pressing research and development need, relative to making better use of shrubs for grazing, was field establishment and livestock management of shrub pastures.

Mesquite (*Prosopis julifora* DC.), as well as other species in this legume genus, ranges in size from 2-m shrubs to 10-m trees. According to Burkart (1976), eight species are found in the arid southwestern United States and northern Mexico. Other species are found in South America, Asia, and Africa. Prime habitat is characterized by warm to hot temperatures with periods of rainfall during the growing season. Deep alluvial soils or areas with groundwater are particularly favorable to *Prosopis* because its long taproot may extend to depths greater than 20 m. *Prosopis* species appear to be tolerant of moisture stress, moderate salinity, and high temperatures. Stout thorns on most biotypes complicate the use of plants by foraging animals or by man collecting pods or cutting fuel wood. Considerable research has been conducted on ways to control mesquite (Scifres, 1973). Even so, the aggressive nature of *Prosopis* species may offer advantages for future use.

Prolific production of indehiscent pods containing protein and energy-rich seeds forms the basis of a major use of *Prosopis* for livestock feed. Felker (1979) reviewed the results of numerous reports on pod production and concluded that from 4000 to 10,000 kg/ha could be obtained from *Prosopis* orchards where there is either groundwater or 250 to 5000 mm of annual rainfall. Yields from 25 trees over a 5-year period in Arizona averaged 14 kg/ha (Parker and Martin, 1952); these results contrast sharply with Felker's (1979) report of 7.3 kg of pods from a 5-m tree and 73 kg of pods from a 8.5-m tree in the Coachella Valley of California. Clearly, much could be done by way of selection and minimum cultural practices to increase the yield of pods, which contain 9–13% protein, 13–36% sucrose, and 45–55% total carbohydrate (Walton, 1923). The high sucrose content (30–45%) of pods and seeds can create problems for rumen bacterial–cellulose action. Greater efficiency of animal use, and better digestion, can be obtained if pods are collected, ground, and mixed with other feed materials.

Another use besides feeding pods to livestock is harvesting the wood of older plants for fuel. Because of its hard wood, *Prosopis* has good heat value. Rapid regrowth after cutting permits regeneration. The high costs of energy have created new interest in *Prosopis* as a potential source of biomass fuel. Although symbiotic nitrogen fixation by *Prosopis* has not been directly reported (Felker, 1979), increased soil nitrogen and organic matter content under mesquite suggests that nitrogen fixation occurs.

Problems created by the invasion and increase of mesquite on rangelands cannot be minimized. *Prosopis* thickets occupy space that would otherwise support good forage species, and thus reduce not only the amount but the accessibility of feed. If these thickets could be thinned to provide fuel and make pod production more available, the problem would not be as critical. Felker (1979) believes that better management and the introduction of noninvasive mesquite lines that are more effective in nitrogen fixation and pod and leaf production would benefit arid land people.

Much research is needed to obtain the benefits available from *Prosopis*. Selection and testing of available genotypes from various populations for livestock forage, nitrogen fixation, fuel biomass, and pod production should receive a high research priority. Other studies should emphasize ways to improve management and utilization. In view of the increasing cost of nitrogen fertilizer and the limited number of leguminous shrubs native to the arid and semiarid southwestern United States, genera like *Prosopis* may be important to the problem of increasing natural inputs to ecosystem nitrogen.

Ramon (*Brosimum alicastrum*) is not a typical arid-land fodder tree but has unique drought-resistant characteristics that make it a good candidate for further study and development. The National Academy of Science (1975a) report, *Underexploited Tropical Plants with Promising Economic Value,* describes ramon foliage as providing valuable fodder for livestock during the dry months, when the branches are lopped off. Further potential for use of this 20- to 30-m tree lies in its milky latex, which can be drunk like milk, or in its seeds, which can be boiled, roasted, or eaten raw.

C. Medicinal Plants

Over a long period of time, man has learned through trial and error that numerous plants have useful medicinal qualities as well as dangerous qualities. Many medicinal plants have been described by Morton (1977). Krochmal *et al.* (1954) identified some useful native plants of the southwestern United States, including those of medicinal value. Some of the main constituents in a plant's medicinal value include alkaloids, saponins, steroids, essential oils, resins, and tannins. These compounds can cause a wide range of physiological responses and many can be used in the formulation of medical products.

Krochmal (1972) pointed out the prohibitive expense of conducting a widespread analysis of plants in order to identify valuable constituents. Instead, he suggested an ethnobotanical approach using folklore and tradition (Fielder, 1975) to uncover plant medicinals that could be isolated and concentrated for more effective use. Whenever a plant species is utilized for a principal constituent, such as rubber or oil, other possibilities should be investigated by the plant products chemist who can develop a chemical flowchart for the extraction of other substances that could be applied during the process of extraction of the major constituent. Table V contains a general description of a few species that have interesting possibilities for medicinal use.

Yucca (*Yucca filifera, Y. decipiens, Y. torreyi, Y. carnerosana, Y. faxoniana,* and other species) grows on alkaline, sandy clay soils throughout the southern United States, and all of Mexico to the temperate forests of Central America. More than 25 species are found in Mexico. Plants of this genus, belonging to the family Agavaceae, have bayonet-shaped leaves and may be acaulescent or possess well-defined stems up to 5 m (as in the Joshua tree, *Y. brevifolia,* of the Sonoran Desert of California, Nevada, and Arizona).

In addition to the fibers that can be recovered from the leaves of some *Yucca* species, the potential for obtaining chemicals and drugs from various plant parts appear to be attractive because of the abundance of the plants, the relatively low

Table V

Selected Plants of Potential Use as New Sources of Medicinal Value

Name					
Common	Scientific	Part used	Medicinal constituent	Medicinal response or use	Potential for development
Yucca	*Yucca* spp. (about 40 in North America)	Extracts from leaves, flowers, and seeds	Steroidal sapogenin, water-soluble extracts, diosgenin, steroid, ascorbic acid	Antibiotic, laxative, steroidal hormone, anti-stress agent in plants and bacteria	Would require an integrated development program involving useful constituents
Mexican yams	*Dioscorea floribunda*	Tubers	Steroidal sapogenins, diosgenin	Steroidal hormone production	Potential is high, can be grown as a crop
Aloe	*Aloe vera*	Succulent leaves	Latex containing anthraquinone glycosides	Used in burn treatment and other skin preparations, also as a laxative	Very high potential because of interest in natural products and easy culture of plants
Juniper	*Juniperus osteosperma* and others	Leaves and wood	Volatile oil	Volatile oil is an aromatic ingredient in ointments	Large populations exist for possible development of volatile oil and wood

value of the land where it grows, and the need for employment of people in the Mexican area of *Yucca* distribution. Numerous papers contained in the state-of-the-art monographs on *Yucca* (Centro de Investigación en Química Aplicada, 1980) include discussions of the kinds of beneficial responses that have been observed in studies of, or experience with, *Yucca*.

Yale (1980) described how *Yucca schidigera* extracts acted as an antistress agent in plant growth, seed germination, and rumen bacteria activity. Small amounts of *Yucca* extract in the human diet appeared to reduce rheumatoid arthritis, hypertension, and cholesterol and triglyceride levels in the blood. *Yucca* seeds may contain up to 12% steroidal sapogenins, although other plant parts, such as leaves, may also contain sapogenin (Wall, 1980). The sapogenin in *Yucca* can serve as steroid hormone precursors, which could be useful in the preparation of cortisone, the sex hormones, and compounds for use in birth control pills. It is also used as a laxative (prepared from the flowers by Indians) and in the biological control of cercaria larvae (*Shistosoma mansoni*).

An integrated development program would be needed to optimize the extraction of all useful constituents from this abundant group of species.

Aloe (*Aloe vera*) is a low-growing succulent plant of the southwestern United States and northern Mexican deserts. It is grown as a pot plant for use as a remedy for numerous ailments, including minor burns and constipation. Recent usage has expanded to include skin lotions, shampoo formulas, dietary additives, and ''health medicines'' (Hylton, 1979). Claims of beneficial use are widespread although some are not substantiated by medical experimentation. There appears to be a high potential for the expanded development of *A. vera* for legitimate purposes.

Mexican yam (*Dioscorea floribunda*) is a climbing, stout-stemmed vine with white-fleshed tubers. The plants are native to southern Mexico in oak forests and into tropical forests. Plantings have been made in the southern United States, but most collections for drug production are from wild populations. The constituents and value of its steroidal sapogenins are similar to those from *Yucca*. Full grown tubers require 3–4 years for development and must be fertilized and grown in deep sandy loam. Potential for development is high, but the possibility of chemical synthesis of the steroidal sapogenin is a deterrent to large investments at the present time (Morton, 1977).

Juniper (*Juniperus osteosperma, J. scopulorum,* or *J. monosperma*) are commonly found in the Colorado Plateau and Great Basin deserts of Colorado, Utah, and Arizona. These slow-growing, drought-tolerant trees are adapted to rocky shallow soils of foothills and plateaus. Short scale-like leaves are evergreen on spreading branches. Trees may attain a height up to 10 m and form a closed ecological community with piñon pine.

In addition to insect- and decay-resistant wood for fence posts, juniper leaves and wood may be useful for their volatile oils. Only in a combined program of wood and product utilization could the constituents be obtained economically.

D. Industrial Plants

Several plants native to North America appear to have a high potential as a source for industrial raw materials (Table VI). These species are well known locally, and some, such as jojoba and guayule, have attracted international notice. However, full-scale development has not been achieved for a number of reasons. As energy costs increase the production costs of conventional crops under standard methods, marginal lands and species adapted to them will undoubtedly receive greater attention for development. A few crops of industrial potential have been selected for discussion.

Jojoba [*Simmondsia chinensis* (Link) Schneid.] is an evergreen shrub native to the Sonoran Desert of North America. Densely branched plants are drought and salinity resistant and grow on deep, sandy, coarse desert soils. Plants are dioecious, although a few monoecious plants have been found. Flowering occurs over the extended period of January to April in North America (Gentry, 1958).

Nut-like fruits about the size of a peanut are ready for harvest in mid to late summer. Five-year-old plants may produce from 125 to 250 g of dry clean seeds, although considerable variation exists within local populations. Fruits contain up to 60% liquid wax, which is useful as a high-temperature-resistant lubricant and also for use in the manufacture of cosmetics, soaps, and other oil-based products (Spadaro and Lambou, 1973). The residual meal that is left after wax extraction can be used for animal feed. The liquid wax is an excellent replacement for sperm whale oil, which has been forbidden for import to the United States since 1970. Considerable interest in jojoba has developed in the United States as a result of the unavailability of sperm whale oil. High worldwide oil prices and shortages have further stimulated interest in jojoba production (Yermanos, 1980).

Development of jojoba has been a dream of a few visionary plant scientists since the first mention of it in 1882, but not until the 1970s have substantial areas been planted. Fifty thousand hectares have been planted world wide (National Academy of Sciences, 1985). Of this total several hundred hectares have been established in Israel, Mexico, Costa Rica, and Australia. Other countries with suitable climates for jojoba anticipate experimental plantings. The National Academy of Sciences (1975b, 1985) recognized the possibilities of jojoba as a promising crop for arid lands. Commercial groups have promoted the potential of jojoba wax, and a wave of interest has resulted in desert land promotional sales, formation of product-development companies, development of management services, and the organization of marketing institutions. The development of jojoba

Table VI

Native Plants with Potentials for Industrial Product Development as Fiber, Fuel, Oil, Wax, Latex, and Resins

Name		Part used	Nature of product	Land type/ adaptation	Season of year best used	Yield	Development potential
Common	Scientific						
Oil							
Jojoba	*Simmondsia chinensis*	Seed	Liquid wax	Sonoran desert	Late fall when seeds ripen	—[a]	High
Bladderpod	*Lesquerella fendleri*	Seed	Oil, seed protein	Arizona, Texas, marginal dry cropland	Late summer	182 to 364 kg/acre	Low
Fiber							
Ixtle	*Agave lecheguilla*	Leaves	Leaf fibers	Warm desert of Mexico	Summer	—	Low
Yucca	*Yucca* spp.	Leaves	Leaf fibers		—	—	Low

Latex							
Guayule	*Parthenim argentatum*	Stems	Latex rubber equal to Hevea rubber	Warm desert of Northern Mexico and SW United States	Late summer	682 kg/acre	High
Rubber rabbitbrush	*Chrysothamnus nauseosus*	Stems	Latex rubber	Intermountain cold deserts	Late summer	—	Low
Milkweed	*Asclepias* spp.	Stems, leaves	Latex hydrocarbon	Marginal agricultural land	Summer, can be sequentially harvested	—	High
Wax							
Candelilla	*Euphorbia antisyphlitica*	Waxy coating on stems	Wax	Mexican desert	Summer	—	Medium

[a] Data not available.

is chronicled in numerous periodicals such as *Jojoba Happenings,* published by the Office of Arid Land Studies, University of Arizona; *Jojoba News,* published by the Jojoba Research Center in San José, Costa Rica; and *Petroculture,* published by the California State University Arboretum at Fullerton, California.

The potential for production of jojoba wax on a large scale appears to be very favorable, especially since it does not compete with traditional crops for agricultural land and has a low requirement for irrigation. Optimal yields of 7000 kg/ha of seeds are projected on the basis of a study that showed a yield of 3.6 kg per 10-year-old plant (King, 1980), grown at a density of 5300 plants per hectare (Yermanos, 1980). However, plant variability and alternate-year production problems must be solved before accurate yield projections can be made.

Jojoba can be grown on a wide range of well-drained, medium-textured soils within a pH range of 5 to 8, but temperatures below −5°C may be critical. Flowers are more susceptible than seeds to low temperature. Vegetative parts may survive temperatures lower than those that cause damage to reproductive parts.

Selection of plants from native populations has emphasized those with a high potential for yield and wax quality, adaptation to stress, erect growth habit suitable for mechanical harvest, and timing of desirable growth response. Because of the dioecious habit, initial seeding has been at high density to permit later removal of excess male plants and plants with less desirable characteristics. Planting of rooted cuttings from superior female and male plants offers a high degree of assurance for the proper male and female proportions and plant characteristics in a plantation. However, the number of rooted cuttings available from individual plants is limited and cannot meet the large numbers required for large plantings. Until plant geneticists are able to produce superior genetic varieties of jojoba, micropropagation or tissue culture of outstanding plants offers a useful alternative (Wochok and Sluis, 1979).

On the basis of costs for land, labor, and container-grown transplants in the southern California area, Beane (1979) projected a cost of 5947 U.S. dollars per hectare for the first 3 years to establish a jojoba plantation.

Cultural practices for jojoba are being developed as experience is gained. Fertility requirements are moderate and field applications of nitrogen and potassium at 56 kg/ha have not resulted in large growth responses. Continued cropping over many years may be expected to create a need for fertilization. Since jojoba grows naturally in areas receiving 75–450 mm of precipitation annually, irrigation requirements are low. The critical period of moisture use is in late winter and spring during flowering and early seed development. Thus, the most effective water-management program must provide adequate moisture during development, plus midsummer irrigation in excessively dry years to ensure seed fill. Ehrler *et al.* (1979) suggested the use of water-harvesting techniques for increasing yields without resorting to the use of scarce irrigation water.

Harvest of seeds is presently performed by hand, but mechanical harvesting methods are being developed that will require appropriate plant row spacing and an upright plant growth form.

Considerable research is needed to sustain the development of jojoba from a plant of natural curiosity to a substantial crop plant. The tempo of public and privately supported research in several countries is accelerating in response to this need. A major opportunity for exchange of technical information is provided by the series of International Conferences on Jojoba sponsored by the International Council of Jojoba.

Bladderpod (*Lesquerella fendleri*) and other species of the genus belong to the family Cruciferae and grow on well-drained soils in a broad area extending from east central Mexico to Alberta and Saskatchewan, Canada. The best growth occurs on basic calcareous soils. Many species are cold tolerant. Bladderpod is a winter annual or short-lived perennial (Theisen *et al.*, 1978).

Lesquerella seeds contain up to 28% oil that is similar to castor oil. Using experimental plot data, yields of 400 to 800 kg/ha appear to be possible, but few agronomic studies have been made, and work on cultural practices is needed (Barklay *et al.*, 1962). The genetic potential for developing taller varieties with improved erect growth habit and better seed retention capabilities appears high. A reduction in seed dormancy is also needed. As with other plants of wide distribution that have moderately low water requirements, bladderpod has a potential to replace conventional crops where depletion of groundwater reserves is expected to result in reduced irrigation.

Yucca baccata, other *Yucca* species, and *Agave lecheguilla* are native to the Mojave, Sonoran, and Chihuahuan deserts of the southwestern United States and north central Mexico. These slow-growing monocotyledonous plants are well adapted to the deserts with their erect bayonet-like leaves and extensive, shallow root systems. Plants range in height from acaulescent forms from 25 cm to treelike forms 10 m tall.

Many traditional uses have been made of *Yucca* species, including the separation of fibers for construction of ropes, sandals, baskets, and clothing. Soap has been made from extracts of *Yucca* roots. Flowers and fruits provide excellent livestock feed, especially in a dry season (Pena-Lujan, 1980).

Traditional means for the collection of leaves, separation of fibers, and manufacture of various articles by hand have limited the potential for increased utilization of this natural resource. If other uses for the *Yucca* develop, such as recovery of saponins, it may be possible to mechanize and expand the use of *Agave* and *Yucca* for their natural fibers as a residual material after the extraction of saponin (Dominguez, 1968).

Guayule (*Parthenium argentatum*) is a low, gray-green shrub native to the Chihuahuan Desert in Mexico and into the Big Bend area of west Texas. Average annual rainfall in the main area of guayule distribution is from 200 to 375 mm, according to Bullard (1946). The bush is often less than 50 cm high and grows rapidly when water is available, but with the onset of drought, rubber production and deposition occur in cellular sacs throughout the plant (Vietmeyer, 1979). In the natural distribution range of guayule, temperature maxima of 46°C and minima of −9.5°C have been recorded. This wide range of temperature tolerance indicates adaptability to a geographical area much larger than its present natural distribution; this area could include the southern portion of the bordering states of the United States and similar world climates (Office of Arid Land Studies, 1979).

Guayule contains from 8 to 20% rubber, which is virtually identical with that from the rubber tree (*Hevea brasiliensis*). Guayule rubber is a polymer of the five-carbon molecule isoprene. The isoprene units are joined together end-to-end in a linear chain and have a molecular weight similar to *Hevea* rubber. Other products from guayule include resins, terpenes, water-soluble polysaccharides, and soluble salts, wax, and up to 55% bagasse.

Interest in guayule as an adjunct to world supplies of natural rubber have stimulated a periodic effort to produce the plant since the early 1900s. A substantial guayule industry thrived in northern Mexico until depletion of native stands and political disruption forced its termination and transfer to the Salinas Valley of California. With the advent of World War II, the U.S. government gave massive support to guayule rubber production research and development.

More than 12,000 ha were planted during World War II, but at the end of the war support was withdrawn in favor of synthetic and natural rubber and the plantings were plowed under (Vietmeyer, 1979).

Guayule is attractive for culture in arid lands because it can be grown as a nonirrigated crop. Supplemental irrigation may be considered for extended dry periods, but too much water can reduce rubber yields. With appropriate plant spacing, land area becomes a trade-off. The plant can be grown as a replacement for traditional crops in areas of declining irrigation water supplies.

Fertilizer requirements are low based on fertilization studies in which no additional yields were produced by various fertilizer rates. Mechanical harvesting of the entire plant will be necessary to obtain the most favorable production/cost ratio (National Academy of Sciences, 1977).

Guayule rubber production ranging from 800 to 1300 kg/ha holds considerable promise for arid lands where other sources of income or land uses are insufficient to sustain human settlements. Mexico has assessed the native population of guayule and plans for a processing facility are well-advanced. Goodyear Tire and Rubber Company has guayule plantings at its Litchfield Research Park in Arizona, and Firestone Tire and Rubber Company also is funding advanced work on production and utilization of guayule rubber in Texas.

Some of the most pressing research needs concern agronomic practices for specific locations, effects of plant growth regulators, genetic improvement, and use of the rubber and by-products. A major funding appropriation for guayule by the U.S. government is expected to accelerate the production of guayule through research and field production trials.

Rubber rabbitbrush (*Chrysothamnus nauseosus*) is a multibranched shrub of the western United States. Plants range in size from 0.5 to 1.5 m and are found in mixed-shrub populations on deep sandy soils. According to an early survey (Hall and Goodspeed, 1919) rabbitbrush contains chrysil, which could be used to make high-quality rubber. Buchanan *et al.* (1977) analyzed more than 100 plants with a potential to produce valuable hydrocarbons and reported rabbitbrush as having 1–2% rubber content. Given its broad range and adaptability to the cold climates of the Great Basin and Colorado Plateau, some scientists expect that rubber rabbitbrush has a greater potential than guayule for rubber production. Large expanses of desert land presently occupied by mixed-shrub populations could be used for spaced plantings of rubber rabbitbrush. An added advantage to growing rabbitbrush for its rubber content would be the potential of using the woody biomass residual for fuel.

Research is needed to identify high-yielding biotypes of rabbitbrush and to develop cultural practices suitable for producing plant material within the constraints of arid systems.

Milkweed (*Asclepias speciosa*) and other closely related species are found as weeds in agricultural fields in humid and irrigated regions. Plants are perennial and branch profusely from rhizomes, thus making possible the use of harvest methods similar to alfalfa.

One of the potential uses for milkweed is in the production of latex that may be used as an industrial hydrocarbon. According to Calvin (1979), latex exudates from many plants have a potential use for fuel and other hydrocarbon-associated applications. Results of field plantings of milkweed by the Plant Resources Institute in Salt Lake City, Utah, indicate a potential per hectare production of the equivalent of 1600 liters of oil, 2500 liters of ethanol, and 4000 liters of methanol. Small amounts of rubber, resins, wax, and a protein residual for livestock feed are also possible (Savage, 1980). However, costs of production and processing milkweed for its hydrocarbon content are in excess of values for petroleum-based fuels at current prices in 1985. Thus milkweed must remain as an alternative hydrocarbon source in the future.

Research on problems of appropriate cultural practices, including irrigation and cutting frequency, is needed to provide a better estimate on the potential of milkweed as an alternate crop for marginal lands.

Candelilla (*Euphorbia antisyphilitica*) plants, with numerous pencil-sized stems, grow in sandy areas of the Chihuahuan Desert. Candelilla is drought and salinity resistant and has been collected for many years for the hard waxy material that can be extracted from its epidermis (Dominguez, 1968). Stems are immersed in hot, acidifed water and the dissolved wax is skimmed off and cooled. A primary use for the wax is in making candles and other wax products. The economics of the collection and extraction operations are marginal.

E. Other Plants

LANDSCAPING AND EROSION CONTROL

An extensive choice of plant species is available for landscaping use in areas where stress conditions are present. Native species may be suitable in areas where normal maintenance such as irrigation, clipping, or fertilization is not available, or is too expensive. Some of the prime locations for extensive landscaping include rights-of-way, revegetation of disturbed areas, industrial sites, and transmission corridors. Numerous guides for choice of species have been published (Natural Vegetation Committee, 1973; U.S. Department of Agriculture, 1972; Stark, 1966). No mention of individual species is considered necessary at this time inasmuch as the conditions of the site, soil, climate, and desired plant characteristics would dictate the kinds of plants chosen for a given landscaping project.

IV. SUMMARY

Development of a native plant species to commercial or economically significant levels cannot be easily accomplished based on current observations of jojoba and guayule. Many constraints must be overcome to satisfactorily develop a native plant.

A. Constraints

1. SCIENTIFIC CONSTRAINTS

A major scientific problem in plant development is lack of technical information. A sufficient amount of general information is needed to identify plants of high potential. Additional species information can help determine feasibility for development and the suitability of products or uses to meet identified needs. Progress toward development may well depend on technical data regarding planting, management, harvest, processing, and conservation. Pilot demonstration programs designed to answer technical problems and work out procedures are essential to guide development.

Particularly needed are scientific studies on ways to establish plants and obtain optimum productivity under arid and semiarid conditions. Problems dealing with microorganisms and plant growth, drought resistance, physiology of stress, and application of plant breeding to improve adaptation present challenges for scientific research on indigenous plants. Where outstanding biotypes exist, tissue culture may be a solution to the multiplication of desirable genotypes.

2. ENVIRONMENTAL CONSTRAINTS

Existing land uses may pose one of the largest constraints to development. Such commitments of land must be seen as a positive value in relation to need for the land uses as they currently exist. For native plants with industrial potential (i.e., guayule), processing may influence air and water quality. Whether or not utilization costs can be internalized in the value of the product or plant use must be determined. In most instances, it must be seen that the environmental benefits of developing native or adapted plants will most likely outweigh the negative impacts. The potential of reducing existing environmental degradation and more sustained land utilization must be considered a positive consequence.

3. CULTURAL CONSTRAINTS

Impact of the development may cause social change, community growth, and increased need for services. Such changes need to be addressed, but at this time little information is available. In general, the cultural impact from developing new crops or practices from native plants should be positive, or at least neutral. Resistance to change may be manifested by the refusal to cooperate or allow project development. Involvement of local leaders and decision makers is a necessity.

4. ECONOMIC CONSTRAINTS

The major economic constraint to development is probably the lack of seed money, venture capital, or government support to conduct pilot scale programs. From the pilot program, cost data can be extrapolated for planting, production, transportation, and marketing. From these preliminary data, decisions can be made toward major financing and long- or short-term commitment of funds, either by the private sector or through government grants and loans. Because of the generally speculative nature of developing high-potential native plants to meet needs that are not clear, private sector funding may have to be government subsidized in areas that are clearly in the public interest.

5. POLITICAL CONSTRAINTS

A major political constraint is the instability and short terms of office of many political leaders. New crops must not appear to compete with existing production systems on public lands but should complement them. Constituents must be convinced that the proposed developments will provide benefits equitably. Adapted crops, which could represent a higher cash return than traditional crops, should receive special attention to sponsor their development and commercialization for the economic benefit of the country.

B. RECOMMENDATIONS

It would be false to assume that only a large commercial-type farm or a family-sized farm would be suitable for native plant development. Much depends on the nature of the plant species and the magnitude of development necessary. Some crops such as buffalo gourd could easily be grown on small plots and collected for commercial markets. In contrast, industrial feedstocks, biomass, and high-volume crops such as guayule would be better grown in large fields and harvested and treated mechanically.

Small field operations would cause little change to socioeconomic structure except to provide an additional income stream to communities. Large operations may disrupt communities by increasing their population or requiring the establishment of new communities. An excellent example in the United States is the 50,000–ha Navajo Irrigation Project near Farmington, New Mexico. The large commercial farm operation has left little opportunity for community development of a traditional native culture, and it has provided no opportunity for family farm or cooperative-group farm development. The project has addressed only the large scale production–economic aspects of development. Socioeconomic problems remain unsolved as illustrated by the attempts being made to resettle Navajo workers in a modern housing subdivision, which is quite foreign to existing patterns of community settlement.

The various options available to develop native plants of high potential could enhance existing social and economic patterns or could disrupt them with large developments, depending on the suitability of the land, the adaptability of native plants to given locations, and the institutional insensitivity that might prevail in their development. Properly designed pilot projects could help provide guidelines for larger scale development and experience in problem solving, both of which are necessary for the commercial utilization of high-potential native species.

REFERENCES

Bailey, L. H. (1975). "Manual of Cultivated Plants." Macmillan, New York.

Barklay, A. S., Gentry, H. S., and Jones, Q. (1962). The search for new industrial crops. II. *Lesquerella* (Cruciferae) as a source of new oil seeds. *Econ. Bot.* **16,** 95–100.

Beane, J. H. (1979). Jojoba development cost analysis. *Avocado Grower,* April, 38–41.

Bemis, W. P., Berry, J. W., and Weber, C. W. (1979). The buffalo gourd, a potential arid crop. *In* "New Agricultural Crops" (G. Ritchie, ed.), pp. 65–87. AAAS Selected Symposium No. 38. Westview Press, Boulder, Colorado.

Bonsma, H. C., and Mare, G. S. (1942). Cactus and oldman saltbush as feed for sheep. Union of South Africa, Dep. Agric. For. Bull., No. 236.

Botkin, C. W., and Shires, L. B. (1948). The composition and value of pinyon nuts. N.M. State Univ., Agric. Exp. Stn. Bull. No. 344.

Buchanan, R. A., Cull, I. M., Otey, F. H., and Russell, C. R. (1977). Hydrocarbon and rubber producing crops: evaluation of 100 U.S. plant species. Paper presented at 11th Great Lakes Regional Meeting, Am. Chem. Soc., Univ. of Wisconsin, Stevens Point.

Bullard, W. E., Jr. (1946). Climate and guayule culture. Emergency rubber project. *U.S. For. Serv.*

Burkart, A. (1976). A monograph of the genus *Prosopis* (Leguminosae, subfam. Mimosoideae). *J. Arnold Arbor.* **57**(3), 217; **57**(4), 450.

Calvin, M. (1979). Petroleum plantations for fuel and materials. *BioScience* **24,** 533–538.

Centro de Investigacion en Quimica Aplicada (1980). "*Yucca.* Serie el Desierto," Vol. 3. Comision Nacional de las Zonas Aridas, Saltillo, Coahuila, Mexico.

Committee on Feed Composition (1958). "Composition of Cereal Grains and Forages." Nat. Acad. Sci., Nat. Res. Counc., Washington, D.C., Publ. 585.

Cook, C. (1972). Comparative nutritive values of forbs, grasses and shrubs. *In* "Wildland shrubs—Their Biology and Utilization" (C. M. McKell, J. P. Blaisdell, and J. R. Goodin, eds.), pp. 303–310. *Gen. Tech. Rep. INT-1. U.S. Dep. Agric. For. Serv. Intermountain For. Range Exp. Stn.*

Curtis, L. C. (1972). "An Attempt to Domesticate a Wild, Perennial, Xerophytic Gourd, *Cucurbita foetidissima.*" Prog. Rep. No. 1. The Ford Foundation, Beirut.

Delury, G. E. (ed.) 1978. Countries in North America. *In* "World Almanac and Book of Facts." Newspaper Enterprises and Newspaper Enterprises Assoc., New York.

Dice, L. R. (1943). "The Biotic Provinces of North America." Univ. of Michigan Press, Ann Arbor.

Dominguez, X. A. (1968). Industrial racional de las plantas nativas de zonas aridas. *In* "International Symposium on Increasing Food Production in Arid Lands" (T. W. Box and P. Rojas-Mendoza, eds.), pp. 225–236. Texas Tech Univ., Lubbock.

Draz, O. (1974). Range management and fodder development report to the Government of the Syrian Arab Republic. *Agric. Serv. Bull. (F.A.O.),* TA 3292.

Ehrler, W. L., Fink, D. H., and Michell, S. T. (1979). Growth and yield of jojoba plants in native stands using runoff-collecting microcatchments. *Agron. J.* **70,** 1005–1009.

Eyre, S. R. (1963). "Vegetation and Soils." Aldine, Chicago.

Felker, P. (1979). Mesquite. *In* "New Agricultural Crops" (G. Ritchie, ed.), pp. 89–132. AAAS Selected Symposium No. 38. Westview Press, Boulder, Colorado.

Fenneman, N. E. (1931). "Physiography of Western United States." McGraw-Hill, New York.

Fielder, M. (1975). "Plant Medicine and Folklore." Winchester Press, New York.

Frischknecht, H. C. (1963). Contrasting effects on big sagebrush and rubber rabbitbrush on production of crested wheatgrass. *J. Range Manage.* **16,** 70–74.

Gasto, J. C., and Contreras, D. (1972). "Analisis del Potencial Pratense de Fanerofitas y Camefitas en Regiones Mediterraneas de Pluviometria Limitada." Congreso Nacionales Interdisciplinarias de Estudios de las Zonas Aridas del Norte de Chile, Arica, Chile.

Gentry, H. S. (1958). The natural history of jojoba (*Simmondsia chinensis*) and its cultural aspects. *Econ. Bot.* **12,** 261–295.

Goodin, J. R. (1979). *Atriplex* as a forage crop for arid lands. *In* "New Agricultural Crops" (G. Ritchie, ed.), pp. 133–148. AAAS Selected Symposium No. 38. Westview Press, Boulder, Colorado.

Goodin, J. R., and McKell, C. M. (1970). *Atriplex* spp. as a potential forage crop in marginal agricultural areas. *In* "Proceedings XI International Grassland Congress," pp. 158–161. Univ. of Queensland Press, Brisbane.

Guinta, B. C., Stevens, R., Jorgenson, R., and Plummer, A. F. (1978). "Antelope Bitterbrush, an Important Wildland Shrub." Utah State Div. Wildl. Resour., Publ. 78-12. Salt Lake City, Utah.

Hall, H. M., and Goodspeed, T. M. (1919). A rubber plant survey of western North America. *Univ. Calif. Publ. Bot.* **7,** 159–278.

Hunt, C. B. (1974). "Natural Regions of the United States and Canada." Freeman, San Francisco.

Hylton, W. H. (ed.) (1979). "The Rodale Herb Book." Rodale Press, Emmaus, Pennsylvania.

Jaeger, E. C. (1957). "The North American Deserts." Stanford Univ. Press, Stanford, California.

King, P. J. (1980). Planting jojoba for future returns. *Venture* (March 1979), 77–79.

Knowles, D., and Wilk, L. (1943). Vitamin C (ascorbic acid) content of the buffalo berry. *Sci.* **97,** 43.

Krochmal, A. S. (1972). Medicinal values. *In* "Wildland Shrubs—Their Biology and Utilization" (C. M. McKell, J. P. Blaisdell, and J. R. Goodin, eds.), pp. 98–100. *Gen. Tech. Rep. INT-1. U.S. Dep. Agric. For. Serv. Intermountain For. Range Exp. Stn.*

Krochmal, A., Paur, S., and Duisberg, P. (1954). Useful native plants in the American southwestern deserts. *Econ. Bot.* **8,** 3–20.

Landsberg, H. E. (ed.) (1974). "World Survey of Climatology: Climates of North America," Vol. 2. Elsevier Sci. Publ., Amsterdam.

Lehouerou, N. H. (1980). Abstracts of proceedings. International symposium on browse in Africa. International Livestock Center for Africa. Addis Ababa, Ethiopia.

MacMahon, J. A. (1979). North American deserts: their floral and faunal components. *In* "Arid-Land Ecosystems Structure, Functioning and Managements," Vol. 1 (R. A. Perry and D. W. Goodall, eds.). IBP 16. Cambridge Univ. Press, London and New York.

Malm, N. R., and Rachie, K. O. (1971). "The *Sitaria* Millets, A Review of the World Literature." S.B. 513. Univ. of Nebraska.

McKell, C. M., Goodin, J. R., and Garcia-Moya, E. (1968). New uses for arid range plants. *In* "International Symposium on Increasing Food Production in Arid Lands" (T. W. Box, ed.), pp. 155–167. International Center for Arid and Semi-arid Land Studies. Texas Tech Univ., Lubbock.

Medin, D. E., and Ferguson, R. B. (1980). High browse yield in a planted stand of bitterbrush. USDA For. Serv. Res. Note INT-279.

Morton, J. F. (1977). "Major Medicinal Plants, Botany, Culture and Uses." Charles C. Thomas, Springfield, Illinois.

Mozafar, A., and Goodin, J. R. (1970). Vesiculated hairs: a mechanism for salt tolerance in *Atriplex halimus* L. *Plant Physiol.* **45,** 62–65.

National Academy of Sciences (1975a). "Underexploited Tropical Plants with Promising Economic Value." Advisory Committee on Technology Innovation, Washington, D.C.

National Academy of Sciences (1975b). "Products From Jojoba: A Promising New Crop for Arid Lands." Board of Science and Technology for International Development, Nat. Acad. Sci., Washington, D.C.

National Academy of Sciences (1977). "Guayule, an Alternative Source of Natural Rubber." Report of ad hoc panel, Advisory Committee on Technology Innovation Board on Agriculture and Renewable Resources. U.S. Gov., Washington, D.C.

National Academy of Sciences (1985). "Jojoba: New Crop for Arid Lands, New Raw Material for Industry." Report of ad hoc panel, Advisory Committee on Technology Innovation. Washington, D.C.

Natural Vegetation Committee (1973). "Landscaping with Native Arizona Plants." Soil Conserv. Soc. Am. (Arizona chapter). Univ. Arizona Press, Tucson.

North Dakota Agricultural Experiment Station (1935). Native fruits of North Dakota and their use. Bull. 281. Fargo, North Dakota.

Office of Arid Land Studies (1979). "A Sociotechnical Survey of Guayule Rubber Commercialization." Nat. Sci. Found. Tucson, Arizona.

Otsyina, R., McKell, C. M., and Van Epps, G. (1980). Fodder shrubs and crested wheatgrass proportions to meet nutrient requirements of livestock for fall and early winter grazing. *Abstr. Annu. Meet. Soc. Range Manage.*, 33rd. p. 18.

Parker, K. W., and Martin, S. C. (1952). The mesquite problem on southern Arizona range. U.S., Dep. Agric., Circ. 968. U.S. Gov. Printing Office, Washington, D.C.

Pena-Lujan, I. (1980). Algunas aspectos sobre las plantas del genero *Yucca*. *In* "Yucca," pp. 13–20. Centro de Investigacion en Quimica Aplicada. Saltillo, Coahuila, Mexico.

Petrov, M. P. (1976). "Deserts of the World." Translated from Russian by Israel Program for Scientific Translations. Wiley, New York.

Plummer, A., Christensen, D. R., and Monson, S. B. (1968). "Restoring Big Game Range in Utah." Utah Div. Fish Game. Publ. 68-3.

Reed, C. M. (1960). Drought feeding suggestions. *Oklahoma Agric. Ext. Serv.*, Leaflet 12.

Revelle, R. (1978). "Flying Beans, Botanical Whales, Jack's Beanstalk and Other Marvels." Nat. Acad. Sci., Advisory Committee on Technology Innovation. Washington, D.C.

Richardson, S. G., and McKell, C. M. (1980). Water relations of *Atriplex canescens* as affected by the salinity and moisture percentage of processed oil shale. *Agron. J.* **72,** 946–950.

Ritchie, G. A. (ed.) (1979). "New Agricultural Crops." AAAS Selected Symposium Series. Westview Press, Boulder, Colorado.

Rojas-Mendoza, P., and Malo, F. J. (1965). "Estudios Agroeconomico del Nopal Forrajero (*Opuntia* spp.) en El Estado de Nuevo Leon," Vol. X. Informe de Investigacion, Escuela de Agricultura y Ganaderia, Institute Technologico de Monterrey, Monterrey.

Rambaugh, M. D., Johnson, D. A., and Van Epps, G. A. (1981). Forage yield and quality in a Great Basin shrub, grass and legume pasture experiment. *Utah Sci.* **42**(3), 114–117.

Sandhu, T. S., Arora, B. S., and Singh, Y. (1974). Interrelationships between yield and yield components in foxtail millet. *Indian J. Agric. Sci.* **44,** 563–566.

Savage, C. (1980). Squeezing oil from acres of milkweed in Utah. *Christian Sci. Monitor.* Dec. 29.

Scifres, C. M. (1973). "Mesquite Research Monograph I." Texas Agric. Exp. Stn., Texas A&M Univ., College Station.

Shelford, V. E. (1963). "The Ecology of North America." Univ. of Illinois Press, Urbana.

Spadaro, J. J., and Lambou, M. G. (1973). Preparation of jojoba products and their potential uses. *In* "Jojoba and Its Uses, An International Conference" (E. F. Haase and W. G. McGinnies, eds.), pp. 47–60. Office of Arid Land studies, Univ. of Arizona, Tucson.

Stark, N. (1966). Review of highway planting information appropriate to Nevada. *Desert Res. Inst., Univ. Nev. Coll. Agr. Bull.*, B-7.

Szarek, S. R., and Woodhouse, R. M. (1977). Ecophysiological studies of Sonoran desert plants. II. Seasonal photosynthesis patterns and primary production of *Ambrosia deltoides* and *Olneya cesota. Oecologia* **28,** 365–375.

Theisen, A. A., Knox, E. G., and Mann, F. L. (1978). "Feasibility of Introducing Food Crops Better Adapted to Environmental Stress," Vols. I and II. Nat. Sci. Found., Washington D.C.

U.S. Department of Agriculture (1972). "Landscape for Living." U.S. Dep. Agric. yearbook of Agriculture. House Document No. 229. U.S. Gov. Printing Office, Washington D.C.

Usher, G. (1974). "A Dictionary of Plants Used by Man." Constable, London.

Van Epps, G. A., and McKell, C. M. (1977). Shrubs plus grass for livestock forage. *Utah Sci.* **33,** 75–78.

Van Epps, G. A., Barker, J., and McKell, C. M. (1982). Energy biomass from large rangeland shrubs of the intermountain United States. *J. Range Manage.* **35,** 22–25.

Vietmeyer, N. D. (1979). Guayule, domestic natural rubber rediscovered. *In* "New Agricultural

Crops'' (G. A. Ritchie, ed.), pp. 167–176. AAAS Selected Symposia Series No. 38. Westview Press, Boulder, Colorado.

Wall, M. E. (1980). *Yucca* and *Agave*—renewable biomaterials for production of steroid hormones. *In* ''*Yucca,*'' pp. 257–278. Centro de Investigacion en Quimica Aplicada, Saltillo, Coahuilo, Mexico.

Walsh, J. (1980). What to do when the well runs dry. *Sci.* **210,** 754–756.

Walton, G. P. (1923). A chemical and structural study of mesquite, carob, and honey locust beans. USDA Dep. Bull. No. 1194.

West, N. E., Cain, D. R., and Gifford, G. F. (1973). ''Biology and Renewable Resource Management of the Pygmy Conifer Woodlands of Western North America, A Bibliography.'' Utah State Univ., Agric. Exp. Stn., Res. Rep. No. 12.

Wilsie, C. P. (1962). ''Crop Adaptation and Distribution.'' Freeman, San Francisco.

Wochok, Z., and Sluis, C. J. (1979). Micropropagation for jojoba improvement programs. *Jojoba Happenings* **26,** 1–5.

Wright, P. H. (1943). A dryland fruit. *Canadian Horticulture and Home Magazine* **66**(4), 79–80.

Yale, J. W. (1980). Anti-stress action of *Yucca* extracts. *In* ''*Yucca,*'' pp. 229–241. Centro de Investigacion en Quimica Aplicada. Saltillo, Coahuilo, Mexico.

Yermanos, D. M. (1980). Personal communication and information contained in a fact paper distributed by the Botany and Plant Sciences Department, Univ. of California, Riverside.

6

PEOPLE'S REPUBLIC OF CHINA

Hsioh-Yu Hou
Laboratory of Plant Ecology
Institute of Botany
Academia Sinica
Beijing, People's Republic of China

INTRODUCTION AND PHYSIOGRAPHY

The People's Republic of China is situated in the southeastern part of Eurasia, the world's biggest continent, with the Pacific, the world's biggest ocean, to its east. The summer southeast monsoon, which comes from the Pacific Ocean, plays a great role in the climate of the eastern half of China, while the southwest monsoon, from the Indian Ocean, influences mainly China's southern and southwestern parts. The annual precipitation of the eastern half of China ranges from 550 to 1000 mm in the north, and is mostly between 1000 and 2200 mm in the south. These parts belong to the humid forest regions. Due to its great distance from the sea, and the mountains and plateaus that obstruct the wet winds blowing in from the southeast, the northwest of China is very dry, mostly receiving an annual precipitation of less than 150 or 200 mm, with some places receiving only ~50 mm. These parts belong to the arid desert regions. Lying between the two areas mentioned above is a land of semiarid steppes where the annual precipitation is about 300–400 mm.

The arid desert regions of China include basins, plateaus, and mountains.

They vary greatly in elevation: the Dzungaria Basin, 300–400 m; the Darim Basin and the Takla-Makan, 800–1300 m; the Tsaidam Basin, 3000 m; the northern Tibetan Plateau, 5000 m; the Ala Shan Plateau, 1200 m; the Ordos Plateau, 1000 m; the Altay, 3000 m; the Tienshan, 3000–5000 m; the Kunlun Mountains, 5000–7000 m; the Chilien Mountains, 3000–5000 m; and the Holan Mountains, 3000 m. The deserts are surrounded by mountains, on which the vegetation is mainly of the steppe, forest, and alpine meadow types.

The climate of the Chinese deserts is of the temperate and warm-temperate type, with the greatest climatic extremes in the world. A pronounced contrast exists between summer and winter. The winter is cold and long, whereas the summer is hot and short. Thus, most plants of the Chinese deserts are brush, low brush, subshrub, and low subshrub with, or without, deciduous small leaves. (In the leafless varieties, photosynthetic processes are assumed by the green stems of the plant.) The grassy vegetation is poorly developed under normal conditions. According to the dryness of the air, the climate of the Chinese deserts may be divided into two types: dry and extremely dry. The dry climate area, occupying the Dzungaria and the eastern portion of the Ala Shan, receives an annual precipitation of 150 to 200 mm. However, the climate of the Dzungaria Basin is somewhat moderate in aridity and continentality. There, the precipitation is caused by the action of cyclones originating in the northern Atlantic and is evenly distributed throughout the year. Spring is wet with melting snows, and some spring ephemerals flourish. In contrast to the Dzungaria Basin, the climate of the Ala Shan Plateau is extremely continental. The rainy season is associated with the arrival of the southeast monsoon in the summer. The spring is dry, cold, and windy. Spring ephemerals are entirely absent, whereas plants having a summer maximum growth are more common. The extremely dry regions, which receive an average annual precipitation of less than 50 mm, and no rainfall at all in some places, comprise the Bei Shan, Takla-Makan, and Tarim and Tsaidam basins, where scanty moisture is supplied by the summer southeast monsoon, and spring ephemerals are also entirely absent.

The annual mean temperature of the Dzungaria Basin is 3–6°C; the Tarim Basin and Ala Shan, 9–12°C; the Tsaidam Basin, 1–14°C; and the northern Tibetan plateau, −8 to −10°C. The soils of deserts are poorly developed and closely connected with the parent materials and physiography. Most of them are characterized by (1) small amounts of clay and loam, (2) surface soil with little humus, (3) abundance of calcium carbonates, (4) certain amounts of calcium sulfate in the surface soil or subsoil, and (5) certain amounts of sodium sulfate and sodium chloride.

Since the climates are severe and soils are poor in the desert regions of China, the flora is poor. Plants belonging to Chenopodiaceae are very common, even though some of them are represented by only a single genus or single species. In addition, plants of Zygophyllaceae, Tamaricaceae, Compositae, Leguminosae,

Polygonaceae, Ephedraceae, and Gramineae are also frequently found. Most of them are succulent, dwarfed, woody, spiny, of harsh texture, and leafless, or with small leaves. The vegetation is generally sparse. The physiognomy, life forms, floristic composition, and elementary chemical composition of the dominant plants in these deserts are closely related to the local climates and soils.

The types of desert vegetation in China may be grouped into two general categories: one is dependent on the local precipitation, and the other is dependent on moisture from sources outside the desert itself, such as rivers or lakes fed by the surrounding mountains. In addition, since the mountains surrounding the desert regions receive much higher amounts of rainfall, the mountain vegetation is entirely different from that found in the deserts. Accordingly, types of vegetation in desert regions may be divided into three general groups: (A) the deserts, (B) the floodplain meadows and gallery forests, and (C) the mountain steppes, forests, and alpine meadows.

A. Deserts

The deserts of China as a whole may be divided into six categories: (1) the sandy deserts, (2) the sandy–pebbly or gravelly deserts, (3) the rocky or hilly deserts, (4) the loessial or loamy deserts, (5) the solonchak or saline deserts, and (6) the cold, high deserts.

1. SANDY DESERTS

Particularly widespread in China, this type of desert features sandy soil, which favors the growth of plants, especially shrubs and subshrubs. However, the shifting sand dunes occupy much of the central parts of the Takla-Makan and the western part of the Ala Shan. The dominant plants in the different sandy deserts have varying mineral compositions (Table I).

The eastern parts of the Ala Shan and Ordos are semidesert regions with an annual precipitation of 200 to 300 mm. The underlying rock of the Ordos region is soft sandstone that has given rise to expanses of sand and dunes. *Artemisia ordosica,* with vegetation cover equaling 30–40%, is widespread on the stabilized sand dunes. On deeper soil where fresh groundwater is very high, there are some leguminous shrubs such as *Oxytropis aciphylla, Caragana microphylla, C. korshinskii, Hedysarum scoparium, Salix flavida,* and *S. mongolica. Artemisia* is a subshrub of about 50 to 70 cm in height. It is a sand-stabilizing plant and can be used as forage during the fall and winter. All leguminous shrubs, which contain 3% nitrogen in the leaves, are excellent camel fodder. In the moist depressions between sand dunes grow *Pugionium cornutum* and *Psammochloa mongolica.* The former, containing very high amounts of calcium (8.28%), is a wild vegetable with a slightly hot taste. The latter is a sand-

Table I

Elementary Composition of Dominant Plants Growing on Sandy Deserts (% Dry Weight)

Plant name, no., and location where collected	Components in dry plant material											Water-extracted	
	Ash	N	P	S	SiO_2	Fe	Al	Mn	K	Na	Ca	Cl	SO_4
Artemisia ordosica 2483, Ala Shan	14.51	2.13	0.080	—	5.56	0.083	0.094	0.000	3.742	0.150	1.112[a]	—	—
Artemisia sphaerocephala 4237, Ala Shan	9.17	—	0.261	0.16	0.98	0.047	0.057	0.000	3.655	0.104	1.295	1.18	0.20
Calligonum mongolicum 2756, Dsungaria	12.82	2.43	0.171	0.32	0.22	0.000	0.059	0.000	3.118	0.972	1.140	—	—
Calligonum leucocladm 2765, Dsungaria	11.06	2.41	0.136	0.23	0.47	0.006	0.054	0.000	1.963	0.635	2.964	—	—
Calligonum rigidum 2624, Dsungaria	12.59	2.53	0.107	0.30	0.74	0.011	0.111	0.000	6.517	1.285	4.669	—	—
Haloxylon persicum													
2633, Dsungaria	11.14	2.86	0.126	0.12	0.54	0.000	0.073	0.003	3.332	1.028	2.462	—	—
2757, Dsungaria	20.66	2.03	0.063	0.32	0.48	0.011	0.095	0.000	3.808	2.069	4.407	—	—
Artemisia terrae-albae 2747, Dsungaria	9.39	1.44	0.094	0.14	4.18	0.190	0.066	0.000	1.101	0.355	0.586	—	—
Artemisia arenaria 2762, Dsungaria	8.27	1.85	0.209	0.17	1.99	0.027	0.059	0.000	2.049	0.296	1.352	—	—
Artemisia santolina 2763, Dsungaria	8.21	2.10	0.188	0.20	2.54	0.015	0.045	0.000	1.522	0.476	0.932	—	—
Aristida pennata 2631, Dsungaria	7.39	2.29	0.021	0.10	2.00	0.007	0.032	0.000	1.911	0.191	1.319	—	—

Agropyron sibiricum 2787, Dsungaria	4.94	0.68	0.062	0.06	3.03	0.038	0.087	0.000	0.513	0.059	0.494	—	—
Ceratoides latens 2783, Dsungaria	13.05	1.62	0.071	0.20	0.63	0.000	0.115	0.000	2.890	0.104	1.969	—	—
Haloxylon ammodrendron													
2403, Dsungaria	28.11	2.42	0.079	1.41	0.46	0.024	0.022	0.000	2.537	9.931	3.987	4.44	4.88
2781, Dsungaria	31.81	1.22	0.078	0.33	1.58	0.017	0.195	0.000	2.451	8.912	3.428	—	—
Horaninowia ulicina													
4171, Dsungaria	23.79	—	0.067	—	2.49	0.061	0.075	0.000	3.238	2.251	2.860	2.96	1.02
4172, Dsungaria	22.81	—	0.067	—	2.02	0.038	0.059	0.000	2.813	2.225	1.931	2.75	0.96
Halogeton glomeratus 2614, Dsungaria	42.69	3.88	0.074	0.98	1.94	0.027	0.112	0.003	3.501	16.608	1.032	—	—
Lepidium latifolium 2625, Dsungaria	12.72	3.96	0.198	1.00	0.74	0.006	0.113	0.002	5.701	0.263	1.500	—	—
Reaumuria soongarica													
2401, Dsungaria	26.45	1.56	0.071	1.84	1.16	0.037	0.073	0.000	0.968	9.831	0.521	8.12	—
2407, Dsungaria	30.30	2.62	0.064	2.02	0.99	0.048	0.074	0.000	1.088	10.579	0.544	8.59	4.74
Tamarix ramosissima 2743, Darim	21.15	—	0.036	1.72	1.41	0.079	0.039	0.000	1.002	5.092	1.402	—	—
Tamarix hispida 2742, Darim	30.48	—	0.069	2.51	2.71	0.114	0.103	0.000	1.430	7.152	2.088	—	—
Tamarix laxa 2718, Darim	26.46	—	0.098	2.81	2.90	0.057	0.102	0.000	1.265	6.516	2.626	—	—
Tamarix sp. 2472, Turfan	23.80	2.23	0.076	2.25	1.17	0.094	0.221	0.000	1.864	4.528	2.492	—	—

[a] Not represented.

stabilizing plant and can be used as fodder for sheep, cattle, horses, and camels. Its leaves can also be used for fibrous materials.

Artemisia sphaerocephala is the pioneer sand binder of the sandy deserts and is widely distributed in the Ala Shan Plateau. On stabilized sand dunes it is replaced by *Calligonum mongolicum,* which has short, filiform leaves that shed quickly. The twigs are assimilative organs. At more than 10 m, the accessory roots of *Calligonum* are very long and are chiefly distributed horizontally in the moist subsoil. Its new fruit is good forage for livestock.

The Tarim Basin is occupied principally by the Takla-Makan sandy desert, which is almost devoid of vegetation. Only in the dune valleys, which have shallow water tables and saline soils, do several species of *Tamarix* grow. The stem of *Tamarix* can grow new roots when buried by sand.

In the Dzungaria Basin there is enough precipitation for the natural colonization of the sand by plants. *Aristida pennata* is the first invader of the sandy deserts, followed by *Calligonum, Haloxylon,* and others. *Aristida* is the most valuable sand-binding grass, forming lateral shoots that extend existing tufts, and forming new tufts when buried by sand. On the stabilized sandy deserts two species of *Haloxylon* are dominant. *Haloxylon persicum* is found on the ridges of dunes, 20–40 m high. Its growth is dependent only on local precipitation. It grows to a height of 3 to 4 m and bears small leaflets. Its heavy wood provides excellent fuel and building material, but the wood is so hard that it is difficult to chop with an axe. *Haloxylon ammodendron* grows on the bottom of sand dunes in moist and slightly saline, sandy soil. It is entirely devoid of leaves and is even larger and taller than *H. persicum,* with a maximum height of 7 m. It is also valuable for binding desert sands. In autumn, when the fruit ripens, it sheds its assimilative twigs.

It is clear that the two species of *Haloxylon* have different ecological properties and mineral composition. *Haloxylon ammodendron* is a saline plant, whereas *H. persicum* is not. The ash, sodium, sulfur, and nitrogen content of the former is clearly higher than that of the latter.

Haloxylon persicum often grows together with *Calligonum leucocladum, Artemisia santolina,* and *A. terrae-albae.* In the spring, on more or less stable sands, there appears an ephemeral vegetation consisting of *Carex physodes* and *Poa bulbosa* var. *vivipara,* among others. *Haloxylon ammodendron,* however, is always associated with many saline plants, such as *Limonium gmelinii, Suaeda altissima, Atriplex micrantha,* and *Horaninovia ulicina.*

2. SANDY–PEBBLY OR GRAVELLY DESERTS

These deserts are widespread in the arid areas of China. Since relative humidity and physiography are important factors in controlling the desert vegetation, the type of vegetation on gravelly deserts varies with different areas and the dominant plants of different gravelly deserts show a different elementary chemical composition (Table II).

Table II

Elementary Composition of Dominant Plants Growing on Gravelly and Rocky Deserts (% Dry Weight)

Plant name, no., and location where collected	Components in dry plant material											Water-extracted	
	Ash	N	P	S	SiO_2	Fe	Al	Mn	K	Na	Ca	Cl	SO_4
Caragana tibetica 4243, Ala Shan	12.05	—[a]	0.104	0.14	5.24	0.199	0.511	0.000	1.448	0.109	1.991	—	—
Caragana stenophylla 4244, Ala Shan	10.46	—	0.153	0.34	1.22	0.101	0.197	0.000	1.228	0.105	2.827	0.98	—
Ceratoides latens 4245, Ala Shan	12.75	—	0.415	0.21	1.33	0.076	0.168	0.000	3.470	0.102	2.392	0.96	0.81
Ammopiptanthus mongolicus 2668, Ala Shan	6.58	3.77	0.186	0.22	0.91	0.000	0.110	0.001	2.222	0.169	0.659	—	—
Reaumuria soongarica													
4287, Ala Shan	23.02	—	0.097	3.22	1.41	0.066	0.119	0.000	1.055	3.648	1.837	4.19	8.15
4291, Ala Shan	32.17	—	0.089	3.44	2.15	0.078	0.167	0.000	1.423	5.265	2.561	7.90	7.16
4342, Ala Shan	27.35	—	0.067	3.63	1.59	0.103	0.045	0.000	1.085	4.496	2.373	5.03	5.53
Salsola passerina													
4330, Ala Shan	33.92	—	0.040	3.39	4.12	0.222	0.256	0.000	2.490	6.359	1.779	3.65	11.39
4341, Ala Shan	31.32	—	0.057	3.18	2.85	0.096	0.171	0.000	2.976	7.615	1.510	7.08	10.25
4289, Ala Shan	39.14	—	0.069	3.90	2.40	0.142	0.296	0.000	2.358	8.734	1.646	5.82	12.25
Ephedra przewalskii 2585, Darim	6.62	—	0.088	0.18	0.23	0.032	0.032	0.000	0.731	0.366	2.460	—	—
Nitraria sphaerocarpa													
2570, Darim	12.67	—	0.127	0.44	0.82	—	0.093	0.003	2.946	2.509	1.527	—	—
2571, Darim	12.16	—	0.048	0.36	1.19	0.071	0.102	0.019	1.965	2.475	1.183	—	—
Calligonum roborowskii													
2592, Darim	7.27	—	0.168	—	0.62	0.002	0.062	0.000	1.882	0.293	0.926	—	—
2600, Darim	10.84	—	0.102	0.14	0.64	0.007	0.000	0.000	1.721	0.241	2.665	—	—
Halogeton glomeratus 2719, Darim	36.01	—	0.105	1.13	0.86	0.047	0.074	0.000	3.076	9.058	0.642	—	—
Anabasis salsa 2415, Dsungaria	34.05	1.90	0.039	0.25	2.08	0.108	0.168	0.000	2.326	11.808	3.936	6.45	2.08
Anabasis aphylla													
2414, Dsungaria	21.09	3.34	0.120	0.34	0.93	0.014	0.058	0.000	2.121	5.832	2.050	—	—
2416, Dsungaria	28.61	1.86	0.045	0.32	0.71	0.015	0.067	0.000	1.593	10.623	3.329	3.19	1.07

(continued)

Table II *(Continued)*

Plant name, no., and location where collected	Ash	Components in dry plant material										Water-extracted	
		N	P	S	SiO_2	Fe	Al	Mn	K	Na	Ca	Cl	SO_4
Sympegma regelii 2746, Turfan	25.76	—	0.051	0.64	—	0.000	0.000	0.000	1.710	5.044	2.373	—	—
Iljnia regelii 2474, Bei Shan	32.78	—	0.208	0.29	1.12	0.031	0.086	0.000	2.045	11.684	1.278	—	—
Capparis spinosa 2469, Darim	24.35	3.03	0.189	2.24	2.26	0.035	0.081	0.000	2.476	2.441	2.929	—	—
Anabasis brevifolia													
2744, Bei Shan	30.62	—	0.070	0.98	0.84	0.049	0.106	0.000	1.748	6.082	0.386	—	—
2745, Bei Shan	26.95	—	0.050	0.61	0.25	0.008	0.031	0.000	1.983	6.073	1.543	—	—
Cerathcarpus turkestanicus 2621, Dsungaria	15.04	3.79	0.294	0.23	1.36	0.000	0.059	0.002	5.786	0.266	3.616	—	—
Ceratoides latens 2767, Dsungaria	13.26	2.77	0.123	0.37	0.99	0.048	0.047	0.000	2.152	0.807	2.094	—	—

[a] Not represented.

In the semidesert area of eastern Ala Shan, where the annual precipitation is ~200 mm, shrubby deserts, with vegetation covering 30–50%, occur on the sandy–gravelly soil of the piedmont plains. The predominant species are *Caragana tibetica* (50 cm tall), *Tetraena mongolica* (50 cm tall), and *Potaninia mongolica* (10–30 cm tall), which grow together with xerophyllous grasses, such as *Stipa glareosa* and *Cleistogenes mutica.*

In some places, the evergreen *Ammopiptanthus mongolicus* (100–150 cm tall) is dominant. In addition, *Caragana microphylla* is frequently seen. All these plants are excellent camel fodder.

The gypseous gravelly deserts, with dominant *Reaumuria soongarica,* are extensively distributed in the Ala Shan, Bei Shan, and Darim. The codominant *Salsola passerina* is confined to the Ala Shan area. These two plants are both sulfur-absorbing, xerohalophytes less than 30 cm tall, and their roots are largely distributed in the topsoil within 40 cm of the surface. They are excellent camel fodder during the fall, winter, and spring, but not in summer.

On the piedmont of the Kunlun Mountains, *Reaumuria trigyria* and *R. kaschgarica* occur.

On the Gobic Plains in the Darim and Bei Shan, the annual precipitation averages 50 mm or less and mainly falls in the summer. The predominant plants are *Ephedra przewalskii, Zygophyllum xanthoxylon, Z. kaschgaricum,* and *Gymnocarpus przewalskii,* and occasionally *Calligonum roborowskii* and *Nitraria sphaerocarpa* in local areas. The sparse shrubby and subshrubby vegetation is found on high foothills of all the mountain systems, growing along the channels of temporary streams. The vegetation cover is less than 5 to 10% in some places. In the Tsaidam, northern Ala Shan, and east of Dzungaria, *Ephedra przewalskii* and *Haloxylon ammodendron* are dominant, but without *Nitraria sphaerocarpa* and *Zygophyllum xanthoxylon.* Except for *Ehpedra,* they make excellent camel fodder. The leaves of some plants are vestigial, their function having been taken over by green branches, which reduce transpiration. Some have roots more than 12 m long for absorbing moisture at that depth. *Ephedra* is a medical plant from which ephedrine can be extracted. The roots of *Nitraria sphaerocarpa* are parasitized by *Orobanche* sp. and *Cynomorium coccineum,* which are vaulable medicinal plants. All the above are sand-stabilizing plants.

The Dzungarian gravelly deserts are dominanted by *Anabasis salsa* and *Nanophyton erinaceum.* The former is found predominantly on saline soils, in shallow depressions. It is a small subshrub 5–15 cm tall with slightly branched annual shoots; these die off yearly, almost to the surface of the ground. The latter plant is a small, procumbent subshrub with small, spiny leaves. Both of them are xerohalophytes containing high amounts of sodium and chlorine. They are excellent camel fodder during the fall and winter. In eastern Dzungaria, where the climate is drier, *Anabasis aphylla* is dominant.

3. ROCKY OR HILLY DESERTS

This desert type is found mainly on the hills and low mountains of the Bei Shan and the Darim Basin, from which all the finer products of weathering have been blown away, and where the exposed rocks have undergone severe wind erosion due to sandblasting. The surface is covered with a pavement of stones. Beneath the pavement there may be a water-repellent, dusty soil that is rich in gypsum and salt, thus preventing the development of a plant cover. The rocky desert areas are cleft by deep erosion valleys with steep, rubble-covered slopes. In the cracks and crevices of the rocks a few low-growing, xerohalophytic species can be found.

The vegetation cover is less than 1%. *Sympegma regelii* and *Iljnia regelii* are frequent. The former provides camel fodder in any season, while the latter provides it only in the fall and winter. On the foothills occur *Ephedra intermedia* and *E. equisetina,* which are an excellent source of ephedrine. On stony soil in the Turfan Depression and Darim Basin, *Capparis spinosa* is frequently found. This plant is a decumbent undershrub with long branches bearing large bright leaves, and carrying two spiny stipules at the base. It is a sulfur-absorbing plant (2.24%), and its seeds are rich in edible oil (Table II).

4. LOESSIAL OR LOAMY DESERTS

This type of desert occurs on loessial soil, which is not excessively saline. It is found mainly in areas of the Dzungaria where rain falls evenly throughout the year. The vegetation is predominantly composed of low subshrub and certain ephemerals, which develop in spring when the soil is moist, but fade quickly with the coming of hot weather. The main low subshrubs are *Artemisia kaschgarica, A. borotalensis,* and *A. terrae-albae,* from which essential oil can be extracted. The common ephemerals are *Poa* and *Carex. Poa bulbosa* var. *vivipara* is a perennial grass that is usually 30–40 cm tall and occurs in small tufts 2–3 cm in diameter. At the base of its stem it forms what appear to be small bulbs, which are capable of surviving drought over very long periods and give rise to new plants. The narrow-leaved sedge, *Carex physodes,* is another perennial ephemeral. The annual ephemerals are *Trigonella arcuata, Meniocus linifolius, Tetracme quadricornis,* and *Lepidium perfoliatum.* In addition, there is a perennial *Ferula* sp., but it dies after its seeds have ripened. This plant yields a foul-smelling, resinous substance used in medicine. The resin is contained in the vigorously developed roots.

Chemical analyses show that the dominant plants of this type of desert contain lower amounts of ash, sodium, and sulfur (Table III).

5. SOLONCHAK OR SALINE DESERTS

Saline deserts with sparse vegetation occur in contour depressions wherever brackish underground water is near the surface in dry lakes, old valleys, and river

Table III

Elementary Composition of Dominant Plants Growing on Loessial Deserts (% Dry Weight)

Plant name, no., and location where collected	Ash	Components in dry plant material									
		N	P	S	SiO_2	Fe	Al	Mn	K	Na	Ca
Artemisia borotalensis 2423, Dsungaria	21.04	—[a]	0.104	0.14		0.559	0.298	0.004	1.565	0.485	2.572
Artemisia terrae-albae											
2778, Dsungaria	5.83	1.49	0.098	0.27		0.008	0.158	0.000	1.032	0.383	0.544
2750, Dsungaria	11.85	2.07	0.210	0.32		0.079	0.90	0.000	2.627	0.225	0.732
Artemisia lessingiana 2784, Dsungaria	16.97	0.73	0.102	0.14		0.295	0.041	0.000	1.094	0.716	2.484
Lipidium perfoliatum 2617, Dsungaria	14.11	4.11	0.017	0.33		0.000	0.025	0.002	2.425	0.334	2.175
Trigonella faenum-graecum 2616, Dsungaria	11.34	3.33	0.216	0.97		0.018	0.069	0.025	3.409	1.284	1.043
Tetracme quadricormis 2626, Dsungaria	18.02	3.06	0.115	1.30		0.006	0.054	0.003	3.765	0.283	3.665

[a] Not represented.

deltas. The soils often have salt crystalizing on the surface, with 10 to 30% sodium chloride or sodium sulfate. The dominant halophytic plants are low subshrubs and shrubs, herbs and grasses, and annual succulent hydrohalophytes.

Around the salt lakes in the Ala Shan and the Tsaidam Basin, *Kalidium foliatum, K. cuspidatum, K. gracile, Nitraria sibirica, Lycium ruthenicum,* and *Tamarix* spp. are dominant. On more moist saline soils occur some annual succulents such as *Salicornia europaea* and *Suaeda* spp. Other halophytic herbaceous plants include *Karelinia caspica* and *Achnatherum splendens.*

On the soils of the Darim, Dzungaria, and Bei Shan, low subshrubs such as *Kalidium caspicum, Suaeda physophora, S. microphylla, Halostachys belangeriana,* and *Halocnemum strobilaceum* occur, in addition to the plants noted above.

Chemical analyses show that plants growing on the saline deserts mostly contain high amounts of ash, sodium, chlorine, and sulfur (Table IV).

Kalidium spp. and *Suaeda* spp. are eaten by camels only in the fall and winter, when they are wilted; however, sheep do not eat them at all. These plants contain high amounts of alkali, from which sodium sulfate and potassium carbonate can be extracted. The lands on which these plants grow are not suitable for agriculture.

6. COLD, HIGH DESERTS

The northwestern corner of Tibet, at 5000 m above sea level, has very sparse vegetation with few varieties. Only short, small subshrubs are found there. Some are creepers; others are cushion plants. Both are adapted to wind and cold, dry conditions. The dominant plants are *Ceratoides compacta, Ajania tibetica,* and *Carex moorcroftii.* They are associated with *Pegaeophyton scapiflorum, Astragalus hedinii, Stipa subsessiliflora* var. *basiplumosa,* and species of woody plants, *Myricaria prostrata* and *Ephedra gerardiana.*

B. Floodplain Meadows, Gallery Forests, and Oases

This group of vegetation types is mostly distributed along river valleys, in oases, and around some lakes. Because of the saline subsoil, the abundance of salt in the river water, and the intense evaporation, the soils in the desert regions, as a rule, are salinized. All the natural plants are mesophytic because there is enough water in the soil, but they are more or less physiologically xerophytic because of the dry air and saline soil. The vegetation may be grouped as (1) the saline meadows and shrubs, (2) the gallery forests, and (3) the oases.

1. SALINE MEADOWS AND SHRUBS

Saline meadows are widespread in the contemporary floodplain of the Darim River and places with a high water table in desert regions. They are composed of

Table IV

Elementary Composition of Dominant Plants Growing on Saline Deserts (% Dry Weight)

Plant name, no., and location where collected	Ash	Components in dry plant material										Water-extracted	
		N	P	S	SiO_2	Fe	Al	Mn	K	Na	Ca	Cl	SO_4
Kalidium gracile 4320, Ala Shan	43.16	—[a]	0.060	—	0.32	0.094	0.000	0.000	1.730	15.882	0.866	19.81	1.03
Kalidium cuspidatum													
4232, Ala Shan	45.20	—	0.195	1.04	0.21	0.040	0.070	0.000	2.504	17.829	0.065	21.01	3.08
2946, Ala Shan	51.28	—	0.113	1.50	1.04	0.050	0.027	0.007	4.154	19.231	0.462	19.84	6.15
Kalidium foliatum													
4233, Ala Shan	38.30	—	0.184	0.79	0.99	0.069	0.051	0.000	1.658	15.288	0.041	16.91	2.47
2468, Turfan	39.33	1.70	0.117	1.45	0.62	0.074	0.158	0.000	1.289	14.368	0.682	—	5.31
Halostachys belangeriana													
2556, Darim	30.08	—	0.144	0.81	0.86	0.018	0.074	0.000	4.914	9.233	0.894	—	—
2591, Darim	30.75	—	0.128	0.78	0.87	0.041	0.093	0.000	2.532	7.062	0.966	—	—
Halocnemum strobilaceum													
2562, Turfan	38.70	—	0.071	0.62	3.56	0.218	0.229	0.000	3.064	12.040	2.030	—	—
2404, Dsungaria	40.06	1.21	—	0.74	2.03	0.027	0.179	0.000	1.710	15.923	1.211	13.39	11.96
Suaeda physophora 2465, Turfan	35.22	2.91	0.029	2.02	0.74	0.032	0.066	0.000	1.502	10.261	0.982	11.30	4.48
Suaeda corniculata 4281, Ala Shan	36.52	—	0.285	1.62	0.10	0.019	0.061	0.000	2.612	11.569	0.052	10.91	4.49
Salicornia europaea													
2745, Darim	44.56	—	0.100	1.15	0.61	0.056	0.048	0.000	2.311	14.679	1.093	—	—
975, Tsaidam	55.00	1.41	0.091	1.57	6.19	0.251	0.488	0.011	0.933	11.656	1.257	20.77	4.52
Nitraria tangutorum 4370, Ala Shan	19.85	—	0.495	0.35	0.36	0.000	0.000	0.000	3.197	5.215	1.149	3.18	0.25
Nitraria sibirica													
4221, Ala Shan	23.71	—	0.118	0.72	0.75	0.071	0.189	0.000	2.740	5.670	1.606	9.57	1.99
4379, Ala Shan	36.21	—	0.069	0.79	1.24	0.054	0.054	0.000	0.784	8.923	3.971	14.71	2.81
Lycium ruthenicum 2464, Turfan	23.69	—	0.064	1.52	0.76	0.068	0.067	0.000	3.022	4.550	2.161	12.53	3.12
Karelinia caspica 2459, Turfan	24.19	2.58	0.079	0.69	0.50	0.025	0.056	0.000	1.541	5.077	2.066	4.73	9.89
Scorzonera mongolica var. putjatae 2557, Darim	19.55	—	0.355	1.51	1.17	0.036	0.006	0.008	3.231	1.146	2.179	—	—

[a] Not represented.

many solonchak grasses and herbs. Frequently occurring grasses are *Aeluropus littoralis, Achnatherum splendens* (*Lasiagrostis splendens*), *Aneurolepidium dasystachys, Cynodon dactylon, Puccinellia distans, Phragmites communis,* and *Calamagrostis pseudaphragmites,* which are adapted to different local climatic and soil conditions. *Aeluropus littoralis,* which belongs to the rhizome group, is characterized by leaves that are covered with grains of salt emitted by the leaf tissues. It has a powerful root system. *Cynodon dactylon* is also a rhizomatous plant. Their rhizomes can be used for matting. *Puccinellia distans* is sod forming. They all are eaten readily by livestock. *Achnatherum splendens* is a tall grass, up to 1.5 m tall, which forms a sod. The height of *Phragmites* decreases with increases in soil salinity. Stems of both *Achnatherum* and *Phragmites* are used for weaving mats. They also yield excellent raw materials for the manufacture of paper and artificial cotton.

Among the herbaceous plants of the saline meadows are many legumes, such as *Glycyrrhiza glabra, G. inflata,* and *G. uralensis.* Their roots have many uses in Chinese medicine, and their stems are used for weaving mats. Other common legumes are *Alhagi pseudalhagi, Sophora alepecuroides, Thermopsis lanceolata, Swainsona salsula,* and *Melilotus* spp. There are many other herbaceous plants. *Paocynum hendersonii* yields a valuable fiber; *Xanthium sibiricum*'s seeds yield oils, which not only are edible but also have industrial uses, and its fruit has medical use.

According to chemical analyses, the ash, SiO_2, and sodium concentrations of the saline grasses found in desert regions are higher than those found in the same species in the semiarid and humid areas. The legumes contain large amounts of sodium and sulfur (Table V).

In some areas of the saline meadows there are a number of shrubs: several species of *Tamarix,* which have pink and violet flowers similar to the color of the landscape, *Lycium ruthenicum,* and *Halimodendron holodendron.* The stems of *Tamarix* can be used to weave baskets. Leaves are eaten by camels and sheep, and the new stems can be used as medicine. The roots are parasitized by *Orobanche;* it and the spiny *Halimodendron,* a legume, are valuable medicinal plants.

2. GALLERY FORESTS (FLOODPLAIN FORESTS)

The tree vegetation along the rivers of the desert region is most unique. It consists of a very small number of species. *Populus euphratica* (*P. diversifolia*) predominates. This tree has narrow and long lower leaves, like *Salix,* while its upper leaves are broad. It is widespread on the floodplain of the Darim River, which is the largest river in Chinese desert regions and is fed by water from the glaciers and snows of the surrounding high mountains. The soil on which it grows is slightly saline, with a water table of about 1.5 to 2.5 m. It grows to a height of 10 to 15 m normally, and 20 to 30 m maximum. However, if the water

Table V

Elementary Composition of Herbaceous Plants Growing on Saline Meadows (% Dry Weight)

Plant name, no., and location where collected	Components in dry plant material											Water-extracted	
	Ash	N	P	S	SiO_2	Fe	Al	Mn	K	Na	Ca	Cl	SO_4
Achnatherum splendens													
2491, Ala Shan	9.80	1.40	0.057	0.50	2.98	0.012	0.027	0.000	2.193	0.330	0.291	—[a]	—
2523, Ala Shan	8.05	1.37	0.153	0.17	3.46	0.000	0.031	0.000	1.480	0.208	0.182	—	—
Aeluropus littoralis 2454, Turfan	13.24	2.22	0.192	0.34	5.86	0.026	0.051	0.000	1.222	1.789	0.524	2.25	0.95
Phragmites communis													
4241, Ala Shan	20.30	—	0.105	0.15	14.81	0.000	0.000	0.000	1.719	1.051	0.014	1.16	1.48
961, Tsaidam	12.57	—	—	0.10	9.53	0.141	0.180	0.000	—	0.487	0.304	—	—
2406, Dsungaria	9.15	—	0.061	0.26	5.62	0.009	0.000	0.000	0.756	1.044	0.198	—	—
Glycyrrhiza inflata 2458, Turfan	7.21	1.97	0.149	0.19	0.55	0.034	0.083	0.000	1.543	0.282	1.731	0.96	0.19
Swainsona salsula 2723, Darim	25.56	—	0.013	2.49	1.52	0.024	0.104	0.000	3.909	1.225	5.040	—	—
Sophora alopecuroides 2739, Darim	8.14	—	—	0.32	0.41	0.021	0.019	0.000	0.798	0.931	2.227	—	—
Melilotus suaveolens 2667, Ala Shan	15.88	2.86	0.173	1.39	0.56	0.041	0.096	0.001	3.760	0.135	8.294	—	—
Melilotus albus 2467, Turfan	12.89	5.05	0.052	0.58	1.00	0.096	0.498	0.000	2.477	0.384	3.294	—	—
Alhagi pseudalhagi 2470, Darim	16.37	2.78	0.111	2.55	0.38	0.034	0.068	0.000	1.198	0.321	4.491	—	1.68
Scorzonera mongolica var. putjatae 2681, Ala Shan	22.71	2.56	0.110	1.41	0.77	0.021	0.099	0.001	6.728	1.755	1.737	—	—
Xanthium sibiricum 2725, Darim	12.22	—	0.259	0.42	0.65	0.018	0.062	0.000	0.704	0.382	4.990	—	—
Poacynum hendersonii 2722, Darim	18.88	—	0.043	1.88	0.94	0.053	0.032	0.000	2.562	0.499	2.883	—	—
Glaux maritimus 974, Tsaidam	14.28	1.53	0.082	1.08	1.22	0.020	0.079	0.004	1.501	1.895	1.001	2.03	2.01
Triglochin striata 970, Tsaidam	23.64	2.93	0.113	0.52	0.59	0.008	0.050	0.005	2.332	3.823	1.682	9.87	1.57

[a] Not represented.

table becomes lower, and the soil more saline, the growth conditions become worse.

Another species, *Populus pruinosa,* is found only on the higher floodplains of the Darim River, where the soil is less saline. The sodium and calcium content of *P. pruinosa* is lower than that of *P. euphratica* (Table VI). Both species of *Populus* are sand- and bank-stabilizing, wind-resistant trees. Their wood is moderate in hardness and can be used as building material in the desert region. The leaves are excellent fodder for sheep in winter. The trunk of *P. euphratica* excretes an edible salt that can also be used as an industrial material.

Apart from *Populus,* the gallery forests contain *Elaeagnus angustifolia,* which grows to a height of 4 to 7 m on slightly saline, sandy soils, and is often found on the edges of *Populus* forests. Its fruits are edible, with a sour–sweet taste, and can be fermented for manufacturing wine. Its seeds produce oil and its leaves are fodder for camels and sheep. The flavorful flowers are the original material of "honey sugar," and the wood is very hard.

3. OASES

In the arid regions of China, the deserts are surrounded by high, snow-clad mountains. A number of streams come from the mountains; irrigation canals, especially along the margins of the basins, lead the water from these rivers to the numerous oases, which form areas of green in an otherwise brown landscape.

In the temperate climate oases, the agricultural crops and the harvesting systems are different from those in the oases of warm-temperate climate. The oases of temperate climate in the Dzungaria Basin produce oats, millet, buckwheat, and beans. Spring wheat and extremely early ripening rice can be raised only under the best irrigation conditions. The economic crops are sugar beets, oil flax, and fiber flax. There is only one harvest a year. In areas with a warm-temperate climate, the main crops are winter wheat, corn, kaoliang, and some rice. The oases in the Turfan Depression and the Tarim Basin are noted for their good yields of long-staple cotton. The seedless white grapes of Turfan and sweet Hami melons have long been well known both at home and abroad. Deciduous fruit trees, including many good varieties of pear, apricot, plum, and walnut, are cultivated in the southern Darim Basin.

Analyses of *Populus* spp., planted along the irrigation canals, and *Morus alba,* grown in the oases, are shown in Table VII.

C. Mountain Steppes, Forests, and Alpine Meadows

The deserts in China are surrounded by high mountains on which the climate is semiarid and humid, and the vegetation is of the steppe and forest types.

Different mountains have different types of steppes and forests, which vary with the geographical location and the climate. On the northern parts of the

Table VI

Elementary Composition of Dominant Trees of Gallery Forests Growing on Saline Meadows (% Dry Weight)

Plant name, no., and location where collected	Components in dry plant material										
	Ash	N	P	S	SiO_2	Fe	Al	Mn	K	Na	Ca
Populus euphratica											
2741, Darim	9.62	—[a]	0.020	0.58	0.23	0.014	0.048	0.000	1.240	1.359	2.174
2705, Darim	8.72	—	0.073	0.28	0.36	0.009	0.026	0.000	1.284	1.862	1.070
2460, Turfan	12.96	2.58	0.139	2.25	0.75	0.069	0.081	0.000	0.953	1.209	3.774
Populus pruinosa											
2596, Darim	11.54	—	0.093	0.84	0.84	0.052	0.136	0.001	3.914	0.147	1.541
2610, Darim	10.77	—	0.011	0.30	0.66	0.008	0.032	0.000	1.990	0.040	0.239
Elaeagnus angustifolia (E. turcomanica)											
2717, Darim	10.37	—	0.070	0.40	1.13	0.046	0.079	0.000	1.617	0.663	1.575
2573, Darim	6.88	—	0.098	0.33	0.84	0.049	0.075	0.000	1.600	0.294	0.846
Elaeagnus moorcrofftii											
2716, Darim	11.90	—	0.058	0.44	0.92	0.033	0.128	0.000	1.781	0.704	2.258
2704, Darim	11.39	—	0.044	0.44	1.13	0.036	0.107	0.000	1.691	0.368	1.563

[a] Not represented.

Table VII

Elementary Composition of Trees Planted in Oases (% Dry Weight)

Plant name, no., and location where collected	Ash	Components in dry plant material									
		N	P	S	SiO_2	Fe	Al	Mn	K	Na	Ca
Populus alba 2733, Darim	13.62	—[a]	0.125	0.50	2.56	0.108	0.067	0.000	0.713	0.107	3.050
Populus bolleana 2735, Darim	17.78	—	0.056	0.27	5.74	0.062	0.096	0.000	0.745	0.140	4.067
Populus balsamifera 2737, Darim	9.56	—	0.085	0.16	1.48	0.190	0.025	0.000	0.935	0.125	2.095
Morus alba 2738, Darim	23.08	—	0.064	0.78	11.26	0.044	0.061	0.000	2.731	0.156	4.846

[a] Not represented.

Chilien (6000 m above sea level) and Holan Mountains (3500 m in elevation), the vegetation at lower latitudes is of the steppe type dominated by *Stipa bungeana* and *Artemisia frigida*. Much higher up, on the north-facing slopes, one first finds *Pinus tabulaeformis* forests, and then *Picea crassifolia* forests, while on the south-facing slopes there are coppices of *Ulmus glacescens* and *Prunus mongolica* in the Holan Mountains, and those of *Sabina salturia* in the Chilien Mountains. The subalpine meadows, admixed with shrubs of *Caragana jubata* and *Potentilla fruticosa,* and the alpine *Kobresia* meadows are found above the forest zone. Everlasting snows and glaciers are seen at the top of the Chilien Mountains.

On the northern parts of the Tienshan (5000 m in elevation), above the semidesert belt of the lower elevation, *Stipa capillata* and *Festuca sulcata* predominate on the grassy steppes. As the elevation increases, *Picea shrenkiana* forests, which grow as tall as 50 m, appear on the north-facing slopes, while steppes are on the south-facing slopes. Above the zone of coniferous forests lie the subalpine and alpine meadows, which are dominated by *Kobresia* spp., some herbaceous plants, having leaves at the base of the plant in the form of a rosette, and some subshrubs, represented by cushion-like forms. These plants include *Saxifraga* spp., *Acantholimon,* spp. and *Androsace* spp., among others. Permanent snows and glaciers are seen on the summits of the mountains.

On the lower slopes in western parts of eastern Tienshan, where the climate is warmer and moister, patchy deciduous forests, including *Malus sieversii* and *Juglans regia,* exist, although they are not very extensive.

The Altay reaches an elevation of ~300 m in China. With increasing elevation, one first finds typical steppes, then expanses of coniferous forests, and finally, vast subalpine and alpine meadows with a slight suggestion of alpine tundra.

The typical mountain steppes are dominated by *Festuca sulcata* and *Stipa capillata. Larix sibirica* forests appear extensively and grow as an admixture with *Pinus sibirica* in the narrower, moist valleys. Coniferous forests dominated by *Picea obovata* and *Abies sibirica* are only distributed in the northwestern parts of the Altay where precipitation is relatively abundant. Subalpine meadows are composed of *Festuca* spp., *Poa* spp., and *Trisetum* spp. mixed with some alpine herbs, such as *Polgonum viviparum* and *Leontopodium sibirium.* The alpine meadows are dominated by *Kobresia filiformis, K. sibirica,* and *Carex* spp. Higher up, they tend to resemble the alpine tundra, with thickets of *Betula rotundifolia* under which mosses prevail.

The climate of southern Tienshan is much drier than that of the northern parts. The desert vegetation reaches an elevation of ~2000 m, or higher. Here the typical mountain steppes are dominated by *Stipa krylovii,* while the alpine (cold-mountain) steppes are dominated by *S. purpurea* and *Festuca ovina.* Patches of *Picea shrenkiana* are occasionally seen in the shadier valleys on north-facing slopes. Immediately above the zone of alpine steppes lie the alpine meadows, and the dominant species is *Kobresia capillifolia*. The permanent snows and glaciers occur at an elevation of 3000 to 4300 m.

The northern slope of the Kunlun Mountains is an extremely arid region of China, where the desert vegetation can reach an elevation of 3000 m above sea level. Only the steppes, with desertic characteristics, are found on the high mountain slopes, and they are composed of *Stipa glareosa, Artemisia* spp., and *Allium* spp. There are no alpine meadows on the top of the mountains.

In the mountains of the desert regions, steppes and forests occupy a vast area. The grasses growing on these mountains, such as *Stipa capillata* and *Festuca sulcata,* are unsuitable for mowing but provide excellent green fodder for sheep and horses. The grassy steppes also contain a rather large number of dicotyledons, many of which are medicinal plants. The coniferous forests, of economic importance, in addition to affording shelter for wildlife, provide a large proportion of the timber, pulpwood, turpentine, firewood, and numerous other commodities in the desert regions.

It is quite evident that the ash, calcium, sodium, and sulfur contents of the trees occurring on the mountains are much lower than those of the plants growing in the deserts (Table VIII).

With the exception of the Dzungaria Basin, the desert regions of China are part of the so-called central Asiatic desert, and are characterized by the following features. As already mentioned, the last traces of the summer monsoons are noticeable in the desert regions. There are extensive areas of gravelly deserts (Gobi) and sandy deserts together with some rocky deserts, and in some places, under extremely arid climates, they are almost devoid of vegetation. The flora is poor. The spring ephemerals and *Haloxylon persicum,* occurring in the Dzungaria, are entirely absent. The shrubby psammophytes, such as *Artemisia ordosica, A. sphaerocephala, Calligonum mongolicum, C. roborowskii, Hedysarum scoparium, Caragana tibetica, Tetraena mongolica, Potaninia mongolica, Ammopiptanthus mongolicus,* and *A. nanus,* predominate among the east Chinese–Mongolian elements. Among the xerohalophytes, *Salsola passerina, Sympegma regelii, Iljnia regelii, Haloxylon ammodendron, Anabasis aphylla, Reaumuria soongarica, Zygophyllum xanthoxylon, Z. kaschgaricum, Nitraria sphaerocarpa,* and *Capparis spinosa* should be mentioned.

Apart from the xerohalophytes, the hydrohalophytes, which grow on wet, saline soil, include shrubby *Kalidium* spp., *Suaeda* spp., *Lycium ruthenicum, Nitraria sibirica,* and *Tamarix* spp. together with a number of herbaceous plants, such as *Salicornia europaea, Suaeda* spp., and *Karelinia caspica. Ephedra przewalskii,* a xerophyte, is widespread on gravelly deserts.

On the floodplains, *Populus euphratica* and *Elaeagnus angustifolia* are common in the gallery forests, while *P. pruinosa* is confined to the upper plains of the Darim River. Plants of the saline meadow comprise *Aeluropus littoralis,*

Table VIII

Elementary Composition of Dominant Trees Growing on Mountains in Desert Regions (% Dry Weight)

Plant name, no., and location where collected		Components in dry plant material										Water-extracted	
	Ash	N	P	S	SiO_2	Fe	Al	Mn	K	Na	Ca	Cl	SO_4
Larix sibirica													
2377, Altay	5.93	2.64	0.215	0.07	1.39	0.000	0.000	0.000	1.394	0.038	0.306	—[a]	—
2371, Altay	6.12	2.63	0.710	0.05	2.51	0.003	0.014	0.033	0.846	0.018	0.529	—	—
2375, Altay	6.51	2.05	0.213	0.13	1.99	0.000	0.000	0.048	1.079	0.019	0.454	—	—
Pinus sibirica 2384, Altay	8.52	4.28	0.172	0.25	2.74	0.046	0.136	0.000	1.478	0.043	0.523	—	—
Picea crassifolia													
4284, Holan	3.72	—	0.137	0.25	0.81	0.033	0.075	0.083	0.828	0.035	0.911	—	—
4285, Holan	3.91	—	0.091	0.26	0.90	0.027	0.068	0.000	0.730	0.046	1.095	—	—
Pinus tabulaeformis 4283, Holan	4.06	—	0.085	—	0.81	0.077	0.080	0.000	0.425	0.053	1.045	—	—

[a] Not represented.

Achnatherum splendens, Cynodon dactylon, Glycyrrhiza spp., *Alhagi pseudalhagi, Poacynum hendersonii,* and others.

Stipa glareosa and *S. krylovia* are the dominant plants of the mountain steppes. *Picea crassifolia* and *Pinus tabulaeformis,* which are found in coniferous forests on the mountains, also should be mentioned.

However, the desert vegetation of the Dzungaria, where precipitation is caused by cyclonic disturbances coming from the northern Atlantic, is somewhat similar to that of middle Asia, especially the deserts of Hazakhstans, in the Soviet Union. The similarity is represented by the presence of sandy deserts with predominant *Haloxylon persicum,* gravelly deserts with dominant *Anabasis salsa* and *Nanophyton erinaceum,* and loessial deserts dominated by *Artemisia* spp. and some ephemerals. On the northern Tienshan, the occurrence of steppes with *Festuca sulcata* and *Stipa capillata,* and coniferous forests with *Picea shrenkiana,* shows the similarity to central Asia.

With regard to the measures undertaken in China to bring the desert under control, those affecting sand stabilization should be mentioned first. The effective methods of fixing active sand dunes is to protect the natural vegetation around farmlands and pastures. It is necessary to choose the ecologically suitable species of trees and shrubs that are to be planted on the periphery of oases. Excellent results have been obtained by planting multiple parallel barriers of trees and shrubs, alternating with canals, in the Turfan Depression, where deflation of the oasis is serious. On the oasis along the western margin of the Takla-Makan Desert, and on the Hosi corridor of the Ala Shan Desert, the work of sand stabilization has been successful as a result of planting trees and shrubs in the hollows between dunes, where water is available. On the wind-eroded area in the southern part of the Dzungaria, planting shelterbelt nets is an effective measure for protecting farmland from shifting sand.

The construction of checkerboard protection is a very important measure for stabilizing shifting sand along communication lines. Within the enclosure of the checkerboard protection, *Artemisia* spp. and various shrubs are planted. By this method large areas of shifting sand have been brought under control in the eastern part of the Ala Shan Desert, where the Paotow–Lanchow railway passes through.

Extensive cultivated areas in arid regions of the world have been transformed into salt deserts because of irrigation without proper drainage. Therefore, utilization of deserts for agriculture is necessarily limited to areas that not only have freshwater irrigation systems, but also have good drainage systems. Melting water from ice and snow in the Tienshan is channeled through the Manas trunk canal on the piedmont plain of the northern slope of the Tienshan to irrigate oases of the Dzungarian Desert. In the gravelly desert, underground canals (*karez*) for irrigating the oasis of the Turfan Depression were constructed a thousand years ago. Another means of providing water is to turn the interdune hollows and lakes into reservoirs in the eastern part of the Takla-Makan Desert.

Finally, planting of the aforementioned wild economic plants, including those for fiber, medicine, fodder, and food, may be extended, although their ecological habitats should be kept in mind.

EDITORS' NOTE

This chapter was prepared by Professor Hou for presentation at the 1978 International Conference on Arid Land Plant Resources. He originally had intended to revise the manuscript, but pressing matters finally forced him to decline the revision. This chapter is, therefore, not in the same format as the others and does not contain text references. We recommend the following sources for additional reading.

REFERENCES

Hou, H.-Y. (1979). "The Vegetation Map of China." Chinese Acad. Sci., Inst. Bot., Lab. Plant Ecol. Geobot., Beijing.

Sung-Chia, Chao (1981). The sandy deserts and the gobi: a preliminary study of their origin and evolution. *In* "ICASALS Pub. No. 81-1," pp. 1–33. International Center for Arid and Semi-Arid Land Studies, Texas Tech Univ., Lubbock.

Sung-Chiao, Chao, and Chin, Han (1981). Large-scale agricultural reclamation in the Tarim Valley and its impact on arid environment. *In* "ICASALS Pub. No. 81-1," pp. 34–47. International Center for Arid and Semi-Arid Land Studies, Texas Tech Univ., Lubbock.

7

SOUTH AMERICA

Juan M. Gastó
Facultad de Agronomía
Universidad Católica de Chile
Santiago, Chile

Alfredo Olivares E.
Facultad de Ciencias Agrarias y Forestales
Universidad de Chile
Santiago, Chile

Rolando H. Braun W.
CONICET
Instituto Argentino de Investigaciones de las Zonas Aridas (IADIZA)
Mendoza, Argentina

I. INTRODUCTION

The objective of this chapter is to present a synthesis of recent advances and research related to plants in the arid and semiarid environments of South America. Because of the extent of the subject, it has not been possible to present an exhaustive and integral paper. To accomplish this work, which includes diverse regions and countries, it was necessary to ask for the cooperation of some of the most outstanding researchers on the continent. Even so, some regions or countries are not included in detail.

The study has been divided into three parts. First, a brief description of the main natural regions is included, giving a general view of the problems and limitations of each one. Then, a synthesis of the recent advances in each of the regions, as well as the general problem of desertification, is presented. The last part is related to the needs and recommendations for future research, according to the interpretation of the authors.

II. PHYSIOGRAPHY

The regions described here experience arid or semiarid conditions during part of the year or season (Figs. 1 and 2). In some regions, the total precipitation may be high, but the combined effect of temperature, soil, and water balance produces the dry environment that characterizes arid regions.

1. DESERT

The South American desert is located on the Pacific coast between 2 and 30°S. The climate is warm and dry, with extreme temperature ranges between day and night, and different seasons. The desert belt is located next to the ocean, and because of the cold Humboldt Current, and air pressure, often has dense fog, which concentrates on leaf surfaces. This moisture stimulates the growth of vegetation, which under certain circumstances is profuse and corresponds to the coastal fog or Lomas desert. The total annual precipitation in the north approaches 100 mm and gradually diminishes toward the south; in Arica and Iquique, Chile, the annual mean is near 0 mm. Further south, it gradually increases, and at 30°S it barely surpasses 30 mm.

The central part of the belt experiences a very low relative humidity, high luminosity, and a minimum of cloudy days, which produce an environment with an almost absolute lack of vegetation. In the ecotone with the high Andean region, 2800 m above sea level toward the east, annual precipitation surpasses 5–10 mm, allowing the growth of ephemeral vegetation and open shrublands.

Next to the Andes, where there is a larger amount of precipitation, extra water comes from canyons, underground rivers, summer surface runoff, and other sources. These sources cause oases, transversal valleys of the desert, salt flats, and pampas; underground water is present at less than 20 m depth. The vegetation and fauna are characteristic of the environment.

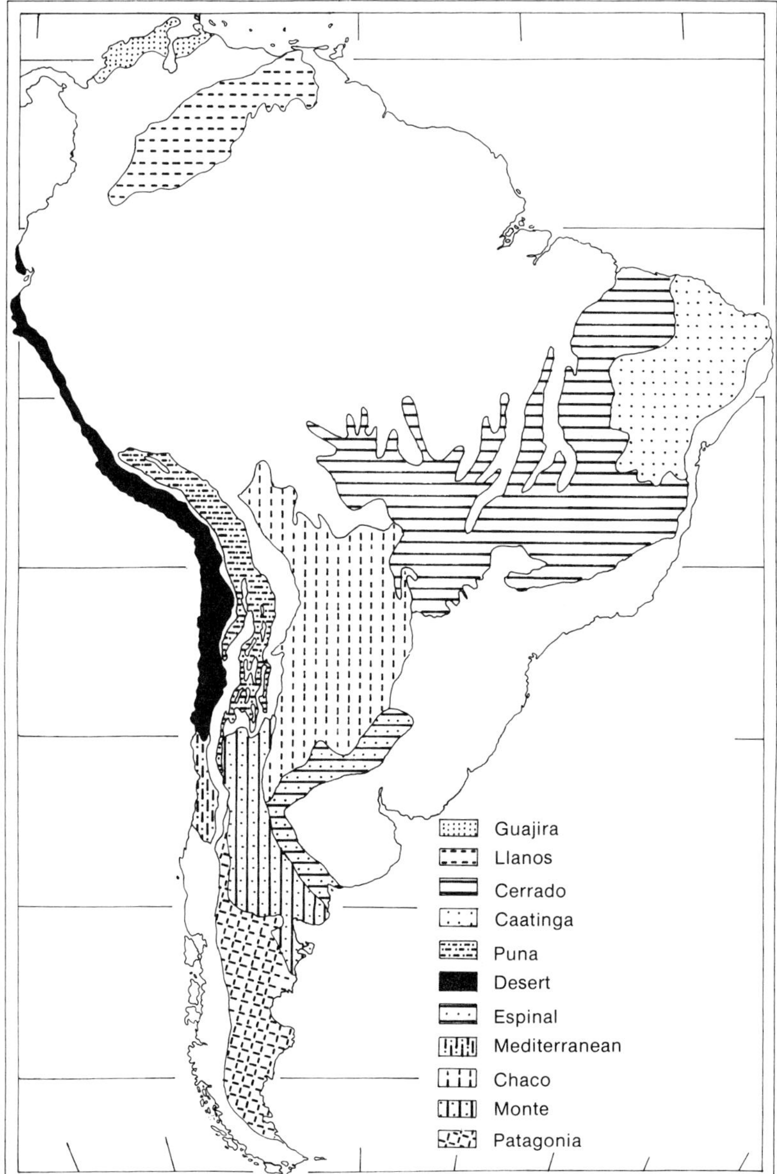

Fig. 1. Natural regions of South America, according to Cabrera and Willink (1973). The regions included, during part of the year or year-round, present arid or semiarid conditions.

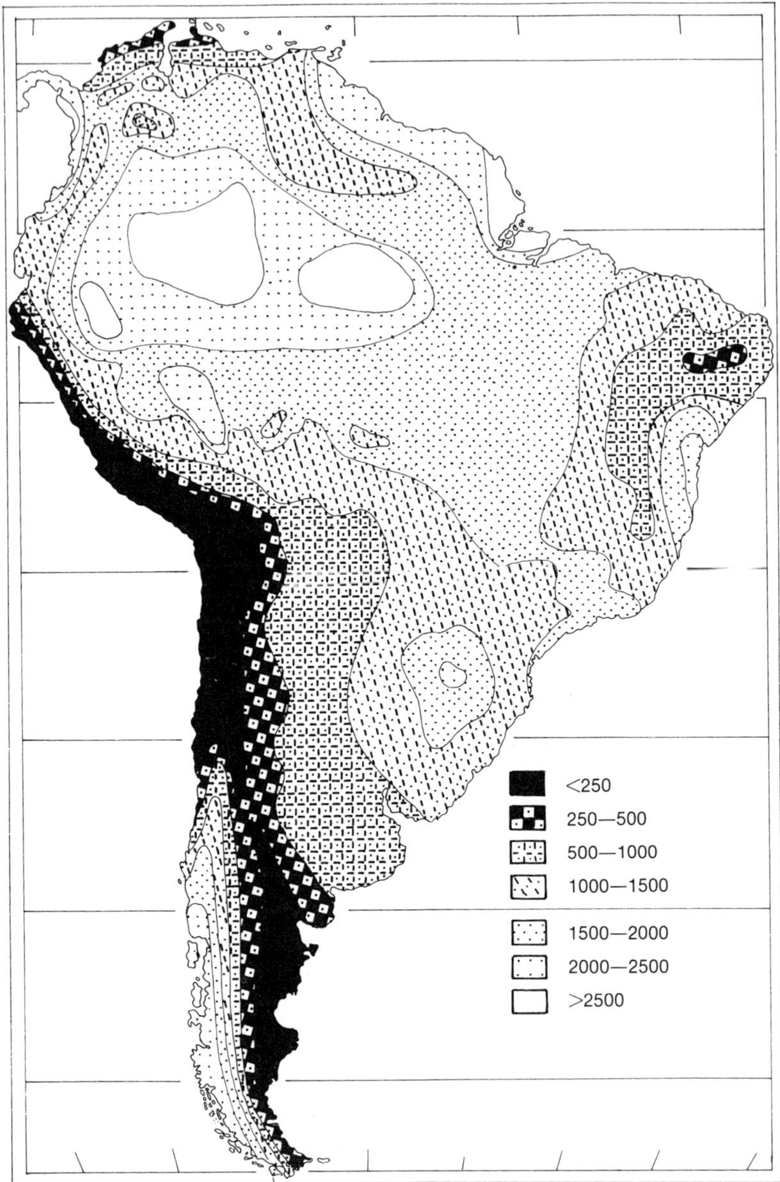

Fig. 2. Total annual precipitation (mm) in South America.

2. PUNA

The Puna is a flat highland located between the Andes and the Cordillera Réal; its altitude ranges from 3200 to 4200 m above sea level. The climate is cold and dry and presents a strong thermic variability between day and night, and between seasons. Mean annual temperature is 9°C. The precipitation is higher in the north, registering up to 800 mm per year and gradually diminishing to 50 mm toward the south. The rainy season occurs during summer, thus ephemeral summer plants are abundant. The remainder of the vegetation corresponds to a steppe whose main vegetation consists of perennial grasses interspersed with low shrubs.

3. PATAGONIA

A temperate–cold climate with strong westerly winds during a great part of the year is the most characteristic trait of this region. Frost, occurring most of the year, and snow, falling during the winter months, gives the area an inhospitable environment. Mean temperature reaches 30°C in the north and 5°C in the south, and precipitation fluctuates between 100 and 200 mm. The physiography corresponds to flatland interspersed with valleys, sierras, and highlands. Soils, in general, are stony and sandy, though with a high range of variability according to the position in the relief. The vegetation is a low shrubby steppe, with grasses interspersed with low shrubs. In the low meadows, grasses predominate over other life forms.

4. MONTE

The monte vegetation corresponds to an open chaparral whose height gets smaller toward the west. The location of this region—between the Andes, chaco, pampean forest and dunes, and Patagonia—produces a wide variety of ecotones. The climate changes from the north, where the average temperature is warmer, to the south. The regional mean is 13–15°C. Precipitation is scarce, ranging from 100 to 250 mm per year. The region has a low carrying capacity, from $\frac{1}{10}$ to $\frac{1}{70}$ animal units per year per hectare, with extensive livestock use. Forage is produced by some herbaceous plants as well as shrubs.

5. CHACO

The predominant vegetation is xerophytic and deciduous, with Cactaceae and Bromeliaceae, and a Gramineae stratum. In addition, there are palm trees and savanna, as well as halophytic, shrubby steppes in localized areas (Cabrera and Willink, 1973). In the eastern part of this area, precipitation is abundant throughout the year, while in the west only the summer season is rainy, with an annual average precipitation between 500 and 1200 mm. Mean temperature is between

20 and 23°C. Adverse environmental conditions are related to the low carrying capacity of the range for livestock, which only reaches $\frac{1}{20}$ to $\frac{1}{33}$ animal units per year per hectare in the north, and $\frac{1}{10}$ in the south.

6. PAMPEAN FOREST AND DUNES

This is a xerophytic open forest, with trees (5–8 m tall) dominated by the endemic legume calden (*Prosopis caldenia*), and associated woody species such as *P. nigra* and *Condalia microphylla.* The pampean forest region forms a belt oriented from northwest to southeast, approximately between the 450 and 600 mm isohyets in the central part of Argentina (Ragonese and Cano, 1971). Annual mean temperature ranges from 15 to 20°C. There exists a certain floristic affinity with the chaco. Edaphic communities with rhizomatous species are frequent on dunes, and there are bushes of fleshy Chenopodiaceae on salty soils (Cabrera and Willink, 1973). Native grassland has great importance. Its carrying capacity ranges between $\frac{1}{10}$ and $\frac{1}{2}$ animal units per year per hectare.

7. MEDITERRANEAN

In the more favorable habitats exposed to the south, the vegetation is dominated by trees, while on northern exposures, or in shallower soils, it is shrubby. Vegetation and stratification are complex, with horizons of pterophytes, geophytes, and phanerophytes at different heights, as well as hemicryptophytes and lianas; all give a high stability to the system.

The climate, as in other Mediterranean regions of the world, is characterized by long summer droughts alternated with a short winter rainy season, which, in the most favorable environments, starts at the end of fall and ends in the middle of spring. In the north, annual precipitation is only 20–25 mm, or even less, whereas toward the south, it increases up to 1400 mm and even 2000 mm. The mean temperature, on the other hand, is higher in the north, reaching 16°C, with a mean of 12°C near the coast and 8°C near the Andes. The average temperature range is high, and nights are cool. The minimum temperature during the coldest months usually does not reach below 0°C. During summer, the maximum temperatures are usually between 32 and 35°C (Di Castri and Mooney, 1973).

The physiographic traits of this zone are hills and terraces cut by transversal canyons and creeks, which carry the winter surface runoff. In winter, the land is covered by annual plants, which dominate the lower strata and ripen and dry during spring. On the other hand, woody species start growing and flowering in spring and stay active during summer. Most species are evergreen.

8. LLANOS

The llanos are located at the extreme north of South America in Venezuela and Colombia. The climate consists of a dry season alternated with an extremely wet season (Cabrera and Willink, 1973). The llanos are characterized by their flatness and are covered by a continuous stratum of Gramineae interspersed with isolated shrubs and trees. During the rainy season, the soil surface is covered in many places by thin, heavy layers of water that reduce the growth rate. During the dry season, the growth rate is kept to a minimum by a water deficit.

9. GUAJIRA

This region is located in the coastal area of northern Colombia and Venezuela, next to the Caribbean Sea. The precipitation concentrates mainly during October and November, when the annual mean reaches only 400 mm. Mean temperature is 28°C (Cabrera and Willink, 1973). Vegetation is dominated by columnar Cactaceae and dwarf shrubs.

10. CERRADO

The cerrado is located in a highland where elevations vary from 500 to 1000 m above sea level. This region covers 200 million ha. Mean precipitation is between 1200 and 2000 mm per year, with a dry season from May to October, and a mean temperature between 21 and 25°C. The vegetation corresponds to open forests of low height, occasionally reaching 8–12 m; shrubby and herbaceous strata develop under them. A high proportion of this region is savanna dominated by a continuous stratum of perennial grasses interspersed with shrubs and small trees (Cabrera and Willink, 1973).

11. CAATINGA

Caatinga is characterized by a shortage of precipitation, averaging between 400 and 700 mm, with high annual variability. In wet years, the surface runoff is lost, while during dry years there is a shortage of water and the seasonal growth is reduced to a minimum. The aridity problem is magnified by high temperatures, with a mean of 27°C. Plant cover is heterogeneous. Trees and shrubs inhabit the best environments and are mixed with open areas dominated by spiny shrubs. A high proportion of the predominant species lose their leaves during the dry season from May to September.

III. PLANT RESOURCES

A. Phytoecological Studies and Management

1. DESERT

Because of the lack of precipitation for long sequences of years and the low relative humidity, few vegetation studies have been conducted in the desert region. Researchers have centered their attention in those areas—oases, valleys, and salt flats—that, due to special circumstances, receive water inputs from other systems, such as surface or subterranean runoff or where the input comes as fog.

Acosta-Solís (1970) described the vegetation and flora of arid lands and lands marginated by drought in Ecuador; these are located in the coastal and inter-Andean region. Ecological conditions and the proposed management were discussed.

In the coastal fog desert, the problems differ according to the area. In the northern savanna, where *Prosopis* and herbaceous plants dominate, the main problem arises from the indiscriminate cutting of *Prosopis* and *Acacia,* which causes eolian erosion. This promotes the advance of sand dunes, which in turn increases the process of desertification and modifies the microclimate (Kummerow, 1966). On the other hand, indiscriminate grazing by goats reduces the importance of herbaceous vegetation, which in turn affects the natural regeneration of shrubby and arboreal vegetation. All of this causes a rupture of the organic-matter cycle, and reduces the potential productivity (MAB-3, 1977).

In the lomas of the coastal region of Peru and Chile, there is a problem of overgrazing, especially of the herbaceous forage plants, which are seasonal and thus are not protected from being consumed by grazing. The lomas vegetation grows mainly during winter on the slopes facing the sea, which are exposed to the south or southwest (Aguilar, 1970; MAB-3, 1977; López Ocaña, 1978). Castaño *et al.* (1978) studied the lomas of the Peruvian coast using a system approach in order to find a methodology to develop and evaluate the natural resource management of the regions in Lachay and Iguanil. This is one of the few integrated studies that has tried to solve the problem with multiple objectives.

The shrubby and herbaceous vegetation of the lomas survives as a result of fog, which increases in intensity from July to October. The process has been carefully studied by several authors who have even quantified the water influx (Kummerow, 1966; Hajek and Saiz, 1976). When vegetation is destroyed, the plant surface, which intercepts the fog, is reduced; this, in turn, causes a reduction or loss of water income to the soil and plants, triggering the desertification process. Because of this cycle, a natural recovery of vegetation does not exist.

The most outstanding section of this phytogeographical unit is the Pampa del Tamrugal in Chile, which is characterized by the presence of a natural open forest dominated by several species of *Prosopis* (Contreras and Gastó, 1978). The density of *P. tamarugo* ranges from 20 to 30 plants per hectare (Pisano, 1966). Since the early 1970s several studies have been done to increase the knowledge necessary to improve man-made irrigation of primary and secondary production, as well as to increase the area planted (Cadahia, 1970). Because of the lack of precipitation and its association with natural forests, this region has received much attention. Several studies analyzing water relations and balance have been conducted in order to state formally a hypothesis leading to the establishment of a cause-and-effect relationship to explain the presence of this association (Botti, 1970; Sudzuki, *et al.*, 1973; Sudzuki, 1975).

The plant community is characterized by its high proportion of support and fruit tissues. A high percentage of the digestible protein is localized in the foliage, even though the highest percentage of digestible energy is located mostly in the fruit (González and Haart, 1960; Lanino, 1966; Muñoz, 1972). Abundant forage that is palatable to sheep, goats, and cattle is produced with 12% crude protein, 30% fiber, and 1.9% ether extract. Litter productivity may reach 74 kg per tree, and that of fruits, 8.2 kg per tree. The carrying capacity of the forest is related to age, reaching up to 10 sheep per hectare per year on a 16-year-old plantation (Elgueta and Calderón, 1970; Elgueta, 1971; Contreras and Gastó, 1978; Lanino, 1966). Under certain circumstances, 30-year-old trees have produced as much as 3.4 kg/m^2 of dry matter, 50% of which is litter, and the rest fruit. Livestock mainly consumes the ground litter, which, because of the dryness of the air and lack of precipitation, is preserved in good condition for many years. Occasionally, they consume the foliage and fruit on the lower branches (Muñoz, 1972; Klein, 1970; Contreras and Gastó, 1978).

The natural trophic chains are mainly of the entomological type; a small number of vertebrates are present, especially lizards and birds (Klein, 1970; León, 1974; Campos, 1968; Klein and Campos, 1977). Because of the low diversity of the system, especially of the phytocoenosis, the stability of herbivorous and carnivorous insects is very low, and a high proportion of the primary production is consumed by them.

In 1963, a research program was started to look for ways to utilize the pampas for animal-raising purposes. The results of these studies were the basis for carrying on a reforestation program; up to 25,000 ha, mainly in the salt flats of Pintados and Zapiga in northern Chile, are now planted with *Prosopis tamarugo*. Improved breeds of livestock also have been introduced (Instituto Forestal, 1971; Lamagdelaine, 1974; Lanino, 1972).

2. PUNA

Ruthsatz (1977) thoroughly studied the plant communities and environmental conditions of the Andean half-deserts of northwestern Argentina.

The agricultural regions of the high Andes have been classified as homogeneous areas with indications of the potential productivity and variety adaptation. This has been done to provide a basis for a plan to grow Andean crops for human consumption. The system stability has its roots in the high diversity of crops and livestock, which characterizes traditional agriculture (Camino, 1977; Earls, 1977). The grasslands of the high Andes also have been considered (Tapia, 1976; Braun, 1964; Flores, 1967; Gómez, 1966; González, 1967, 1969). Great emphasis has been placed on introducing new species and varieties, even though the greatest potential lies in the cultivation of natural grasslands (Segura, 1976). Some studies have been done using an ecological approach (Cárdenas, 1971; Posnasky, 1971). There is abundant information derived from agrostologic studies (Tapia, 1976; Lara, 1972; Johnson *et al.*, 1968).

3. PATAGONIA

Ruiz Leal (1972b) performed studies in the northern portion of the Patagonian plateau and characterized the phytogeographical transition between the monte and Patagonia according to floristic evidence.

Among the programs of vegetation inventories and cartographical studies of semiarid Patagonia, several studies have been conducted to describe and classify those communities that are dominated by *Stipa speciosa* and *S. humilis* (Soriano and Brun, 1973). At the same time, three areas have been described using a geological, geomorphological, edaphic, vegetational, and meteorological approach (Brun *et al.*, 1971). In certain areas, plant communities have been described in large scale on a cartographic basis (De Anchorena, 1973). The flora and vegetation of Patagonia also have been studied (Latour, 1971).

Livestock exclusion from rangelands causes changes in the productivity and botanical composition, even when it happens for short periods of time. Plant succession induced by grazing expresses itself from the first year of use (Soriano, 1973). In the analysis of both existing and potential vegetation, Soriano *et al.* (1978) studied the structure of the climax community before and after utilization by sheep. In another study, range dynamics were analyzed through the measurement of diseminules density, both inside and outside the exclosures.

The natural psammophytes of the Bahía Blanca region were studied by Bishop *et al.* (1968) from a taxonomic, palatability, and bromatologic point of view. In some studies, biomass production of some species and range types has also been considered through the periodic harvest of herbaceous dominant components, especially *Stipa speciosa* and *S. humilis* (Soriano, 1974). The nutritive characteristics of the main range species of a broad Patagonian area have been described by Wernli *et al.* (1976). The structure of natural communities dominated by low, palatable grasses and shrubs can be modified by seeding with improved herbaceous range plants (Serra, 1970). Specifically, Abadie (1967) studied the native range dominated by *Festuca pallescens,* which has a high forage value, in order to increase its natural regeneration.

4. MONTE

Roig (1972) outlined a physiognomic vegetation map of the Mendoza Province, in the central western area of Argentina, in which he described formations having trees (in flatlands and mountains), shrubs (of halophytic, psammophytic, and mountainous swamps and *lacunes*), and mountain grasslands. Guevara *et al.* (1978) prepared an inventory of the natural, renewable resources of Mendoza. One of their conclusions is that the regional vegetation includes valuable forages, with regard to distribution, density, and fodder value.

Ruiz Leal (1965) referred to 16 different species not previously known in Mendoza. Several other floristic studies have been carried out, among which was one related to the species of the genus *Stipa*.

Ambrosetti (1971) studied some of the important species used in the management of regional watersheds in the central west of Argentina. The range changes originating from the exclusion of livestock from a piedmont alluvial basin were studied by Pedrani and Rodriguez (1978). The ecosystem structure of shrubby species, which characterize the monte, was studied through a process of breaking the equilibrium through biological procedures (Zuccardi, 1974). Vegetation dynamics have been studied where changes were triggered by fire (Roig *et al.*, 1972).

In the Mendozinan plain, primary production of a monte community dominated by *Prosopis flexuosa* was analyzed by Braun *et al.* (1978). The energy and nitrogen content of species of this community were studied by Braun and Candia (1980), while other woody and herbaceous species of western Argentina were also studied in detail by Wainstein *et al.* (1974, 1979).

5. CHACO

Large vegetational and environmental units were studied with the use of aerial photographs, as well as measurements and observations on the ground; soil, geomorphology, and vegetation were analyzed (Adamoli, 1972; Morello *et al.*, 1971, 1972, 1974). Vegetation dynamics were studied using remote sensors to detect changes in vegetation structure, and to evaluate invasion of woody species (Morello and Gomez, 1971). Morello and Adamoli (1973) described and defined the ecological subregions of the chaco and prepared the cartography. The variation of some characteristics of plant communities, which developed under extreme pluviometric conditions, also were analyzed (Sejzer, 1973).

A large number of arboreal, shrubby, and succulent species were analyzed from a structural point of view (Neumann, 1973). This has also been done for communities dominated by *Prosopis ruscifolia,* to gain data on their function, as well as their effect on the other components (Gomez *et al.*, 1973). The regional flora has been described in detail (Digilio *et al.*, 1971), including its spatial arrangement (Adamoli, 1972).

The indiscriminate harvest and exploitation of the natural resources of the chaco area, especially during the last century, produced an intensive retrodegradation over most of the area (Ledesma, 1978). One of the most outstanding colonizers is *Prosopis ruscifolia,* whose invasion mechanism has been described by Morello (1970). Gomez *et al.* (1974) described its selective control with herbicides to promote successional changes in the community. Other species also have been studied (Lagomarsino *et al.*, 1974). Some improved species were established in the northwestern area (Díaz, 1972a).

Díaz *et al.* (1972b) studied the digestible value of the most important species in this semiarid region. Aerial and subterranean biomass of some woody species of the occidental (dry) chaco forest, as well as their contribution to the nitrogen balance, have also been studied. Del Aguila *et al.* (1969) studied the animal carrying capacity of rangelands of Los Llanos in La Rioja.

The interstrata interference of some invader species, such as *Prosopis ruscifolia,* is due to the exudate production, which is able to inhibit or delay the seed germination of some subordinated species (Souto and Eilberg, 1972).

6. PAMPEAN FOREST AND DUNES

Several studies have been dedicated to the description of vegetation and productivity (Adler *et al.*, 1974; Cano and Fernández, 1974; Anderson *et al.*, 1970; Cano and Olmos, 1968; Lewis, 1973; León, 1973). Alliney *et al.* (1978) studied the size and number of samples in the range evaluation. The flora and vegetation of the semiarid pampean forest have been studied by Cano (1977), Ragonese and Cano (1971), and Covas (1971), while Anderson *et al.* (1974) studied relics on sandy soils.

Vegetation dynamics have been studied where changes were triggered by fire (Braun and Lamberto, 1974; Lamberto and Braun, 1974; Orionte and Anderson, 1976). In the same way, postfire successions have been studied by Cano and Holgado (1971). The range changes originating from the exclusion of livestock and wildlife were studied by Cano (1969). Recovery time of the desirable species was studied under exclosure conditions by Santo *et al.* (no date).

Feldman (1972) described the management and replacement of woody plant cover in the forest area. Range conditions and trends were studied by Anderson *et al.* (1971) in order to propose the proper management strategy and to optimize plant production. In other studies, ecological successions have been considered in a broader context, where, besides the ecological aspects, the genetic, biochemical, and evolutionary aspects have been considered in such a way as to analyze the ecological convergence when comparing analogous regions (Lowe *et al.*, 1972).

The primary production of ecosystems has been analyzed by several authors. Cano (1977) estimated a production (dry weight) of 2500 kg per hectare per year for shrubs and grasses. The range production that resulted after the controlled burning of pampean forest vegetation was studied by Lutz and Graff (1974) and Braun *et al.* (1974); it was found that during the dry years, the lower strata production was considerably reduced.

7. MEDITERRANEAN

Range associations were determined through phytological analyses (Etienne *et al.*, 1978). The botanical composition, nutritive value, and range production in an area with an annual precipitation gradient from 50 to 2000 mm has been estimated, both for granitic soils as well as marine terraces (Wernli, *et al.*, 1976). Quantitative relationships have been established between the ecological and climatic characteristics and the range structure (Gastó and Contreras, 1979; Parilo, 1978; Rojas, 1978).

The geographic-distribution studies have permitted a selection of native plant species that are adapted to improved environments. Some of the species cultivated at present were collected in natural environments and selected to be included on improved grasslands (Gastó and Contreras, 1972a, 1972b).

The absolute growth rate through the year and the relative growth rate were studied to obtain values for the natural ranges under normal grazing conditions, as well as with fertilizer (Parilo, 1978; Schenckel *et al.*, 1971; Segarra, 1979; Gastó and Contreras, 1979). The greatest increase in the growth rate of the range takes place at the end of winter and the beginning of spring, from August to October, when the best moisture and temperature conditions occur. The precipitation variability and the relationship between precipitation and primary production were studied in detail by Acuña (1978) and Gastó and Contreras (1979).

The ecological influence of fire on the natural shrubland was studied by Altieri and Rodriguez (1974).

Season and frequency of range utilization affect productivity and botanical composition. Experimental results indicate that when a range is utilized for short periods of time, it is not affected, even when heavily grazed. Continued intensive use causes range degradation, which ends in the destruction and change of its botanical composition (Olivares and Riveros, 1979; Olivares and Gastó, 1978; Contreras and Gastó, 1978). Animal nutrition under different range conditions has been studied by Riveros (1977) and Avendaño *et al.* (1978). The effect of livestock on the range, and vice versa, has been considered by Olivares and Riveras (1979) and Grez and Fernandez (1978). The supplementation of livestock feed under grazing conditions allows a better range utilization (Manterola *et al.*, 1979; Hechenleitner, 1973; Catalán, 1973). The influence of seasonal frequency and intensity of range utilization by livestock has been studied by Rodriguez (1978), Trivelli (1973), Di Marco (1973), and Concha (1975).

The range organization in fields and its utilization with rotation and grazing has been studied by Avendaño *et al.* (1978). The study is based on experimental results accumulated over a long period of years in the Mediterranean region of central Chile. Aranda (1971) studied the complementation of winter and summer ranges, as well as the routes and procedures followed in the process of moving the livestock to complementary areas. A general view of livestock migration over a long period was done by Cosio and Vicens (1969).

Range reseeding studies of herbaceous species have continued, even though less intensively than in previous years, since there is a clear knowledge of the adaptations of the main groups of plants (Uslar, 1972).

The study of natural phytocoenosis has increased the understanding of the basic structural elements of the matorral, which characterizes the Mediterranean environments of Chile. Based on its natural structure, a new community, similar to the original one in its morphology and stability, but with a larger rate of man-made irrigation, is created (Gastó and Contreras, 1972a,b) with the changing environment.

Most of the emphasis has been placed on species adaptation to environments that record 50–300 mm of precipitation per year (Gastó and Contreras, 1972a,b), even though moister ranges also have been studied (Cosio, 1970; Vicens and Cosio, 1968, 1970). The adaptations of shrubby species to different soil conditions (Lailhacar, 1976), as well as the natural population distribution (Badilla, 1975), have been studied.

Intraspecific interference is one of the areas that has received most of the attention during the last few years. This knowledge is considered basic for the development of bistratified phytocoenosis, which is required for ecosystem improvement in arid and semiarid Mediterrancan regions of central Chile (Zuñiga, 1973; Gastó and Caviedes, 1976).

B. Food Plants

The botanical, genetic, and agronomic characteristics of some native species of the Andean region have been studied. Among these are *Chenopodium quinua, C. pallidicaule, Lupinus mutabiles, Oxalis tuberosa, Ollucus tuberosum, Tropaeolum tuberosum,* and *Amaranthus caudatus* (Tapia, 1977). Most of the studies are being conducted in Peru and Bolivia (Vorano and García, 1976; Telleria, 1976; Lanino, 1976; Cardozo *et al.*, 1976; and Narrea, 1976). Basic regional studies have permitted the formation of germ-plasm banks in which a high proportion of the local varieties are included. At present, there are adequate facilities to preserve these banks in good condition. The exchange of genetic material with other scientists began in 1979.

Anatomical, morphological, and physiological studies have been done on the above-mentioned species and varieties, giving more information on the development of the areas where they occur (Cornejo, 1976; Lara, 1976; Ignacio *et al.*, 1976; Cortez, 1977; Canales and Baldomero, 1977). The most promising varieties have been selected from a nutritional, agronomic, and economic point of view (Lescano and Palomino, 1976; Lescano, 1976; Telleria and Ballón, 1976). At present, a basic knowledge of the growth of these species (especially *Chenopodium quinua, C. pallidicaule,* and *Lupinus mutabilis*) is available concerning fertilization, cultivation, and pest control (Gandarillas and Tapia, 1976; Chaquilla, 1976; Morales, 1976; Christiansen, 1977; De La Puente, 1977).

Growth, productivity, digestibility, and utilization have been studied in general (Arze and Sotelo, 1976; Arze and Alencastre, 1976; Negron *et al.*, 1976; Cornejo, 1976).

New varieties have been developed by local experiment stations, especially in Peru and Bolivia; some of these varieties, in addition to having better agronomic characteristics, contain a lower percentage of saponin (Gandarillas and Tapia, 1976; Cano and Rosas, 1976). The productivity of one of these varieties, which is adapted to environments with frost-free periods longer than 140 days, reaches 2300 kg/ha, with 14% protein and 6% lysine.

Brücher (1962, 1965, 1973) and Viirsoo (1967) have studied wild potatoes in northwestern Argentina, which are a valuable source of germ plasm for the improvement of cultivated potatoes. In Argentina, *Ximenia americana* and *Geoffroea decorticans,* present in the monte (Ruiz Leal, 1972a), and *Celtis spinosa,* existing in the chaco (Schulz, 1963), have edible fruits of value. *Prosopis flexuosa* and *P. alba,* in the monte, and *P. torquata,* in the chaco, possess edible seeds. The fruits of other species of the monte also are indicated as possible sustenance foods (Ruiz Leal, 1972a): *Amaranthus viridis, Berberis* spp., *Cortesia cuneifolia, Ephedra triandra,* and *E. ochreata.* Additionally, *Portulaca oleracea* and *Arjona patagonica,* in Patagonia and the monte, are edible raw or cooked (Ruiz Leal, 1972a). In the chaco, people eat fruits of *Zizyphus mistol.*

Ortiz Garmendia (1969) has worked on Chilean wild plants with edible fruits; many are found in the arid zones. Among these are *Aristotelia chinensis, Beilschmedia berteroana, Cordia decandra, Cryptocaria rubra, Eugenai chequén, Eulychnia* spp., *Gourliea chilensis, Jubaea chilensis, Krameria cistoidea, Lithraea caustica, Lucuma valparadiceae, Margiricarpus seteous, Mühlenbeckia chilensis, Peumus boldus,* and *Prosopis chilensis.*

Acosta-Solís (1970) pointed out the edible cacti (*Opuntia* spp.) and several herbaceous legumes, such as *Crotolaria* spp., among the native species from Ecuador. Also, several woody legumes, such as *Prosopis pallida* and *P. juliflora,* are known to have edible parts. In Peru, fruits of *Acacia macracantha* and *Dioscorea chancayensis* are eaten.

C. Forage Plants

The inventory of herbaceous and woody plants, which can serve to feed domestic animals, is very large. In arid zones, woody species attain singular importance as a permanent forage resource because of their resistance to the water deficit. Tinto (1974) has indicated the importance of several Argentine species:

Foliage and fruit: *Acacia aroma, Caesalpinia paraguariensis, Condalia microphylla, Ephedra triandra, E. ochreata, Geoffroea decorticans, Lycium chilense, Prosopis alba,* and *Zizyphus mistol*

Only foliage: *Acacia caven, A. praecox, Atriplex lampa, A. platensis, A.*

semibaccata, A. undulatum, Capparis tweediana, Celtis tala, Gradowskia obtusa, Maytenus vitisidaea, Porlieria microphylla, and *Trichomaria usillo*

Foraging fruits: *Prosopis alpataco, P. caldenia, P. denudans, P. nigra, P. patagonica, P. ruscifolia, P. sericantha, P. stricta,* and *P. torquata* (*Larrea* spp. also can be turned into forage)

It is frequently possible to use the above-mentioned plant resources for other purposes, giving a complementary usage and allowing new activities. Many of these species can be used as firewood or have wood that can be used for industrial purposes.

In Chile, according to several authors, there exist various important woody forage species: *Prosopis chilensis, P. tamarugo, P. alpataco, Geoffroea decorticans, Berberis* spp., and *Acacia caven.*

Acosta-Solís (1970) mentioned the following as forage species from Ecuador: *Mimosa quitensis* and several cacti of the genera *Opuntia, Cereus,* and *Trichocereus.*

The protein content and digestibility of some native plants of Argentina, such as *Condalia microphylla, Ephedra ochreata, E. triandra, Larrea* spp., *Lycium chilense,* and *Prosopis caldenia,* as well as native grasses of the pampean forest (*Poa ligularis, Stipa tenuis,* and *Piptochaetium napostaense*) have been studied by Abiusso (1974) and Magoja (1975). The phytochemical characteristics of the species of this region were analyzed in several other works (Rondina *et al.*, 1971; Mendiondo *et al.*, 1973; Hnatyszyn *et al.*, 1974). Wainstein (1974) discussed the nutritional value of some forage species of western Argentina: *Sphaeralcea miniata, Verbena mendocian, V. aspera, Adesmia filipes, Amaranthus crispus, Prosopidastrum globosum, Bougainvillea spinosa, Lycium chilense,* and *L. tenuispinosum.*

The Mediterranean species of Chile have been studied as a raw-material source to be used in the manufacturing of protein extracts. Soluble protein extracts from *Atriplex repanda* and *A. nummularia* leaves, according to Silva and Pereira (1976a,b), contain 19.6% and 26.2%, respectively, of the total protein present in the leaves. Amino acid analyses of the extracted proteins show a balanced composition with the exception of methionine, which is present in lower than usual amounts. Lysine and methionine amounts depend on the phenological stage of the plant (Ferrer *et al.*, 1977). Protein concentrates could be used as valuable supplements for other foods that are low in the essential amino acids (Silva and Pereira, 1976a).

D. Medicinal Plants

Montes (1965) gathered knowledge on the composition of essences of Argentina. Among these are *Acacia caven, Acantholippia seriphioides, A. hastulata, Aloysia polystachia, Artemisia mendozana, Azorella* spp., *Chenopodium rigidum, Hedeoma multiflorum, Lippia trubinata, L. integrifolia, Myroxylon peruiferum, Schinus molle,* and *Trixis antimenorrhea.*

Mateu Amengual and Villa Carenzo (1971) published an extensive phytochemical bibliography; it includes references to Argentine species and those from other South American countries. The natural products of several composites of central western Argentina have been studied (Giordano, 1979): *Tessaria* spp., *Flourensia oolepis, Baccharis* spp., *Senecio* spp., and *Grindelia* spp.

In his work on the flora of Mendoza, Ruiz Leal (1972a) included an extensive list of plants that are popularly used and even sold, generally as infusions, because of their pharmacological value.

A list of herbs with which medicaments are prepared in Peru has been prepared by Vargas (1978). Among the native ones, the following can be cited: *Ambrosia peruviana, Caesalpinia* sp., *Cassia loretana, Chenopodium pallidicaule, Ephedra* sp., *Equisetum* sp., *Persea americana, Portulaca* sp., and *Schinus molle.* The essence of *Peumus boldus,* present in Chile from Coquimbo to the south, has been studied by Schimmel (cited by Fester *et al.,* 1960).

E. Industrial Plants

1. FIBER PLANTS

As wood sources, one can mention (according to Tinto, 1974, and several other authors) *Acacia aroma, A. caven, A. praecox, Caesalpinia paraguariensis, Celtis tala, Geoffroea decorticans, Myroxylon peruiferum, Prosopis caldenia, P. alba, P. chilensis, P. ruscifolia, P. torquata,* and *Zizyphus mistol,* all of which occur in western Argentina. In Ecuador, Acosta-Solís (1970) reported *Cordia alliodora, Geoffroea spinosa, Libidibia corymbosa, Loxopterygium huasango, Prosopis juliflora,* and *Zizyphus thirsiflorus.*

Lippia turbinata, in Argentina, and *Sporobolus rigens,* in Argentina and Chile, are used for basketry; the latter species also is used for making brooms.

Typha dominguensis is used in western Argentina for binding vines, calking barrels, manufacturing of seats, covering greenhouses, protecting glass containers, and so on.

In Argentina, *Condalia microphylla, Porlieria microphylla,* and *Zizyphus mistol* are used for turnery.

The following species, belonging to arid zones of Argentina, are used as wood for industrial purposes: *Caesalpinia paraguariensis, Celtis tala, Condalia microphylla, Geoffroea decorticans, Prosopis alba, P. caldenia, P. chilensis, P. nigra, P. ruscifolia,* and *Zizyphus mistol.*

Tinto (1974) referred to the following species as firewood: *Acacia* spp., *Atriplex lampa, Celtis tala, Condalia microphylla, Geoffroea decorticans, Gradowskia obtusa, Prosopis* spp., and *Zizyphus mistol. Monttea aphylla* also is known for that purpose. In Chile, in addition to some of the above mentioned, *P. tamarugo* and *P. chilensis* are used for firewood, *Cordia decandra* for making charcoal (Ortiz Garmendia, 1969). In Peru, *P. pallida* and *P. chilensis* are used as firewood. Acosta-Solís (1970) mentioned *Mimosa quitensis* and *Baccharis* spp. for the same purpose in Ecuador.

2. OIL PLANTS

Acacia aroma and *A. caven,* found in Argentina, have flowers with essential oils for perfumery (Tinto, 1974).

3. WAX, RESIN, AND LATEX PLANTS

Tinto (1974) indicated that *Acacia caven* has a gummy substance, similar to gum arabic; *Prosopis alba, P. nigra, Porlieria microphylla,* and *Larrea* spp. also have resins. Rique (1974a,b) commented on *Gleditsia amorphoides* from the chaco area as an interesting species, which has a gelatinous gum in its seeds. Ruiz Leal (1972a,b) discussed *Cercidium australe,* from which a kind of glue is obtained, as well as *Bulnesia retama,* also from the monte, which gives wax similar to carnauba. In Peru, the glue of *Cercidium praecox* is used.

4. TANNIN PLANTS

Caesalpinia paraguariensis from the chaco area has been mentioned by Tinto (1974).

F. Other Plants

1. TINCTOREAL PLANTS

Plants native to Chilean arid zones, some of which are also present in border countries, have been discussed by Ortiz Garmendia (1968). Among these, he includes *Balsamocarpon brevifolium, Sphacele campanulata, Berberis glomerata, Caesalpinia spinosa, Oxalis gigantea, Geoffroea decorticans, Tagetes glandulifera, Salix chilensis, Colliguaya adorifera, Bulnesia chilensis, Krameria cistoidea, Schinus molle, Ephedra andina, Lithraea caustica, Schinus polygamus, Berberis* spp., *Maytenus boaria, Cassia* spp., *Zuccagnia punctata, Budleia globosa, Phrygilanthus* spp., *Psittacanthus* spp., *Acacia caven, Prosopis chilensis, P. tamarugo, P. alpataco, P. strombulifera, Larrea* spp., *Bulnesia chilensis,* and *Porlieria chilensis.*

Ruiz Leal (1972a,b) included *Berberis* spp. and *Relbunium richardianum* in this group, the latter for the substances contained in its roots. Acosta-Solís (1970) cited *Coulteria tinctoria.*

2. WASHING SUBSTANCES

Solanum eleagnifolium, with saponins for washing clothes, and *Cestrum parqui,* for washing hair, are found in the monte area (Ruiz Leal, 1972a,b).

3. ANTIOXIDANTS

Larrea spp. and *Succagnia punctata* contain nordihydroguayaretic acid, which is an antioxidant for oils and fats (Ruiz Leal, 1972a,b).

Alkaline ashes are produced by *Allenrolfea vaginata,* a species present in the monte, and are used in the production of soap and the preparation of olives.

4. SWEETENING SUBSTANCES

Olaechea *et al.* (1964) studied the chemical composition of *Glycyrrhiza astragalina,* a species that exists in the Patagonian and western regions of Argentina, and contains glycyrrhizin.

5. SAND-BINDING PLANTS

Hyalis argentea, Panicum urvilleanum, Plectrocarpa tetracantha, and *Prosopis strombulifera* serve as sand binders in the monte and Patagonia. In Peru, *Krameria illucca* and *Pennisetum chinense* can be found performing the same function.

IV. DESERTIFICATION

The natural ecosystems of the arid zones, which are in advanced successional stages, can reach an equilibrium between organizational and disorganizational forces.

Humans, who are socially, culturally, and politically organized, are capable of manipulating the topological arrangement of plant architecture through inputs to the ecosystem in the form of energy, matter, and information. This produces an ecological convergence that is different from the one in the natural state, where the anthropic component is absent (Fig. 3).

The technological capacity of organized humans enables them to transform arid and semiarid ecosystems; based on ecological laws, and through functioning and systemogenic processes, it also allows humans to channel these ecosystems to a new state of equilibrium. These processes are the basis for the management of natural ecosystems from an ecological and agronomical point of view.

A case study of desertification in the arid Mediterranean region of Chile was done at the request of UNESCO. The integrated study included specialists on

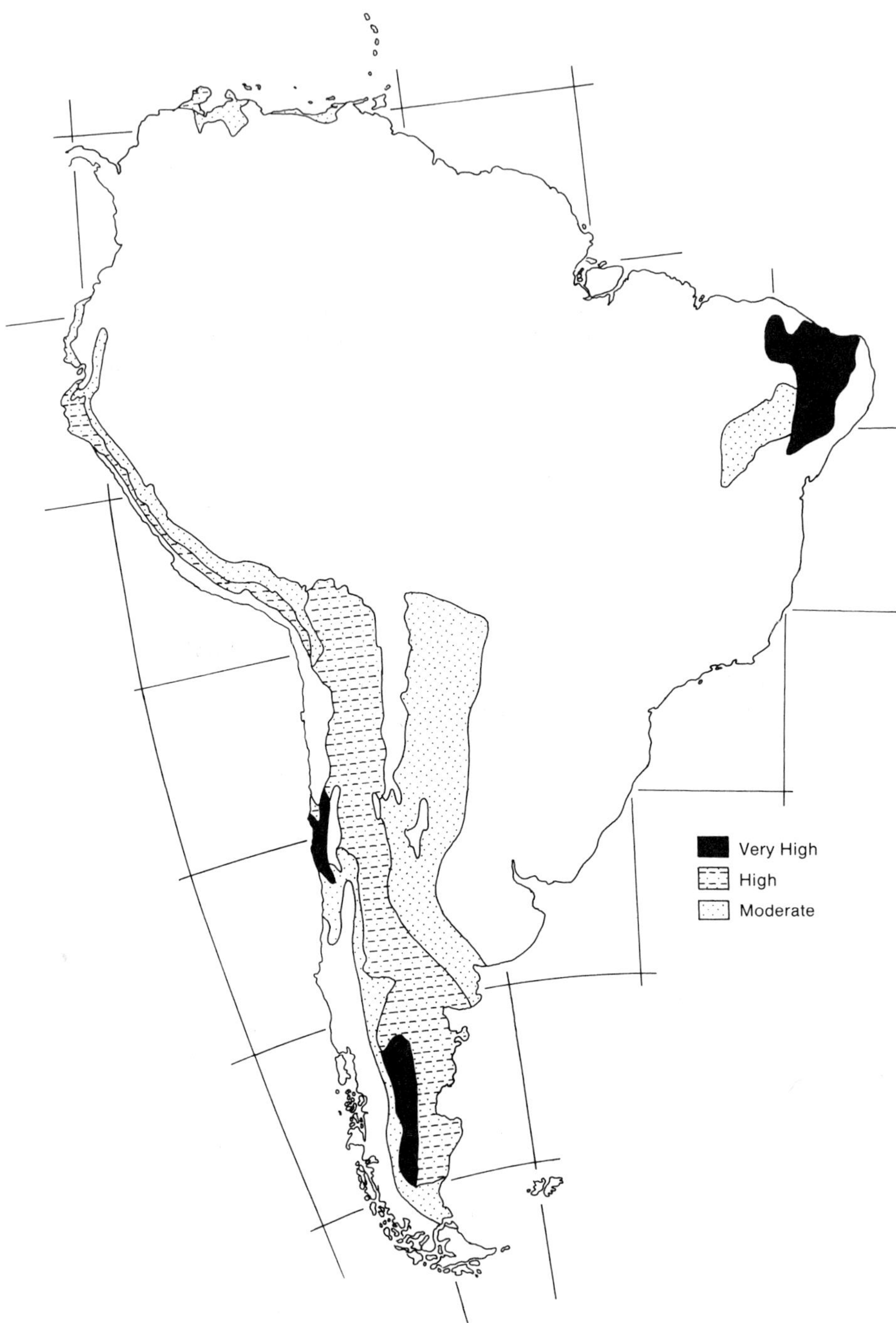

Fig. 3. Degree of desertification hazards in South America, according to the United Nations Conference on Desertification (1977).

geomorphology, geology, soils, bioclimatology, climatology, ecosystems, phytocoenosis, terrestrial and marine zoocoenosis, anthropology, range, and crops (INIA, 1977). The study presented a general scheme of the process of systemogenic degradation that ends in ecosystem desertification. In international meetings, representatives from several South American countries presented studies related to desertification. Also included were regions where precipitation is beyond the limits of aridity. The destruction of ecosystem components initiates a process of aridization, especially at the microclimatic level (Instituto Ecuatoriano de Recursos Hidráulicos, 1977). The United Nations has prepared a world map of desertification, where several studies conducted by the countries in these continents were included (United Nations, 1977; Peralta, 1977; Suarez de Castro, 1977).

V. SUMMARY

Research to be conducted in the arid South American environments should be placed in a general framework whose purpose is to solve general problems in a global way, and in which sociostructure, technostructure, and incident systems are closely related with biogeostructure. Very commonly, there exists an excessive emphasis on detailed local studies, which are weakly connected with the general problems of each natural region. There is often an obvious gap in the scientific and technological knowledge, which is available and applicable to the South American continent. Frequently, the research being accomplished refers to problems that have already been solved, or to those that easily could be solved, using the available scientific and technological knowledge. There is an obvious lack of interest in conducting research projects that produce information of general value for solving specific local problems.

The material resources invested in arid-land research have not been scarce, especially in relation to the necessities and economic capacity of the region. Human resources, available for research, are also abundant. With good organization and integration in each region, and among homologous regions, it would be possible in a short period of time to reach results corresponding to the necessities of the area. The experience obtained in South American arid-land research and the adaptation of techniques developed in homologous regions of the world, has led to the development of well-qualified groups that are able to organize and carry out research to solve the regional problems of arid lands.

As time passes, the necessity of conducting integrated studies to solve problems on an ecosystem basis becomes obvious. The integration of research efforts among countries that share natural regions is a must, in order to optimize the utilization of human and material resources, and to achieve research goals.

A general ecosystem approach toward the elucidation of arid-land problems should give better results than sophisticated studies centered on detailed subjects of an ecosystem's components, which are unrelated to one another. Because of existing limitations in arid zones, a transformation based solely on technology is not possible. The natural ecological processes, especially ecological successions, combined simultaneously with the application of adequate levels of technology, should provide an optimum combination to allow arid-land ecodevelopment.

The efforts made in South America to search for solutions to arid-land problems is proof of the interest that most of these countries have in their arid lands. The effort to overcome human and material problems, which are often found in these regions, should be recognized as a demonstration of the intellectual and material possibilities to solve them.

ACKNOWLEDGMENTS

The authors wish to acknowledge the contribution of regional information provided by the following persons: Dr. Mario Tapia, IICA, La Paz, Bolivia; Roberto J. Candia, IADIZA, Mendoza, Argentina; Prof. David Contreras, Universidad de Chile, Santiago; and Prof. Alfredo Gargano, Universidad del Sur, Bahía Blanca, Argentina. Some of the information presented in this chapter comes from summaries written by them.

REFERENCES

Abadie, C. A. (1967). La regeneración de *Festuca pallescens* (St. Ives) Parodi en los pastizales del oeste del Chubut. *IDIA* **239,** 58–64. INTA. Buenos Aires, Argentina.

*Abiusso, N. G. (1962). Composición química y valor alimenticio de algunas plantas indígenas y cultivadas en la Rep. Argentina. *Rev. Invest. Agropecu.* **16**(2), 92–247. Buenos Aires.

*Abiusso, N. G. (1964). Composición química y valor alimenticio de algunas plantas indígenas y cultivadas en la Rep. Argentina. II. *Rev. Invest. Agropecu., Ser. 2* **1**(13), 311–338. INTA, Buenos Aires.

*Abiusso, N. G. (1973). Estudio químico de cuatro gramíneas forrajeras: *Eragrostis curvula, Agropyron intermedium, A. elongatum* y *A. cristatum. Rev. Invest. Agropecu., Ser. 2: Biologia y Producción Vegetal* **3,** 89–110. INTA, Buenos Aires.

Abiusso, N. G. (1974). "Niveles Proteicos y Valores de Digestibilidad (KIUMS) de Tres Gramíneas Invernales Nativas de La Pampa," Vol. 5. Reunión Nacional para el Estudio de las Regiones Aridas y Semiáridas, Mendoza, Argentina.

Acosta-Solís, M. (1970). "Geografía y Ecología de las Tierras Áridas del Ecuador." Instituto Ecuatoriano de Ciencias Naturales, Contribución No. 72., Quito, Ecuador.

Acuña, H. (1978). "Influencia de la Pluviometría en la Producción Primaria de Una Pradera Anual de Clima Mediterraneo." Facultad de Agronomía, Univ. de Chile. Tesis Mg. Sc. Santiago.

Adamoli, J. (1972). "Frecuencia, Confinamiento y Transgresividad en Especies del Chaco Argentino." I Reunión Argentina de Ecología, Vaquerías, Córdoba, Argentina.

*Adamoli, J. (1973). "Manejo y Recuperación de Recursos Forrajeros Degradados en el Chaco Salteño." II Reunión Argentina de Ecología, Salta, Argentina.

*General references for further information.

*Adamoli, J. *et al.* (1972). El chaco aluvial salteño. *Rev. Invest. Agropecu., Ser. 3* **9,** 165–237. INTA, Buenos Aires, Argentina.

Adler, M. *et al.* (1974). "Relevamiento Ecológico del Borde Occidental de los Llanos de la Rioja." III Reunión Argentina de Ecología, Puerto Madryn, Chubut, Argentina.

Aguilar, F. (1970). "Las Lomas Costeras del Perú." Univ. Nac. Agraria La Molina, Lima. Congreso Internacional de Zonas Aridas, Arica, Chile.

Alliney, J. E. *et al.* (1978). "Efecto del Tamaño y del Número de las Muestras para Medir Producción en el Pastizal Natural de San Luis." IV Reunión Argentina de Ecología, Corrientes, Argentina.

Altieri S., M. A., and Rodriguez M., J. A. (1974). "Acción Ecológica del Fuego en el Matorral Natural Mediterráneo de Chile, en Rinconada de Maipu." V Reunión Nacional para el Estudio de las Regiones Aridas y Semiáridas, Mendoza, Argentina.

Ambrosetti, J. A. (1971). Especies interesantes en la ordenación de la cuenca Papagayos. I. *Deserta* **2,** 207–237. IADIZA, Mendoza, Argentina.

Anderson, D. L., del Aguila, J. A., and Bernardon, A. E. (1970). Las formaciones vegetales en la provincia de San Luis. *Rev. Invest. Agropecu., Ser. 2: Biología y Producción Vegetal* **7**(3), 81–87. INTA, Buenos Aires, Argentina.

Anderson, D. L. *et al.* (1971). "La Dinámica y el Manejo de la Pastura Natural en el Centro-Sur de San Luis." Reunión de Programación sobre Pastizales de la Región Semiárida, Villa Mercedes, San Luis, Argentina.

Anderson, D. L., Orionte, E. L., and Vera, J. C. (1974). "Un Relicto del Pastizal de San Luis." IV Reunión Argentina de Ecología, Río Cuarto, Córdoba, Argentina.

Aranda, X. (1971). "Un Tipo de Ganadería Tradicional en el Norte Chico, la Trashumancia." Centro Demostrativo Corral de Julio y Dep. de Geografía, Univ. de Chile, Santiago.

Arze, J. A., and Alencastra, S. J. (1976). "Digestibilidad de los Granos de Quinua (*Chenopodium quinua* Wild) y Canihua (*Chenopodium pallidicaule* Aellen) en Ovinos," pp. 166–171. II Convención Inter. Quenopodiáceas, IICA. Inf., Conf., Cursos Reuniones, No. 96., Potosí, Bolivia.

Arze, J. A., and Sotelo, G. (1976). "Epoca Óptima de Corte de Canihua (*Chenopodium pallidicaule* Aellen) para su Uso como forraje," pp. 155–162. II Convención Inter. Quenopodiáceas, IICA. Inf., Conf., Cursos Reuniones, No. 96., Potosí, Bolivia.

Avendaño, J., Ovalle, C., and Franco, I. (1978). "Un Sistema Ovino Propuesto para la Zona Centro-Sur del Secano-Interior. Descripción y análisis económico." Estación Experimental Quilamapu, Sub-Estación Cauquenes, Inst. Invest. Agropec., Bol. Tec. 13., Chile.

*Badilla, S. I. (1975). "Características Ecológicas y Fitosociológicas de *Atriplex repanda* Phil." Facultad de Agronomía, Univ. de Chile. Tesis Ing. Agrónomo. Santiago.

Bishop, J. P. *et al.* (1968). "Estudio de las Especies Espontáneas Psamofilas de la Región de Bahía Blanca en la Alimentación de Ovinos: Análisis Taxonómico y Nutritivo." III Reunión Nacional para el Estudio de las Regiones Aridas y Semiáridas, Trelew, Chubut, Argentina.

Botti, C. (1970). "Relaciones Hídricas del Tamarugo (*Prosopis tamarugo* Phil.) en la Localidad de Canchones." Facultad de Agronomía, Univ. de Chile. Tesis Ingeniero Agrónomo. Santiago.

Braun, O. (1964). "Forrajeras del Altiplano. Bolivia." Servicio Agrícola Interamericano, Boletín Experimental 30.

*Braun W., R. H., and Candia, R. J. (1980). Poder calorífico y contenidos de nitrógeno y carbono de componentes del algarrobal de Ñacuñán (Mendoza). *Deserta* **6,** 91–99. Mendoza, Argentina.

Braun W., R. H., and Lamberto, S. A. (1974). Modificaciones producidas por incendios en la integración de los componentes leñosos de un monte natural. *Rev. Invest. Agropecu. Ser. 2: Biología y Producción Vegetal* **9**(2), 11–24. INTA, Buenos Aires, Argentina.

*Braun, W., R. H., Verettoni, N. H., Gonzalez, M., and Lamberto, S. (1971). "Unidades de Vegetación en la Región Semiarida del Sur Bonaerense." IV Reunión Nacional para el Estudio de las Regiones Aridas y Semiáridas, Santa Rosa, La Pampa, Argentina.

Braun, W., R. H., McKell, C. M., and Lucero, J. C. (1974). "Eficiencia en el Uso de la Humedad Edáfica y Producción de Forraje en un Monte Quemado." IV Reunión Nacional para el Estudio de las Regiones Aridas y Semiáridas, Mendoza, Argentina.

Braun, W., R. H., Candia, R. J., and Leiva, R. (1978). Productividad primaria aérea neta del algarrobal de Ñacuñán (Mendoza). *Deserta* **5**, 7–43. Mendoza, Argentina.

Braun, W., R. H., and Candia, R. J. (1980). Poder calorífico y contenidos de nitrógeno y carbono de componentes del algarrobal de Ñacuñán (Mendoza). *Deserta* **6**, 91–99. Mendoza, Argentina.

Brücher, E. (1962). Nuevas especies de *Solanum* (Tuberarium) de la zona semiárida del NW Argentino. *Rev. Fac. Cienc. Agrar., Univ. Nac. Cuyo* **9**, 7–14. Mendoza, Argentina.

Brücher, E. (1965). Las papas indígenas en rápida desaparición en la Argentina. *Cienc. Invest.* **21**, 212–215. Buenos Aires.

*Brücher, H. (1968). Südamerika als Herkunftsraum, *In* E. J. Fittkau *et al.* "Biogeography and Ecology in South America," pp. 251–301. The Hage.

Brücher, E. (1973). Las especies tuberíferas de *Solanum* (papas silvestres) como elemento ecológico-florístico de la vegetación semiárida de Mendoza. *Deserta* **4**, 147–159. IADIZA, Mendoza, Argentina.

Brun, J. M., *et al.* (1971). "Relevamiento Ecológico en Áreas Piloto de la Patagonia." Reunión de Programación sobre Pastizales de la Región Semiárida, Villa Mercedes, San Luis, Argentina.

Cabrera, A. L., and Willink, A. (1973). "Biogeografía de América Latina." Programa Regional de Desarrollo Ceintífico y Techológico, Dep. Asuntos Cientificos, OEA, Washington, D.C.

Cadahia, D. (1970). "Informe Sobre el Plan Forestal-Ganadero. Pampa del Tamarugal." Estudio FAO para BID. Corporación de fomento. [Mimeographic report].

*Cairnie, A., and Monesiglio, J. C. (1967). Composición química de especies forrajeras nativas e introducidas en la región semiárida pampeana. *Rev. Invest. Agropecu., Ser. 2* **4**(11), 207–221. INTA, Buenos Aires, Argentina.

Camino, A. (1977). "Monocultivo y Policultivo en las Montañas Tropicales: Un Estudio Preliminar en el Distrito de Cuyo-Cuyo (Provincia de Sandía, Puno)," pp. 44–51. Primer Congreso Inter. Cultivos Andinos, IICA. Reuniones, Cursos Conferencias, No. 178, Ayacucho, Perú.

Campos, S. L. (1968). "Proyecto estudio entomológico del tamarugo. Insectos asociados con el tamarugo en la Pampa del Tamarugal," pp. 180–185. Facultad de Agronomía, Dep. Biol., Univ. de Chile. Memoria Anual, 1967. Santiago.

Canales, G., and Baldomero, O. (1977). "Características Morfológicas Asociadas al Rendimiento en Oca (*Oxalis tuberosa* Mol.) Bajo Condiciones de Allpachaka (3.600 m.s.m.) Ayacucho," pp. 244–247. Primer Congreso Inter. Cultivos Andinos, IICA. Reuniones, Cursos Conferencias, No. 178, Ayacucho, Perú.

Cano, E. (1969). Dinámica de la vegetación de un pastizal de planicie de La Pampa. *Rev. Invest. Agropecu., Ser. 2: Biología y Producción Vegetal* **6**(12), 193–223. INTA, Buenos Aires, Argentina.

Cano, E. (1977). Pastizales en la Región Central de la provincia de La Pampa. *IDIA* **331–333,** 1–15. INTA, Buenos Aires, Argentina.

Cano, E., and Fernández, B. (1974). "Mapa de Vegetación de la Provincia de La Pampa." V Reunión Nacional para el Estudio de las Regiones Aridas y Semiáridas, Mendoza, Argentina.

Cano, E., and Holgado, H. O. (1971). "Cambios en la Vegetación de un Caldenal Despues de un Fuego." IV Reunión Nacional para el Estudio de las Regiones Aridas y Semiáridas, Santa Rosa, La Pampa, Argentina.

Cano, E., and Olmos, A. (1968). "Mapa de Vegetación del Departamento de Loventue, Provincia de la Pampa." III Reunión Nacional para el Estudio de las Regiones Aridas y Semiáridas, Trelew, Chubut, Argentina.

Cano, J., and Rosas, M. (1976). "Evaluación Cualitativa y Cuantitativa del Comportamiento de Siete Variedades de Quinua," pp. 106–112. II Convención Inter. Quenopodiáceas, IICA. Inf., Conf., Cursos Reuniones, No. 96, Potosí, Bolivia.

*Cano, J. *et al.* (1971). "Cambios en la Vegetación de un Caldenal Despues de un Fuego." IV

Reunion Nacional para el Estudio de las Regiones Arida y Semiáridas, Santa Rosa, La Pampa, Argentina.
*Cardenas, M. (1969). "Manual de las plantas económicas de Bolivia." Cochabamba.
*Cardenas, M. (1971). "El Altiplano Como un Sistema Ecológico," pp. 56–65. Primera Reunión sobre Pastura de los Andes Altos, La Paz, Bolivia.
*Cardozo, A. (1970). "Bibliografía Boliviana de Pastos y Forrajes." Ecuador. Inst. Interamericano de Ciencias Agricolas, Programa de Investigacion, Bol. Informativo 1(1), Suplemento Técnico 2, Turrialba, Costa Rica.
Cardozo, G., Romero, A., Sotomayor, E., and de Navarro, O. T. (1976). "El Cultivo de la Quinua en el Ecuador y Colombia," pp. 26–30. II Convención Inter. Quenopodiáceas, IICA. Infor., Conf., Cursos Reuniones, No. 96, Potosí, Bolivia.
Castaño, E. *et al.* (1978). "Diseño de un Sistema para el Desarrollo de los Recursos Naturales de las Lomas de la Costa Peruana." Proyecto Cooperativo en Ejecución de Oficina de Investigaciones de Tierras Aridas, Univ. of Arizona y Univ. Agraria La Molina, Lima, Perú.
Catalán, N. (1973). "Invernada de Novillos Hereford en Pradera Natural de Secano con Suplementación y Termino de Engorda en Dry-Lot con Urea." Facultad de Agronomia, Univ. de Chile. Tesis Ing. Agronomo.
Chaquilla, O. (1976). "Exploración de Herbicidas en el Cultivo de Quinua (*Chenopodium quinoa* Wild.)," pp. 127–129. II Convención Inter Quenopodiaceas, IICA. Inf., Conf., No. 178, Ayacucho, Perú.
Christiansen, J. (1977). "Las Papas Amargas: Fuente de Calorías y Proteinas en los Andes," pp. 201–203. Primer Congr. Inter. Cultivos Andinos, IICA. Reuniones, Cursos Conferencias, No. 178, Ayacucho, Perú.
Concha, R. (1975). "Consumo y Ganancia de Peso Ovino Durante el Período Primavera-Verano en una Pradera Natural Biestratificada con *Atriplex repanda*." Facultad de Agronomía, Univ. de Chile. Tesis M.S. Santiago.
Contreras, D., and Gastó, J. (1978). "Comparación de Arquitecturas Fitocenósicas en el Secano Mediterráneo de Chile." Facultad de Agronomía, Univ. de Chile, Santiago. [Mimeographic report].
de Cornejo, Z. (1976). "Hojas de Quinua (*Chenopodium quinoa* W.) Fuente de Proteina," pp. 177–189. II Convención Inter. Quenopodiácea, IICA. Inf., Conf., Cursos Reuniones, No. 96, Potosí, Bolivia.
Cortez, H. (1977). "Avances en la Investigación en Oca," pp. 227–243. Primer Congr. Inter. Cultivos Andinos, IICA. Reuniones, Cursos Conferencias, No. 178, Ayacucho, Perú.
Cosio, F. (1970). "Informe Técnico Anual. Proyecto de Praderas de Secano y Producción Animal, Area Sur." INIA, Est. Exp. Quilamapu, Chillán, Chile.
Cosio, F., and Vicens, J. (1969). "Efectos de la Fertilización de Fósforo de Mantención en Pradera Mejorada sobre la Producción de Lana." XX Jornadas Agronómicas, Chillán, Chile.
Covas, G. (1971). "Arboles y Arbustos Forrajeros Nativos en la Provincia de La Pampa." IV Reunión Nacional para el Estudio de las Regiones Aridas y Semiáridas, Santa Rosa, La Pampa, Argentina.
*Dawson, G. (1960). Los alimentos vegetales que América dio al mundo. *Mus. La Plata, Ser. Tecn. Didact.* **8,** 1–68. La Plata.
De Anchorena, J. (1973). "Relevamiento de vegetación en el Sudoeste del Chubut." II Reunión Argentina de Ecología, Salta, Argentina.
De La Puente, F. (1977). "Desarrollo Tecnológico del Cultivo de la Papa en el Perú," pp. 183–200. Primer Congreso Inter. Cultivos Andinos, IICA. Reuniones, Cursos Conferencias, No. 178, Ayacucho, Perú.
Del Aguila, J. A., Bernardon, A. E., and Anderson, D. L. (1969). Contribución al estudio de los pastizales naturales de Los Llanos de La Rioja. *Rev. Invest. Agropecu., Ser. 1: Biología y Producción Animal* **6**(7), 81–87. INTA, Buenos Aires, Argentina.
Diaz, H. B. *et al.* (1972a). Estudio de las pasturas naturales e implantación de forrajeras cultivadas

en zonas ganaderas del noroeste Argentino (región semiárida). *Rev. Agrón. Noroeste Argent.* **9**(1), 32–53. Tucumán, Argentina.

Diaz, H. B. *et al.* (1972b). Determinación de la digestibilidad de las especies forrajeras naturales mas comunes y de algunas cultivadas en la zona semiárida del noroeste Argentino. *Rev. Agrón. Noroeste Argent.* **9**(1), 55–68. Tucumán, Argentina.

Digilio, A. P. *et al.* (1971). "Flora Chaqueña." Reunión de Programación sobre Pastizales de la Región Semiárida, Villa Mercedes, San Luis, Argentina.

Di Castri, F., and Mooney, H. A. (1973). Mediterranean type ecosystems. Origin and structure. *In* "Ecological Studies," Vol. 7. Springer, New York.

Di Marco, O. N. (1973). "Consumo y Preferencia Ovina Estacional al Aumentar la Intensidad de Pastoreo de un Bioma Biestratificado con *Atriplex repanda.*" Facultad de Agronomía, Univ. de Chile. Tesis M.S. Santiago.

*Duplancic, A. M., Gonzalez, S., Rey, E., and Avellaneda, M. (1981). Valor nutritivo de las especies forrajeras comunes de la zona precordillerana del centro de Mendoza. *IV Jornadas de Invest., Univ. Nac. Cuyo, Mendoza.*

Earls, J. (1977). "La Coordinación en la Producción Agrícola en el Tawantinsuyo," pp. 52–78. Primer Congreso Inter. Cultivos Andinos, IICA. Reuniones, Cursos Conferencias, No. 178, Ayacucho, Perú.

Elgueta, S. (1971). "Estudio del Tamarugo (*Prosopis tamarugo* Phil.) como Productor de Alimento para el Ganado Lanar en la Pampa del Tamarugal." Instituto Forestal, Informe Técnico 38, Santiago, Chile.

Elgueta, S., and Calderón, S. (1970). "Estudio del Tamarugo como Productor de Alimento de Ganado Lanar en la Pampa del Tamarugal Informe Técnico." Julio, 1970. Instituto, Forestal, Santiago, Chile.

Etienne, G., de la Caviedes, R., and Contreras, T. (1978). "Nuevo Enfoque en la Evaluación de la Productividad de las Praderas." Facultad de Agronomía, Univ. de Chile, Santiago. [Mimeographic report].

Feldman, I. (1972). "Ecología de la Pampa Central o Distrito del Caldenal. Manejo y Reemplazo del Tapiz Vegetal Leñoso." Ia. Reunión Argentina de Ecología, Vaquerías, Córdoba, Argentina.

Ferrer, I., Silva, E., and Barriga, R. (1977). "Evolución de la Distribución Proteica en una Especie con Mecanismo de Fotosíntesis C-4: *Atriplex repanda.*" IX Jornadas Chilenas de Química, Soc. Chilena de Química y Univ. de Chile, Jahuel, Chile.

Fester, G. A. *et al.* (1960). Variedades fitoquímicas en plantas aromáticas Argentinas. *Rev. Fac. Cienc. Agrar.* **8**(2), 45–49. Univ. Nac. Cuyo, Mendoza, Argentina.

Flores, M. A. (1967). "Contribución al Estudio de los Pastos del Altiplano Peruano, Perú." Oficina Nacional de Reforma Agraria, Oficina de Catastros y D.R., Lima, Perú.

Gandarillas, H., and Tapia, G. (1976). "Requerimientos de Fertilizantes en la Quinua; Altiplano Central de Bolivia," pp. 116–123. II Convención Inter. Quenopodiáceas, IICA. Inf., Conf., Cursos Reuniones, No. 96, Potosí, Bolivia.

Gastó, J., and Caviedes, E. (1976). Interferencia intraespecífica de *Atriplex repanda* en el secano Mediterráneo de Chile. Facultad de Agronomía, Univ. de Chile. *Bol. Tec.* **41,** 3–18. Santiago.

Gastó, J., and Contreras, D. (1972a). Análisis del potencial pratense de fanerófitas y caméfitas en regiones Mediterráneas de pulviometría limitada. Facultad de Agronomía, Univ. de Chile. *Bol. Tec.* **35,** 30–61. Santiago.

Gastó, C., and Contreras, D. (1972b). Bioma pratense de la región Mediterránea de pulviometría limitada. Facultad de Agronomía, Univ. de Chile. *Bol. Tec.* **35,** 3–29. Santiago.

Gastó, J., and Contreras, D. (1979). Fertilización nitrogenada y precipitación en la productividad de la pradera anual natural del secano Mediterraneo de Chile central. *Avances en Producción Animal* **4**(2), 111–128.

*Gastó, J., and Olivares, A. (1979). Análisis cuantitativo de la arquitectura de *Atriplex repanda* Phil. *Ciencia e Investigación Agrícola* **6**(2), 105–113. Santiago, Chile.

Giordano, O. S. (1979). "Productos Naturales de Compuestas de las Región de Cuyo." VIII Seminario Latinoamericano de Química, Buenos Aires, Argentina.

*Gomez, I., and Morello, J. (1972). "Incremento de la Vegetación Leñosa en el Chaco. El Papel del Vinal (*Prosopis ruscifolia*)." I Reunión Argentina de Ecología, Vaquerías, Córdoba, Argentina.

Gomez, I., Malverez, I., Morello, J., and Gazia, N. (1974). "Efectos de un Herbicida sobre la Problación de *Prosopis ruscifolia.*" III Reunión Argentina de Ecología, Puerto Madryn, Chubut, Argentina.

Gomez, I. *et al.* (1973). Caracterización estructural de poblaciones de Vinal (*Prosopis ruscifolia* Gris.). *Rev. Invest. Agropecu., Ser. 2: Biología y Producción Vegetal* **4,** 143–150. INTA, Buenos Aires, Argentina.

Gomez, V. (1966). "Aspectos Generales sobre los Pastos Naturales en las Zonas Alto-Andinas de la Sierra Peruana," pp. 99–101. Inst. Veter. Inv. Tropic. y de Altura, Bol. Extraordinario, Lima, Peru.

González, B. G. (1967). Algunos aspectos sobre la situación de pastos y forrajes en Ecuador. *In* "Sexta Conferencia Interamericana de FAO sobre Producción y Salud Animal." Gainesville, Florida.

González, B. G. (1969). "Recomendaciones Generales para el Cultivo de Pastos en la Sierra." Inst. Nac. Invest. Agropecu., Bol. Divulgativo SC-69-12, Colombia.

*González, N., and Haart, E. (1960). "Análisis de Muestras Diversificadas de *Prosopis tamarugo.*" Facultad de Medicina Veterinaria, Univ. de Chile, Informe interno, Santiago, Chile.

Grez, P., and Fernandez, M. (1978). "Encaste Prematuro de Borregas Merino Precoz Francés, Sometidas a Diferentes Planos Nutricionales." Facultad de Agronomía, Univ. de Chile. Tesis Ing. Agr. Santiago.

*Guevara, J. C., Candia, R. J., Mendez, E., and Roig, F. A. (1973). Modificaciones florísticas y producción forrajera invernal del estrato herbáceo de Ñacuñán en un año anormalmente lluvioso. *Deserta* **4,** 125–139. IADIZA, Mendoza, Argentina.

Guevara, J. C. *et al.* (1978). "Inventario de los Recursos Pastorales de la Provincia de Mendoza." VI Reunión Argentina de Ecologia, Corrientes, Argentina.

Hajek, E., and Saiz, F. (1976). Aplicación del método de la sacarosa a la caracterización microclimática del Parque Nacional Fray Jorge, Chile. *An. Mus. Hist. Nat., Santiago, Chile* **9.**

Hechenleitner, K. M. (1973). "Suplementación Proteica a Vaquillas de Carne en Pradera Natural de Secano Durante la Estación Estival," p. 75. Facultad de Agronomía, Univ. de Chile. Tesis Ing. Agr. Santiago.

Hnatyszyn, D. *et al.* (1974). Estudio fitoquímico de plantas indígenas Argentinas. *Rev. Invest. Agropecu. Ser. 2: Biología y Producción Vegetal* **11**(1), 15–23. INTA, Buenos Aires, Argentina.

Ignacio, Q., Fernandez, A., and Cortes, G. (1976). "Contribución al Estudio Morfológico del Grano de Quinua," pp. 58–68. II Convención Inter. Quenopodiáceas, IICA. Inf., Conf., Cursos Reuniones, No. 96, Potosí, Bolivia.

INIA (1977). "Lucha Contra la Desertificación: la Experiencia Chilena." Informe preparado por INIA para UNESCO, Santiago, Chile.

Instituto Ecuatoriano de Recursos Hidráulicos (1977). "El Problema de la Desertificación en la Provincia de Manabi, Ecuador," p. 19. Ministerio de Agricultura, INERHI, Quito, Ecuador.

Instituto Forestal (1971). "Estudio del Tamarugo como Productor de Alimento del Ganado Lanar en la Pampa del Tamarugal." Instituto Forestal, Bol. 38, Santiago, Chile.

Johnson, W. L., Zeppilli, R., and Delgado, D. (1968). "Respuesta a Diferentes Niveles de Alimentación de Vacas Pardo Suizas en Altura." Segunda Reunión Latinoamericana de Producción Animal, Lima, Perú.

*Johnston, M., and Olivares, A. (1976). "Condiciones de Cama de Semilla y Profundidad de Siembra en la Germinación de *Atriplex repanda.*" XXVII Jornadas Agronómicas, Santiago, Chile.

*Josifovich, J., Maddaloni, J., Serrano, H., and Echeverria, I. (1982). Areas forrajeras y de producción animal en la Argentina. *Inf. Técnico* No. 169. EERA Pergamino; INTA, Pergamino.

Klein, K. (1970). "Evaluación de la Producción de Frutos en Tamarugos Adultos del Bosque Junoy, Temporada 1969–1970," p. 12. CORFO, Tarapaca. Laboratorio de Entomología, IN-CONOR, Canchones, Chile.

Klein, K., and Campos, L. (1977). "Biocenosis del Tamarugo (*Prosopis tamarugo* Phil.) con Especial Referencia a los Artrópodos y Fitófagos y Sus Enemigos Naturales," pp. 86–108. *Zeitschrift für Angewandte Entomologie,* Sonderdruck aus Bol. 85 (1977). Verlag Paul Parey, Hamburg, Germany.

Kummerow, J. (1966). Aporte al conocimiento de las condiciones climáticas del bosque Fray Jorge. Facultad de Agronomía, Univ. de Chile. *Bol. Tec.* **24.** Santiago.

Lagomarsino, E. D. *et al.* (1974). Control químico de un renoval en zonas ganaderas del N.W. Argentino. *Rev. Agron. Noroeste Argent.* **11**(3–4), 241–256. Tucuman, Argentina.

Lailhacar, K. (1976). "Effect of Soil Parameters on the Components of Biomass Production in *Atriplex polycarpa* (Torn.) Wats, and *Atriplex repanda* Phil." Univ. of California, Davis. Ph.D. Thesis.

Lamagdelaine, L. (1974). "Antecedentes sobre Forestación y Ganadería en la Pampa del Tamarugal y en el Altiplano de Tarapacá." Depto de Ciencias Sociales, Univ. de Chile, Iquique.

Lamberto, S. A., and Braun, R. H. (1974). Cambio en el estrato bajo de un monte natural inducidos por incendios. *Cienc. Invest.* **30,** 327–333. Buenos Aires, Argentina.

Lanino, R. I. (1966). "Comparación de Tres Razas Ovinas Alimentadas con Tamarugo (*Prosopis tamarugo* Phil.) en la Pampa del Tamarugal." Facultad de Agronomia, Universidad de Chile. Tesis Ingeniero Agrónomo. Santiago, Chile.

Lanino, R. I. (1972). "Instrucciones Generales para la Siemba y Plantación de Tamarugo (*Prosopis tamarugo* Phil.)." [Mimeographic report].

Lanino, R. I. (1976). "Informe Preliminar Sobre el Cultivo de Quinua en el Altiplano Chileno, Zona de Isluga," pp. 24–25. II Convención Inter. Quenopodiaceas, IICA. Inf., Conf., Cursos Reuniones, No. 96, Potosí, Bolivia.

Lara, M. (1972). "Pastos Naturales del Altiplano de Bolivia." LBTA, Bol. 13, La Paz, Bolivia.

*Lara, R. R. (1976). "Algunos Caracteres Sistemáticos Exomorfológicos de Importancia para una Clasificación de Tipos Agronómicos de Quinuas Cultivades." pp. 51–57. II Convención Inter. Quenopodiáceas, IICA. Inf., Conf., Cursos Reuniones No. 96, Potosí, Bolivia.

Latour, C. M. (1971). "Nuevo Estudio sobre Gramineas Forrajeras en la Patagonia." IDIA 283. INTA, Buenos Aires, Argentina.

Ledesma, N. R. (1978). "La Degradación Ecológica del Ecosistema en el Chaco." VI Reunión Argentina de Ecología, Corrientes, Argentina.

León, J. C. (1974). "Estudio Comparativo de Espolvoreos y Pulverizaciones para el Control de Insectos del Tamarugo (*Prosopis tamarugo* Phil.)" Facultad de Agronomia, Univ. de Chile. Tesis Ingeniero Agrónomo. Santiago.

Leon, J. C. (1973). "Delimitacion de Comunidades en el Pastizal Puntano y Sus Relaciones con el Pastores." II Reunión Argentina de Ecología, Salta, Argentina.

Lescano, J. L. (1976). "Cariotipo y Poliploidia en Canihua," pp. 81–88. II Convención Inter. Quenopodiáceas, IICA. Inf., Conf., Cursos Reuniones, No. 96, Potosi, Bolivia.

Lescano, J. L., and Palomino, C. (1976). "Metodología de Cruzamiento en Quinua." pp. 78–80. II Conv. Inter. Quenopodiáceas, IICA. Inf., Conf., Cursos Reuniones, No. 96, Poposí, Bolivia.

*López Ocaña, C. E. (1978). 'Informe del Proyecto Especial "Desarrollo de Tierras Aridas y Semiáridas" en el Perú.' IIIa. Reunión de Cordinación, OEA, Washington, DC.

Lewis, J. P. (1973). ''El Espinal Periestépico.'' II Reunión Argentina de Ecología, Salta, Argentina.

Lowe, Ch., Morello, J., and Goldstein, G. (1972). ''Estructura y Variación de la Vegetación en los Desiertos del Monte y de Sonora. Proyecto: Origen y Estructura de Ecosistemas.'' Programa Biológico Internacional, I Reunión Argentina de Ecología, Vaquerías, Córdoba, Argentina.

*Luti, R., Galera, F. M., and Bertran, M. A. (1974). ''Vegetacion de las Provincia de Cordoba: Una Primera Aproximacion.'' IIIa. Reunión Argentina de Ecología, Puerto Madryn, Chubut, Argentina.

Lutz, E. E., and Graff, A. B. (1974). ''Evaluación de las Pasturas Naturales Resultantes de la Quema Controlada de Monte en la Región Semiárida Pampeana.'' V Reunión Argentina para el Estudio de las Regiones Aridas y Semiáridas, Mendoza, Argentina.

MAB-3 (1977). ''Descripción, Funcionamiento y Transformación de los Ecosistemas Agropastorales en Bolivia, Chile y Perú para el Benficio del Hombre.'' Proyecto de Cooperación presentado a UNESCO por Comité MAB-3 de Bolivia, Chile y Perú.

Magoja, J. L. (1975). La composición mineral y su variación estacional de tres forrajeras nativas en La Pampa: *Stipa tenuis* Phil., *Poa ligularis* Nees *ex* Stendel y *Piptochaetium napostaense* Speg. Hacker ag. Stuckert. *Rev. Invest. Agropecu. Ser. 2: Biología y Producción Vegetal* **12**(1), 27–48. INTA, Buenos Aires, Argentina.

Manterola, H., Olivares, A., and Borquez, F. (1979). Efecto de la suplementación con nitrógeno no proteico y la presencia del arbusto *Acacia caven* en la utilización de la pradera natural Mediterránea. *Avances en Produccion Animal* **4**(2), 129–134. Santiago, Chile.

Mateu Amengual, B., and Villa Carenzo, M. (1971). ''Catálogo Bibliográfico Fitoquímico Argentino,'' Vol. II. Miscelanea No. 36. Univ. Nac. de Tucumán, Tucumán, Argentina.

Mediondo, M. E. *et al.* (1973). Estudio fitoquímico de plantas indígenas Argentinas. *Rev. Invest. Agropecu. Ser. 2: Biología y Producción Vegetal* **10**(3), 137–141. INTA, Buenos Aires, Argentina.

Montes, A. L. (1965). ''Produccion de plantas aromaticas nativas,'' pp. 103–111. Reunion de Programacion de Plantas Aromaticas, IDIA, No. 211. INTA, Buenos Aires, Argentina.

Morales, R. A. (1976). ''Control de Plagas en Cultivo de Quinua,'' pp. 131–133. II Convención Inter. Quenopodiáceas, IICA. Inf., Conf., Cursos Reuniones, No. 96, Potosí, Bolivia.

Morello, J. H. (1970). Modelo de relaciones en tres pastizales y leñosas colonizadoras en el Chaco Argentino. *IDIA* **276,** 31–52. Buenos Aires, Argentina.

Morello, J. H., and Adamoli, J. (1972). ''Vegetacion y Ambiente del Chaco (Argentina).'' Ia. Reunion Argentina de Ecología, Vaquerias, Cordoba, Argentina.

Morello, J. H., and Adamoli, J. (1973). Subregiones ecológicas de la provincia del Chaco. *Ecología* **1**(1), 29–33. Buenos Aires, Argentina.

Morello, J. H. *et al.* (1971). ''Vegetación y Ambiente en la Provincia de Salta.'' Reunión de Programación sobre Pastizales de la Región Semiárida, Villa Mercedes, San Luis, Argentina.

Morello, J. H., and Gomez, I. (1971). ''Uso de Sensores en el Inventario de Recursos y Problemas de Leñosas en el Norte Argentino.'' Reunión de Programación sobre Pastizales de la Región Semiárida, Villa Mercedes, San Luis, Argentina.

Morello, J. H., Lowe, Ch., Goldstein, G., and Luti, R. (1972). ''Un Chaparral de Sensibilidad Estival en el Centro de la Argentina.'' I Reunión Argentina de Ecología, Vaquerías, Córdoba, Argentina.

Morello, J. H., Sancholuz, L. A., and Blanco, C. A. (1974). ''Estudio Macroecológico de Los Llanos de La Rioja.'' V Reunión Nacional para el Estudio de las Regiones Aridas y Semiáridas, Mendoza, Argentina.

Muñoz, S. (1972). ''Evaluación de la Proporción de Fruto en Tamarugos Adultos.'' CORFO, Informe Interno, Canchones, Chile.

Narrea, R. A. (1976). "La Producción de Quinua en el Perú," pp. 31–34. II Convención Inter. Quenopodiáceas, IICA. Inf., Conf., Cursos Reuniones, No. 96, Potosí, Bolivia.

*Nava, C., Armijo, R., and Gastó, J. (1979). "Ecosistema. La Unidad de la Naturaleza y el Hombre." Serie Recursos Naturales, Univ. Auton, Agraria "A. Narro," Saltillo, México.

Negron, A. A., Alvarez, E., and Calmet, E. (1976). "La Quinua y la Canihua en Raciones de Pollos Parrilleros en Puno, Perú," pp. 172–176. II Convención Inter. Quenopodiáceas, IICA. Inf., Conf., Cursos Reuniones, No. 96, Potosí, Bolivia.

*Neumann, E. (1978). "Variación Estacional del Contenido de Caroteno en la Pradera Mediterranea y Su Efecto en el Consumo de Forraje." Facultad de Agronomía, Univ. de Chile. Tesis M.S. Santiago, Chile.

*Neumann, R. (1972). "Productividad en el Chaco Seco, Provincia de Salta." Ia. Reunión Argentina de Ecología, Vaquerias, Cordoba, Argentina.

Neumann, R. (1973). "Perfiles Estructurales de Leñosas y Suculentas del Chaco Semiárido, Provincia de Salta." II Reunión Argentina de Ecología, Salta, Argentina.

Olaechea, V., Nico, R., and Escalante, M. (1964). Sustitución de "regaliz" por raíz y rizomas de *Glycyrrhiza astragalina*. *Rev. Farm.* **106**(5–6), 1–7. La Plata, Argentina.

*Olivares E., A., and Gasto, J. (1971). "Comunidades de Terofitas en Subseres Postaradura y en Exclusion en la Estepa de *Acacia caven* (Mol.) Hook et Arn." Facultad de Agronomía, Univ. de Chile. *Bol. Tec.* **34,** 3–24. Santiago, Chile.

Olivares E., A., and Gasto, J. (1978). Función de cosecha por ovinos de la pradera anual Mediterránea de Chile. *Avances en Producción Animal* **4**(1), 45–54.

Olivares E., A., and Johnston, B. M. (1978). Alternativas de mejoramiento en la emergencia de *Atriplex repanda* (Phil.). Revista de Botanica Internacional, *Øyton* **36**(2), 129–137.

Olivares E., A., and Riveros, E. (1979). Cambios en la composición botánica de la estrata de terófitas residentes de una pradera Mediterránea natural sometida a diferentes épocas y frecuencias de talajeo. *Avances en Produccion Animal* **4**(1), 35–44. Santiago, Chile.

Orionte, Y., and Anderson, D. L. (1976). "Influencia del Fuego en un Area Relicto del Sorgastral." IV Reunión Argentina de Ecología, Río IV, Córdoba, Argentina.

Ortiz Garmendia, J. (1968). "Plantas Tintoreas de las Zonas del Desierto y la Estepa Septentrional Chilenas." Contrib. Arqueol., No. 7. Museo de la Serena, Chile.

Ortiz Garmendia, J. (1969). "Plantas Silvestres Chilenas de Frutos Comestibles por el Hombre." Contribución Arqueol., No. 8. Museo de la Serena, Chile..

Parilo B., J. (1978). "Relaciones Sinecólogicas del Bioma Pratense en Suelos Graníticos de Chile y Productividad Potencial del Ecosistema." Facultad de Agronomía, Univ. de Chile. Tesis M.S. Santiago, Chile.

Pedrani, A. R., and Rodriquez, I. A. (1978). "Siete Años de Evolución en Una Reserva del Piedemonte Árido de Mendoza." VII Reunión Argentina de Ecología, Mendoza, Argentina.

Peralta, P. M. (1977). "Mapa Preliminar de los Procesos y Areas de Desertificación en Chile Continental." Facultad de Ciencias Forestales, Departamento de Silvicultura, Univ. de Chile, Informe Interno.

Pisano, E. (1966). Zonas biogeográficas de Chile. *In* "Geografía Económica de Chile." *Tomo* I, 62–72. CORFO, Chile.

Posnansky, M. (1971). "Aspectos Ecológicos sobre Pastos Nativos del Altiplano," pp. 24–31. Primera Reunión sobre Pasturas de los Andes Altos, La Paz, Bolivia.

Ragonese, A., and Cano, E. (1971). "La Vegetacion de la Provincia de La Pampa." Reunión de Programación sobre Pastizales de la Región Semiárida, Villa Mercedes, San Luis, Argentina.

*Ratera, E. L., and Ratera, M. O. (1980). "Plantas de la Flora Argentina Empleadas en Medicina Popular." Hemisferio Sur, Buenos Aires.

Rique, T. (1974a). "Aplicaciones Industriales de Extractivos de Especies Forestales Indígenas de las Zonas Áridas y Semiáridas del País." V Reunión Nacional para el Estudio de las Regiones Áridas y Semiáridas, Mendoza, Argentina.

Rique, T. (1974b). "Recursos Adicionales para Pobladores de Zonas Áridas y Semiáridas, Aprovechando Extractivos de Leguminosas Forestales." V Reunion Nacional para el Estudio de las Regiones Áridas y Semiáridas, Mendoza, Argentina.

*Riveros, E., Neumann, E., Olivares, A., Manterola, H., and Ramirez, R. (1978). Variaciones estacionales en el contenido de caroteno y proteina de la pradera natural y del forraje consumido por ovinos en ecosistemas semiáridos. *Avances en Produccion Animal* **3**(1), 23–30. Chile.

*Riveros, F. A., Avila, G., Aljaro, M. E., Araya, S., Hoffman, A. E., and Montenegro, G. (1976). Comparative morphological and ecophysiological aspects of two sclerophylous Chilean shrubs. *Flora* **165,** 223–234. Chile.

Riveros, V. E. (1977). "Influencia de las Variaciones Estacionales de Caroteno de la Pradera sobre el Contenido Plasmático de Caroteno y Vitamina "A" y sobre la Producción de Lana y Fertilidad de Machos Ovinos en Ecosistemas Semiáridos." Facultad de Agronomía, Univ. de Chile. Teses M.S. Santiago, Chile.

Rodríguez, D. (1978). "Influencia del Momento de Utilizacion de la Pradera Natural de la Zona Mediterranea Central de Chile en el Consumo Ovino." Facultad de Agronomía, Univ. de Chile. Tesis M.S. Santiago, Chile.

*Roig, F. A. (1964). Las gramineas mendocians del genero *Stipax. Rev. Fac. Cienc. Agrar.* **11**(1–2), 3–110. Univ. Nac. de Cuyo, Mendoza, Argentina.

*Roig, F. A. (1970). Flora y vegetacion de la Reserva Forestal de Nacunan. *Deserta* **1,** 25–232. IADIZA, Mendoza, Argentina.

Roig, F. A. (1972). Investigaciones climaxicas. II. Los pastizales disclimaxicos del Melocoton (Mendoza) y nuevas observaciones sobre la biologia de *Schinus polygamus. Deserta* **4,** 173–184. IADIZA, Mendoza, Argentina.

*Roig, F. A. (1976). Las comunidades vegetales del piedemonte de la precordillera de Mendoza. *Ecosur* **3**(5), 1–45. Buenos Aires, Argentina.

*Roig, F. A., and Ambrosetti, J. A. (1971). Investigaciones climaxicas. I. Restos de un estrato arboreo bajo de *Schinus polygamus* en la precordillera de Mendoza. *Deserta* **2,** 115–130. IADIZA, Mendoza, Argentina.

Roig, F. A. (1972). Investigaciones climáxicas. II. Los pastizales disclimáxicos del Melocotón (Mendoza) y nuevas observaciones sobre la biología de *Schinus polygamus. Deserta* **4,** 173–184. IADIZA, Mendoza, Argentina.

*Roig, F. A. *et al.* (1972). "Investigaciones Climáxicas. II. Los Pastizales disclimácicos del Melocotón, Mendoza y Nuevas Observaciones sobre la Biología de *Schinus polyganus.*" II Reunión Argentina de Ecologia, Salta, Argentina.

Rojas, W. (1978). "Relaciones Sinecológicas del Bioma Pratense en Suelos de Terraza Marina de Chile y Productividad Potencial del Ecosistema." Facultad de Agronomía, Univ. de Chile, Tesis M.S., Santiago, Chile.

Rondina, R. V. D. *et al.* (1971). Estudio fitoquímico de plantas indígenas Argentinas. III. *Rev. Invest. Agropecu. Ser. 2: Biología y Producción Vegetal* **8**(1), 29–33. INTA, Buenos Aires, Argentina.

Ruiz Leal, A. (1965). Notas fanerogámicas mendocinas. II. *Rev. Fac. Cienc. Agrar.* **12**(2), 181–200. Univ. Nac. Cuyo, Mendoza, Argentina.

Ruiz Leal, A. (1972a). Flora Popular Mendocina. *Deserta* **3,** 3–296. IADIZA, Mendoza, Argentina.

Ruiz Leal, A. (1972b). Los confines boreal y austral de las provincias patagónica y central respectivamente. *Reedición Especial del Suppl. Bol. Soc. Arg. Bot.* **13,** 89–118. IADIZA, Mendoza, Argentina.

Ruthsatz, B. (1977). Pflanzengesellschaften und ihre Lebensbedingungen in den Andinen Halbwusten Nordwest-Argentiniens. *Dissertationes Botanicae* **39.** J. Cramer, Vaduz, Liechtenstein.

Santo, R., *et al.* (no date). "Comportamiento de Especies Forrajeras Nativas e Introducidas en el Oeste de La Pampa." Hoja informativa, INTA, Rama Caida, San Rafael, Mendoza, Argentina.

*Saravia, T. C. (1973). "Informacion Ecologica Cuantitativa del Efecto de Pastoreo y Explotacion Forestal en Bosques de Quebracho en la Provincia de Salta." II Reunion Argentina de Ecología, Salta, Argentina.

Schenckel, G. *et al.* (1971). Exploración de deficiencias nutritivas con suelos en macetas. II. Cálculo de las lineas de fertilidad sobre el diagrama de fertilidad. *Agric. Tec.* **31,** 106–115. Chile.

Schulz, A. G. (1963). Plantas y frutos comestibles de la región chaqueña. *Rev. Agron. Noroeste Argent.* **4**(1), 57–83. Tucumán, Argentina.

Segarra, F. (1979). "Caracterización de la Curva de Crecimiento de Pradera Natural en el Secano Interior Mediteraneo." Facultad de Agronomía, Univ. de Chile. Tesis Ing. Agr., Santiago, Chile.

Segura, B. M. (1976). Métodos de evaluación de pasturas naturales. *In* "Programación de la Investigación," pp. 33–47. Programa Andes Altos, IICA–OEA, La Paz, Bolivia.

Sejzer, D. (1973). Variación del caracteres estructurales y funcionales en comunidades vegetales chaqueñas. *Ecología* **1**(1), 25–28. Buenos Aires, Argentina.

Serra, J. (1970). Pasturas en Tierra del Fuego. *IDIA* **272,** 7–14. INTA, Buenos Aires, Argentina.

Silva S., E., and Pereira, C. (1976a). Aislación y composición de las proteinas de hojas de *Atriplex nummularia* y *Atriplex repanda. Cienc. Invest. Agrar.* **3,** 169–174. Santiago, Chile.

Silva S., E., and Pereira, C. (1976b). Concentrados proteicos de hojas de *Atriplex nummularia. Cienc. Invest. Agrar.* **3,** 153–157. Santiago, Chile.

Soriano, A. (1973). "Efecto de Diecinueve Años de Clausura sobre el Pastizal de *Stipa* del Oeste de la Provincia de Chubut." II Reunión Argentina de Ecología, Salta, Argentina.

Soriano, A. (1974). "La Producción de Fitomasa en el Pastizal de *Stipa speciosa* del Sudoeste del Chubut." V Reunión Nacional para el Estudio de las Regiones Áridas y Semiáridas, Mendoza, Argentina.

Soriano, A. *et al.* (1978). "Vegetación Actual y Vegetación Potencial en el Pastizal de Coiron Amargo en el S.W. de Chubut. Tendencias Registradas en Hipótesis Propuestas." VI Reunión Argentina de Ecología, Corrientes, Argentina.

Soriano, A., and Brun, J. (1973). Valorización de campos en el centro-oeste de la Patagonia: desarrollo de una escala de puntaje. *Rev. Invest. Agropecu., Ser. 2: Biología y Producción Vegetal* **10**(5), 173–186. INTA, Buenos Aires, Argentina.

Souto, J., and Eilberg, B. A. (1972). Efecto de extractos de hojas de vinal (*Prosopis ruscinfolia* Griseb.) sobre la germinación y el crecimiento de plántulas de varias especies cultivadas. *Rev. Invest. Agropecu., Ser. 2: Biología y Producción Vegetal* **9**(1), 19–27. INTA, Buenos Aires.

Suarez de Castro, F. (1977). "La Desertificación en América Latina desde una Perspectiva Ecológica y Agrícola." Reunión Preparativa Regional para América, IICA, Febrero (1977), Santiago, Chile.

*Sudzuki, H. F. (1969). "Absorcion Foliar de Humedad Atmosférica en Tamarugos (*Prosopis tamarugo* Phil.)." Estación Experimental Agronómica, Bol. Tec. 30, Facultad de Agronomía, Univ. de Chile, Chile.

Sudzuki, H. F. (1975). "Captación y Economía del Agua en Plantas que Viven en Ambientes de Desierto." Estación Experimental Agronómica, Bol. Tec. 40, Facultad de Agronomía, Univ. de Chile, Santiago.

Sudzuki, H. F., Botti, C., and Acevedo, E. (1973). "Relaciones Hídricas del Tamarugo (*Prosopis tamarugo* Phil.) en la Localidad de Canchones." Estación Experimental Agronómica, Bol. Tec. 37, Facultad de Agronomía, Univ. de Chile, Santiago.

Tapia, M. E. (1976). "Sistema de Producción Ganadera en Base a Pastizales y *Avena*-Alfalfa, en el Altiplano de Puno," pp. 16–32. Programación de la Investigación, Programa Andes Altos, IICA–OEZ, La Paz, Bolivia.

Tapia, M. E. (1977). "Características de los Sistemas Agrícolas Andinos." pp. 90–103. Primer Congreso Internacional de Cultivos Andinos, Univ. Nacional San Cristóbal-Huamanga, Ayacucho, Perú.

Telleria, W. (1976). "La Quinua y Cañihua en Bolivia," Vol. II, pp. 21–23. Convención Inter. Quenopodiáceas, IICA. Inf., Conf., Cursos Reuniones, No. 96, Potosí, Bolivia.
Telleria, W., and Ballón, E. (1976). "Componentes de Rendimiento en Quinua." Vol. II, pp. 96–101. Convención Inter. Quenopodiáceas, IICA. Inf., Conf., Cursos Reuniones, No. 96, Potosí, Bolivia.
Tinto, J. C. (1974). "Recursos Forrajeros Leñosos para Zonas Áridas y Semiáridas." V Reunión Nacional para el Estudio de las Regiones Aridas y Semiáridas, Mendoza, Argentina.
*Toro, J. M. (1967). "Desarrollo Radical del Tamarugo." Facultad de Agronomía, Univ. de Chile. Informe Interno. Santiago.
*Toursarkissian, M. (1980). "Plantas medicinales de la Argentina." Hemisferio Sur, Buenos Aires.
Trivelli R., A. (1973). "Fertilización, Época y Frecuencia de Utilización en la Bioma Biestratificada de *Atriplex repanda* Phil. y Terófitas Residentes." Facultad de Agronomía, Univ. de Chile. Tesis M.S. Santiago.
United Nations (1977). "World Map on Desertification." FAO, UNESCO, OMM.
Uslar, R. (1972). "Estudio de la Efectividad de Algunas Cepas Nativas de *Rhizobium trifolii* en Simbiosis con *Trifolium subterraneum*." Facultad de Agronomía, Univ. de Concepción. Tesis Ing. Agrónomo. Chillán, Chile.
Vargas, V. (1978). "Manual para Preparar Medicamentos a Base de Hierbas." Ier. Seminario Nacional sobre Tecnologías Adecuadas, Ayacucho, Perú.
Vicens, O. J., and Cosio, F. (1968). "Resumen de Investigación en Praderas de Secano en el Campo Experimental Collipulli." INIA, Chile. [Inédito].
Vicens, O. J., and Cosio, F. (1970). "Productividad Potencial de Gramíneas con Distintos Niveles de Nitrógeno en el Secano Central." INIA, Chile. [Unpublished].
Viirsoo, E. V. (1967). Comportamiento de especies silvestres tuberíferas y de formas autóctonas del género *Solanum* en las Montañas de Catamarca. *IDIA* **231,** 16–31. INTA, Buenos Aires, Argentina.
*de Vizcarra, I. A. (1972). "Bibliografia de Pastos y Forrajes de los Andes Altos." IICA, La Paz, Bolivia.
Vorano, A. E., and García, R. C. A. (1976). "La Quinua (*Chenopodium quinoa*) en la Provincia de Jujuy. Informe Preliminar de Su Situación Actual y Perspectivas Futuras." pp. 19–20. II Convención Inter. Quenopodiáceas, IICA. Inf., Conf., Cursos Reuniones, No. 96, Potosí, Bolivia.
*Wainstein, P. (1974). "Valor Nutritivo de Plantas Forrajeras del este de la Provincia de Mendoza," Vol. III. V. Reunión Nacional para el Estudio de Zonas Aridas y Semiáridas. I Encuentro de la Zona Arida Latinoamericana. Mendoza, Argentina.
*Wainstein, P., and Gonzalez, S. (1962). Valor forrajero de trece especies de *Stipa* de Mendoza. *Rev. Fac. Cienc. Agrar.* **9**(1), 3–18. Univ. Nac. de Cuyo, Mendoza, Argentina.
Wainstein, P. *et al.* (1974). "Valor Nutrilivo de Plantas Forrajeras del Oeste de la Provincia de Mendoza," Vol. III. V Reunión Nacional para el Estudio de Zonas Áridas y Semiáridas. I Encuentro de la Zona Árida Latinoamericana, Mendoza, Argentina.
Wainstein, P., Gonzalez, S., and Rey, E. (1979). Valor nutritivo de plantas forrajeras de la provincia de Mendoza. III. *Cuaderno Técnico* **2,** 97–108. IADIZA, Mendoza, Argentina.
Wernli, K., Silva, C. M., and Trucco, A. (1976). "The Nutritive Value of Range Pastures in Chile," pp. 149–157. I Mediterranean Grasslands of Central Chile. Proc. First International Symposium Feed Consumption, Utah State Univ., Logan.
Zuccardi, R. (1974). "El Ecosistema de Los Valles Calchaquies." III Reunión Argentina de Ecología, Puerto Madryn, Chubut, Argentina.
Zuñiga F., M. (1973). "Determinación de Curvas de Crecimiento en *Atriplex repanda* Phil. en Función de las Densidad Poblacional." Facultad de Agronomía, Univ. Católica de Chile. Tesis Ing. Agr. Santiago.

8

UNION OF SOVIET SOCIALIST REPUBLICS

N. T. Nechayeva
Desert Institute
Ashkhabad, USSR

I. INTRODUCTION AND PHYSIOGRAPHY

The arid area of the Union of Soviet Socialist Republics occurs primarily in the southern part of the country and consists of two zones: the plains and the mountains. The plains include the northern desert, southern desert, and foothill desert subzones; the mountain zone, characterized by vertical relief, consists of mountains of low, medium, and high altitude. River valleys are intrazonal. These subzones will be discussed later in more detail. [For locations, summaries, and further information concerning these areas, see Fig. 1, Table I, and Nikolaev *et al.* (1977).]

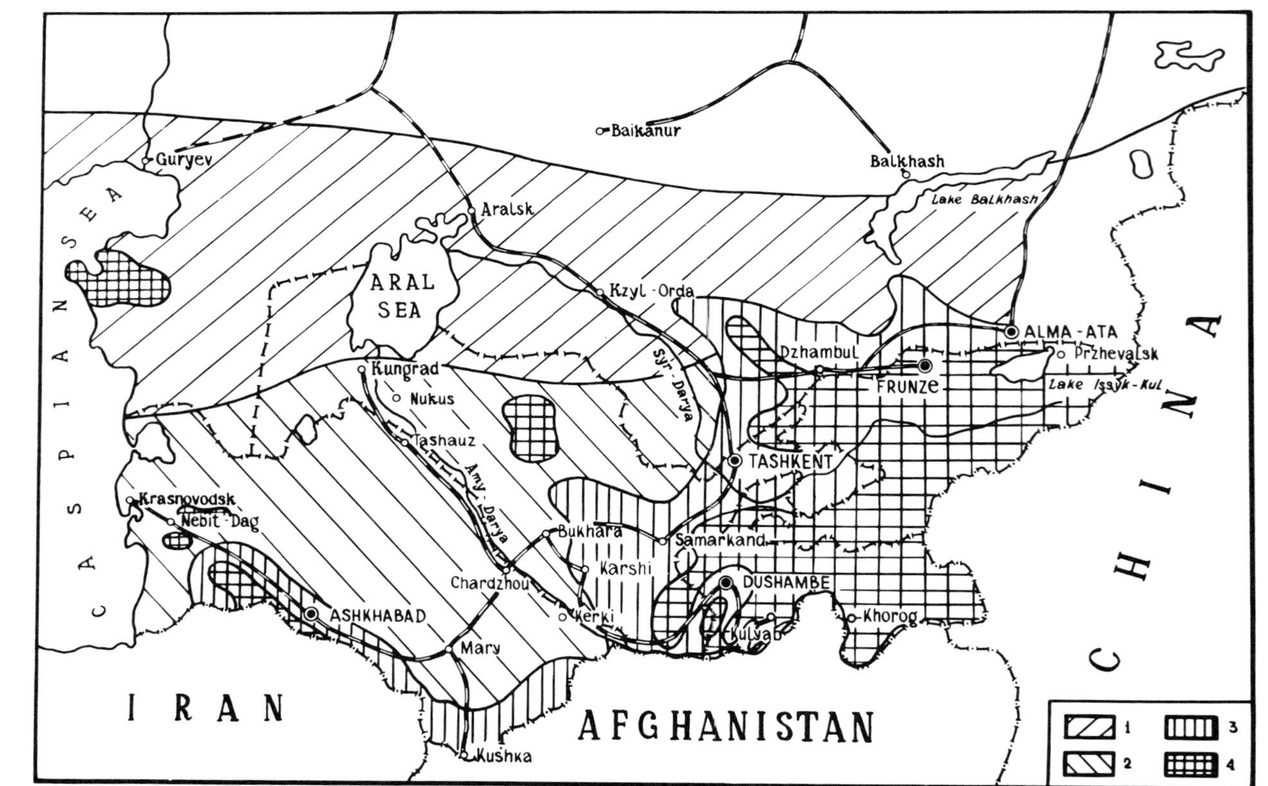

Fig. 1. Ecological profile of ridges and takyr-like plains in the central part of the Kara-Kum Desert. Association No. 2: *Calligonum rubens*, *Mausolea eriocarpa*, and *Carex physodes* on the slopes of high ridges. Association No. 3: *Salsola arbuscula*, *Artemisia kemrudica*, and *Carex physodes* on hummocky sands. Association No. 4: *Salsola gemmascens*, *Artemisia kemrudica*, and *Gamanthus gamocarpus* on takyr-like plains.

Table I

Areas (in Thousand Square Kilometers) of Principal Desert Types in Middle Asia and Southern Kazakhstan[a]

	Desert types						
Subzones	Sandy	Gypsum	Clay	Stony	Loess	Type combinations	Total
Northern desert	147.1	216.3	140.7	—[b]	—	24.7	528.8
Southern desert	301.0	163.8	45.9	—	—	55.3	566.0
Foothill desert	—	—	—	10.3	94.0	—	104.3
River and lake valleys	—	—	—	—	—	—	29.4
	448.1	380.1	186.6	10.3	94.0	80.0	1228.5

[a] Nikolaev *et al.* (1977).
[b] Not represented.

A. Plains Zone

1. NORTHERN DESERT SUBZONE

The altitude above sea level does not exceed 600 m. The mean annual temperature ranges from 7.6°C in the town of Guriev to 7°C at the Aral sea station. The mean monthly air temperature ranges from 25.4 to 26.3°C in July, and from −10.4 to 13.5°C in January. The total mean monthly temperature accumulation above 10°C is 3000 to 3700°C per year.

The mean annual precipitation in Guriev is 16.3 cm, and in the Aral sea region it is 10.3 cm. The precipitation is distributed regularly throughout the different seasons. Evaporation is high, exceeding the total annual precipitation 9- to 10-fold.

The main soil types are the sandy, gray-brown, and shallow clays, which are poor in humus and frequently highly alkalized. Ground waters are not deep.

2. SOUTHERN DESERT SUBZONE

The elevations of areas in this subzone are 400 m or less. The mean annual air temperature ranges from 13.2 to 16.1°C; the mean temperature for the hottest month (July) ranges from 29 to 30°C. Summers are long, hot, cloudless, and without precipitation. Winters are short and unstable, with a wide variety of temperatures. The mean annual January temperature ranges from 2.4 to −4.1°C. The total monthly temperature accumulation above 10°C ranges from 4700 to 5162°C.

Mean annual precipitation ranges from 9.7 to 11.3 cm, with the maximum precipitation recorded in winter and early spring. This is beneficial to plants, particularly the abundance of annual and perennial herbs with a winter–spring growth. A deep penetration of precipitation sustains shrubs and semishrubs despite a very high evaporation rate, which exceeds the annual precipitation value 26-fold.

Both saline and fresh groundwaters are found at varying depths, most frequently deeper than 10 m, making them unavailable to plants.

The soils are sandy, gray-brown alkalized, and white alkalized with gypsum intercalations, heavy and takyr-like, and are typical takyrs containing considerable concentrations of chlorides and sulfates (Lobova, 1960). All the above soil types are slightly humic, structureless, and salted, except the sands. Takyrs with a compact surface ensure rainfall runoff. Runoff from large takyrs is used to freshen water in wells and to construct small freshwater ponds for livestock watering sites. On the borders of takyrs, favorable conditions are created for the cultivation of melons.

3. FOOTHILL DESERT SUBZONE

This subzone is situated at an altitude of 400 to 1000 m above sea level and is hilly, with valley-like depressions. Considerable elongation of the foothills accounts for a wide range of climatic indices. Generally, the climate is more favorable here than in typical deserts. The mean annual temperature is 14°C; the mean monthly temperature of July is 26°C and that of January, ~0°C. The cumulative total of mean monthly temperatures higher than 10°C ranges from 2500 to 4560°C.

The mean annual precipitation of 16 to 26 cm occurs primarily in early spring and winter. A considerable amount of moisture is conducive to the formation of a thick turf of perennials, preventing the spread of shrubs. Evaporation exceeds the total yearly precipitation five- to eightfold.

The thin soil cover of the highest hills is slightly developed and stony. Lower gentle ranges are covered with unsalted, light, and typical sierozem. Groundwaters are largely salted and deep.

B. Mountain Zone

1. LOW MOUNTAIN SUBZONE

This subzone ranges from 600 to 800 m above sea level. Climatically, these are arid, sharply continental areas with a mean annual precipitation between 40 and 60 cm, timed largely to winter and spring. The mean annual temperature is 14.8°C (Dushanbe). The warmest month is July (28°C), and the coldest is January (1.4°C).

Groundwater occurs at varying depths, and livestock water comes mainly from brooks, small rivers, which often dry up in summer, and wells.

Soil cover is diverse. On light chestnut and chestnut soils, gramineous ranges with *Bothriochloa ishaemum, Stipa, Festuca,* and *Poa* predominate. Shallow, brown, and sometimes stony soils underlie fruticulose (*Salsola laricifolia*) and dwarf semishrub (*Artemisia, Convolvulus, Ceratoides*) ranges. The ranges are used in spring, summer, and fall.

2. MIDMOUNTAIN SUBZONE

The altitude of this area is 1000 to 2000 m above sea level. It is moderately dry with a mean annual precipitation of ~80 cm. The bulk of the precipitation falls in winter and spring; summer is thus characterized by a moisture shortage. The winter is unstable and warm.

Soil cover diversity, associated with mountain slope exposure, accounts for the heterogeneity of the vegetative cover. Dark sierozems and brown soils underlie gramineous ranges of *Elytrigia, Festuca,* and *Bromus;* dark chestnut and chernozem-like soils underlie stands of *Dactylis, Stipa, Poa, Festuca,* and *Elytrigia.* These grass openings occur in open woodlands with *Juniperus, Acer, Juglans, Cotoneaster,* and others. Dry slopes typically are covered with shibliak of *Amygdalus, Rhus, Celtis, Zizyphus,* and *Pistacia.* These ranges are used in spring, summer, and fall.

3. HIGH MOUNTAIN SUBZONE

This subzone is situated at a height of 2200 to 3300 m above sea level. The mean annual temperature ranges from 2 to 7°C. Winter is long and cold, with persistent snow cover; summer is cold. The mean annual precipitation is 120 cm, and the vegetation is herbaceous. The high mountain meadows, steppes, and heathlands that make up this subzone are used as summer ranges.

II. DEMOGRAPHY

The population density of the Soviet Union arid zone is highly heterogeneous. Around large oases, the rural population density is more than 100 people per square kilometer, but generally it is from 50 to 100 people per square kilometer. In the deserts, density ranges from 1 to 10 people per square kilometer, but more frequently it is less than 1 person per square kilometer.

III. SOCIOECONOMIC FACTORS

With many different peoples inhabiting the Soviet Union's arid zone, the following Soviet socialist republics were created: the Kazakh SSR in 1920, and the Kirghiz SSR, Tajik SSR, Turkmen SSR, and Uzbek SSR in 1924. These republics are independent, with each having its own plan of the multibranch economy. The republics are in turn united by the Soviet Union's integrated plan of economic development.

The organization of these independent republics created favorable conditions for a rapid advance in the economy and culture of each. Their natural resources, industry, and agriculture have been developed extensively. The existing agricultural units are state and collective farms. The state farms largely specialize in either plant growing or husbandry; the collective farms, as a rule, combine the two. Range husbandry—astrakhan and wool sheep breeding, and camel breeding—is carried out primarily in arid territories; plant breeding of cotton, cereals, and forage plants occurs on irrigated lands in oases. Cereals are also grown on unirrigated lands in the mountains and in the foothills. The construction of large irrigation systems, for example, Golodnaya steppe in the Soviet Union, Nurek Dam in the Tajik SSR, and Karakum Canal in Turkmenia, has increased the irrigated land area. The Karakum Canal created favorable conditions for organizing specialized gardening and vine-growing farms and also for cultivating medicinal and other economically important plants. The development of large-scale irrigation projects continues. The Karakum Canal is being extended; projects are being developed for the construction of several dams and for the transfer of part of the runoff of some Siberian rivers. However, even if these projects are completed, only 15% of the heathlands could be irrigated in the near future. Hence, the focus of attention is toward a fuller and more rational utilization of desert plant resources, specifically the plants that can be used for forage, range, firewood, and such.

IV. PLANT RESOURCES

The Soviet Union arid zone flora (middle Asia and southern Kazakhstan) features ~7000 plant species, of which 1600 are desert species and 5400 are mountain species (Korovin, 1962). This flora is rich in a variety of plants with diverse applications. In some respects, arid zone plants are of specific interest; for instance, many are richer in essential oils than the same or similar species found in other natural zones. There also occurs in the arid region a number of plants containing a considerable amount of alkaloids. There is also a very large number of tinctorial plants, yielding a wide range of dyestuffs. It was not long ago that carpet and colored felt (*koshma*) manufacture was dependent on resistant dyestuffs of plant origin. Additionally, the river valleys contain large amounts of raw material for the paper–cellulose industry.

The favorable climatic conditions found in a number of mountain ranges with mesic gorges are ideal for thickets of fruit trees of many species and forms. These areas are also ideal for improving garden species and varieties (Blinovsky and Mizgireva, 1971).

Vast plains areas, which support arid vegetation, are ranges for sheep and camels. Desert forage resources are the basis for valuable products of husbandry, such as wool, astrakhans, and meat. These products are thus cheaper in these areas compared to other natural zones.

A. Food Plants

About 150 plant species in arid middle Asia and Kazakhstan are used, without processing, for food.

1. TREES AND SHRUBS

About 50 species of wild fruit trees and shrubs have been recorded; most are found in mountain areas.

Juglans regia L., a large tree with a life span of a few hundred years, is distributed in mountain valleys. Wild forms produce thin-walled fruits up to 5 cm in diameter. These plants are equal to the best cultivated varieties and can be used without additional breeding; the kernel constitutes up to 52% of the fruit, with an oil content of 73%.

The fruits are used directly for food and for the production of oil, which is used for food, in perfumery, and for manufacturing painting colors. The green pericarp is rich in tannins—up to 25%—and dyestuffs.

Pistacia vera L. is a tree 5–6 m tall, with a thick, short trunk and a wide crown. These trees form open woodlands reminiscent of savannas. This highly drought resistant plant is found in low mountains at an elevation of 600 to 800 m.

Pistacia vera is a valuable caryocarpous plant. The nuts contain 57% fat, 22% protein, 3% cellulose, and 3% ash; its weight constitutes 50% of the fruit weight, exclusive of the pericarp. The nuts are used for food and in confectioneries. Nut crop yields are low and variable from year to year. Natural *Pistachia* woodlands yield moderate crops every other year and abundant crops once in 3 to 5 years, with large trees each yielding ~45 kg of fruit.

The leaves and galls contain tannin, which is used in medicine. The resin is used in the lacquer industry, and the timber, which is hard, is used in manufacturing turning and joining wares.

Cultivation of *Pistacia vera* began in the 1880s. At present, terrace plantations of *Pistachia* are widely distributed; these play an antierosion, slope-stabilizing role (Klyushkin, 1958).

The genus *Amigdalus* contains several species that occur naturally: *A. communis* L., *A. bucharica* Korsh., *A. turcomanica* Lincz., *A. Spinosissima* Bge., *A. scoparia* Spach., and *A. vavilovii* M. Pop. These species are trees and, less frequently, shrubs and form large groves in mountain gorges and on slopes in the lower and middle zones.

The fruits are bitter; sweet fruit and thin cortical forms are rare. The nucleus accounts for 14 to 55% of the fruit weight, with an oil yield of 49 to 52%. The oil and extraction wastes (oil cake) are used in the pharmaceutical and perfumery industries. Highly drought resistant, it is of interest for hybridization purposes and is used as a rootstock for cultivated almond trees and other nutshell species (Blinovsky and Mizgireva, 1959).

Punica granatum is a large shrub that grows in mountain gorges, on dry slopes, and also is cultivated. Varying in taste, from acid to sweet, and size, up to 10 cm in diameter, the fruits are used directly for food, or the juice is extracted for the preparation of soft drinks and liqueurs. The bark of the trunk, branches, and roots, and flowers contain alkaloids and have medicinal use. The pericarp, and the bark of the trunk, branches, and roots yield a tanning agent as well as citric and acetic acids. Young leaves are used for the preparation of a tea substitute. *Punica granatum* often gets frostbitten, but is soon restored.

Berberis turcomanica Karel. is distributed in mountain regions and frequently forms undershrubs. The fruits are used to prepare jam, sweets, fruit flour, a coffee substitute, and soft drinks. Leaf infusion is used as a tonic. *Berberis turcomanica* yields dyestuffs.

Pyrus L. grows in the middle part of mountain ranges on slopes and in gorges. There are several species: *P. boissieriana* Bushe., *P. communis* L., *P. turcomanica* Maleev., and *P. bucharica* Litv. The fruits are invariably small and yellow with a large number of grit cells. Cultivation and selective breeding produces larger fruits with juicy, palatable flesh. Some cultivated species, such as *P. turcomanica,* yield good-quality dried products. *Pyrus* serves as a drought-resistant rootstock for cultivated varieties, and some species, such as *P. bucharica,* are used as ornamental plants.

Malus rutcomanica Juz. is a small tree 3–4 m high, with numerous shoots, which form extensive underbrush. The fruits are small, although there also are large-fruited trees that can be cultivated. Seedlings of wild apple trees with large fruits yield a wide range of fruits with respect to shape, color, and taste; this holds much promise for future breeding.

Prunus divaricata L. is a shrub ~2 m high, occasionally becoming a tree 3–4 m high, with abundant trunk shoots. The fruits are dark red, rosy, or yellow. They contain 5–7% sugars, 4–7% citric acid, and vitamin A. The fruits are used for food and for preparation of jams and vitamin-rich extractions. *Prunus divaricata* L. is a good drought-resistant rootstock for peach and plum trees.

Ficus carica L. occurs as a bush and sometimes as a tree 5–6 m high. The fruits vary in quality from nearly inedible to economically important: the yellow-fruited fig, by dry weight, contains 50% glucose and fructose; the green-fruited fig contains 48% fructose and sucrose. Economically important varieties are cultivated. *Ficus carica* L. is easily frostbitten, but frost-resistant forms do exist.

Crataegus pontica C. Koch. is distributed in the middle and lower zones. The fruits are juicy and palatable, with an apple odor, range in shape from ribbed and flat to rounded, and are up to 3 cm in diameter. Crop capacity is up to 20 kg per tree.

Crataegus pseudoazarolus M. Pop. has large red fruits. This and a number of other species, including *C. hissarica* Pojark, and *C. turkestanica* Pojark., are of importance in the future breeding of large-fruited garden forms.

Vitis silvestris Gmel. is distributed in mountain valleys. Often, it grows with escaped cultivated vines of *V. vinifera* L., and numerous crosses are produced. Many authors have reported the presence of both wild (*V. silvestris* Gmel. and *V. hissarica* Vass.) and escaped (*V. vinifera*) vines (Popov, 1929). The berries from wild vines are highly variable in shape, size, coloration, and taste. Along with small-fruited, slightly acid forms with low sugar content, there also occur valuable forms that are suitable for cultivation. The grape cluster weight of these forms ranges from 25 to 485 g; berry weight, from 1 to 3.9 g; sugar content, from 17 to 22% and, rarely, up to 28%; and acidity, from 5 to 20% (Solovyev, 1955). These forms are characterized by drought and cold resistance and show promise for breeding and hybridization. These plants can also serve as excellent stabilizers of steep slopes.

Rubus sanguineus Friv. and *R. caesius* L. occur on the banks of rivers and brooks and form brush. The fruits can be eaten both fresh and dried, and are used for preparing jams and drinks. The leaves yield a good tea substitute.

In addition to the above-described plants, there occur, in both wild and cultivated forms, other useful food-producing shrubs and trees, such as *Eleagnus angustifolia* L., *E. orientalis* L., *Zizyphus jujuba* Mill., *Cerasus verrucosa* (Franch.) Nevski., and *Sorbus persica* Hedl.

2. HERBS

More than 100 wild herb species are known to be used for food, either fresh or cooked. Only the most popular species are listed below.

Eremurus olgae Rgl. and other species of this genus are most frequently served as boiled and seasoned rosettes of young leaves.

Various species of *Allium* are used, including both the bulbs and above-ground parts; these include *A. cepa* L., *A. vavilovi* M. Pop., *A. sabulosum* Stev., and *A. sativum* L.

Some herb species are used in various soups, sauces, and marinades; this group includes *Rumex tuberosus* L., *R. acetosa* L., *Spinacea oleracea* L., and *Capparis spinosa* L. Used as seasoning are *Apium graveolens* L., *Mentha longifolia* (L.) Huds., *M. regarica* Juz., *Cichorium intybus* L., *Spirorhynchus sabulosus* Kar. and Kir., and others. Regarded as a delicacy in some remote desert regions are baked young stalks of *Ferula assa-foetida* L. In addition, when cooked on a slow fire, peeled stalks yield a dark, sweet material similar to honey.

Table II

Vegetative Cover of the Karakum Desert in the Southern Desert Subzone

Biomorph	Number of biomorphs	%	Number of families	%	Number of species	%
Trees	4	6.9	3	5.2	5	1.8
Shrubs	9	15.5	5	8.6	28	10.1
Dwarf shrubs	4	6.9	3	5.2	4	1.4
Semishrubs	5	8.6	4	6.9	14	5.1
Dwarf semishrubs	10	17.2	6	10.3	25	9.0
Perennial herbs	19	32.9	14	24.2	60	21.6
Biannual herbs	1	1.7	1	1.7	2	0.8
Annual herbs	6	10.3	22	37.9	139	50.2
	58	100.0	58	100.0	277	100.0

B. Forage Plants and Forage Resources

About 1600 species of vascular plants exist in the desert zone, but depending on geomorphology, climate, and soils, the number of species in certain regions may be much smaller, ranging from 300 to 500. Because most of the species in a given area are present in low numbers or are not favored by livestock, the range diet normally consists of ~100 plants. Of greatest significance for grazing are the following families: Chenopodiaceae, Compositae (Asteraceae), Leguminosae (Fabaceae), Cruciferae (Brassicaceae), Gramineae (Poaceae), Cyperaceae, Polygonaceae, and Umbelliferae (Apiaceae). Within this group is a total of 50 genera, which are important as forage plants. The majority of plants are consumed on ranges, both during active growth and when dry. Many species, however, are consumed only in a dry state.

The vegetation cover of desert ranges includes diverse life forms, which range from trees to annuals; their numbers and ratios are presented in Table II.

Although the herbs rank highest with respect to the number of species, they are much less evident than shrubs and semishrubs, which dominate the desert vegetation cover.

The highest amount of ecological differentiation within a biomorph is shown by arborescent and semiarborescent plants, perennials rank second (60 species are represented by 19 biomorphs), and annuals are the least differentiated (139 species are represented by only 6 biomorphs).

1. IMPORTANT FORAGE PLANTS

a. Shrubs. Important components in the diet of sheep and camels, shrubs are most widely distributed in the southern subzone, where they grow almost year-round. Shrubs are consumed by livestock both in fall and winter and have high nutritional value. Shrubs such as *Haloxylon persicum, H. aphyllum, Salsola richteri,* and *Ephedra strobilacea* are important forages for the fall–winter season.

Shrubs with spring–summer growth, such as *Astragalus paucijugus* C. A. Mey, and the polymorphous genus *Calligonum* (~50 species), enrich the spring and early summer ranges. Because different species of *Calligonum* differ in ecological requirements, they are widely distributed. A number of species dwell on poorly stabilized sands (*Calligonum rubens, C. microcarpum* Borszcz., *C. caputmedusae*); others, such as *C. setosum, C. alatum,* and *C. aphyllum* (Pall.) Gurke, occur on well-stabilized, compact sands. One species, *C. junceum* (Fisch. and C. A. Mey.) Litv., even grows on heavy, takyr-like soils. Heavy, partly biannual woody branches bearing fruit are consumed by animals, as are fallen fruits.

In addition to serving as valuable livestock forage, shrubs provide firewood and aid in sand stabilization.

b. Semishrubs. Widely distributed in desert ranges, these biomorphs dominate the range vegetation on gypsum and clay deserts. In the southern subzone these consist of a number of sagebrush species, such as *Artemisia arenaria, A. deserti, A. kemrudica,* and *A. terrae-albae,* and saltworts such as *Salsola gemmascens, S. orientale,* and *Anabasis salsa.*

The role of semishrubs, such as *Astragalus turcomanicus, A. ammodendendron,* and *A. villosima,* is notable in sandy, breakstone plains.

Semishrubs are more widely distributed in the northern desert subzone, where monotonous ranges are created by the domination of two to three species of semishrub. The most important species are *Artemisia terrae-albae, A. uzbekistanica, Anabasis aphylla, A. salsa, Salsola gemmascens,* and *Nanophyton erinaceum.*

In the foothill zone, semishrub vegetation is found in stony areas with outcrops of parent rocks. These are predominantly sagebrushes such as *Artemisia badhysi* and *A. turanica.*

c. Perennial Herbs. These frequently are dominant in the plant cover of many range types. In the sandy desert, *Carex physodes* grows everywhere on stabilized sands. This highly nutritive sedge is favored by livestock and forms the basis of the diet of sheep. *Carex physodes* contributes much to sand stabilization (Karycum, Kyzylkum) due to its powerful system of rhizomes and roots. On the sands of the northern subzone, *Agropyrum sibiricum* (Willd.) P.B. is found.

In the foothill deserts, the vegetation cover is dominated by the perennials *Carex pachystylis* and *Poa bulbosa;* together with the annuals they form a continuous carpet.

d. Annual Herbs. Although annuals do not dominate the ranges, they are readily adapted to various ecological conditions, and the high number of species serves to enrich the ranges and diversify the diet of livestock.

Table III

Forage Evaluation of Principal Plants

Plant	Consumption by sheep[a]				Nutritive value of 10 kg of forage in Soviet forage units[b]			
	Spring	Summer	Fall	Winter	Spring	Summer	Fall	Winter
Shrubs								
Aellenia subaphylla var. *arenaria* (C. A. Mey.) Allen	2	3	5	4	44	82	68	51
Calligonum caput-medusae Schrenk.	4	5	2 (fall)	3 (fall)	80	47	—[c]	—
C. rubens Mattei.	4	5	2	3	76	75	58	51
C. setosum Litv.	4	5	2	3	79	63	50	33
Ephedra strobilacea Bge.	3	1	3	5	86	61	46	53
Haloxylon aphyllum (Minkw.) Iljin.	1	0	4	3	59	47	64	59
H. persicum Bge.	1–2	2	5	4	71	72	71	50
Salsola arbuscula Pall.	2	4	3	2	51	53	56	55
S. richteri Kar.	2	0	4	3	69	60	78	46
Semishrubs								
Astragalus unifoliolatus Bge.	5	2	2	3	92	63	53	45
Convolvulus divaricatus Rgl. and Schmalh.	4	4	2	3	69	34	41	34
C. korolkovii Rgl. and Schmalh.	3	3	2	3	72	61	—	35
Mausolea eriocarpa (Bge.) Poljak.	4	5	3	3	62	53	53	45
Salsola gemmascens Pall.	0	0	4	5	80	67	75	54
S. orientalis S. G. Gmel.	2	3	4	5	—	38	41	53
Artemisia kemrudica Krash.	3	2	4	4	72	74	35	45
Smirnovia turkestanica Bge.	4	3	2	3	55	53	44	45
Perennial herbs (summer–fall vegetative period)								
Aeluropus litoralis (Gouan.) Parl.	3	4	3	4	55	53	48	45
Alhagi persarum Boiss. and Buhse.	3	3	4	3	95	65	63	47
Aristida karelinii (Trin. and Rupr.) Roshev.	0	0	2	3	28	68	40	33
A. pennata Trin.	1	1	3	3	58	71	36	29
Heliotropium arguzioides Kar. and Kir.	1	3	3	3	80	86	63	40

H. dasycarpum Ledeb.	1	3	2	2	92	89	—	—
Tournefortia sogdiana (Bge.) M. Pop.	1	3	3	—	79	83	64	—
Perennial herbs (winter–spring vegetative period: ephemerals)								
Crex physodes Bieb.	5	5	4	4	98	52	52	38
Cousinia schistoptera Yuz.	3	4	2	2	93	41	36	34
Iris longiscapa Ledeb.	4	—	—	—	97	—	—	—
Annual herbs (summer–fall vegetative period)								
Agriophyllum latifolium Fisch. and Mey.	0	0	3	3	73	82	40	32
A. minus Fisch. and Mey.	0	0	3	3	82	84	43	34
Climacoptera lanata (Pall.) Botsch.	0	0	4	5	81	42	57	48
Euphorbia ceirolepis Fisch. and Mey.	3	5	4	2	90	93	86	53
Gamanthus gamocarpus (Moq.) Bge.	0	0	5	5	97	58	47	46
Salsola paulseni Litv.	1	2	3	4	72	77	52	56
S. sclerantha C. A. Mey.	0	1	4	4	70	76	70	49
Annual herbs (winter–spring vegetative period: ephemerals)								
Acantholepsis orientalis Less.	4	2	2	3	52	36	33	16
Alyssum desertorum Stapf.	5	2	—	—	91	72	—	—
Arnebia decumbens (Vent.) Coss. and Kral.	5	3	3	—	58	42	47	—
Bromus tectorum L.	5	4	3	3	95	41	43	33
Delphinium camptocarpum Fisch. and Mey.	5	4	3	—	92	64	42	—
Eremopyrum bonaepartis (Spreng.) Nevski., E. orientale (L.) Jaub. and Spach., E. distans (C. Koch.) Nevski	5	4	3	4	90	68	53	41
Isatis violascens Bge.	5	4	3	—	94	46	31	—
Lappula caspia (Fisch. and Mey.) M. Pop.	5	—	—	—	63	—	—	—
Senecio subdentatus Ledeb.	5	4	—	—	73	66	—	—
Spirorrhinchus sabulosus Kar. and Kir.	5	3	—	—	92	71	—	—
Strigosella (Malcolmia) grandiflora (Bge.) Botsch.	5	3	3	—	100	79	63	—
Tetracme recurvata Bge.	5	4	2	—	94	51	56	—

[a] 5, excellent; 4, very good; 3, satisfactory; 2, poor; 1, very poor; 0, not consumed at all.
[b] 1 Soviet forage unit is equal to 1 kg oats (*Avena sativa*) in nutritive value.
[c] Not represented.

The sandy desert is characterized by numerous annuals. The winter–spring vegetation includes the following genera: *Bromus, Eremopyrum, Hypecoum, Tetracme, Malcolmia, Isatis,* and *Koelpinia. Salsola sclerantha, Euphorbia cheirolepis,* and *Agriophyllum latifolium* vegetate in summer and fall.

The gypsum desert is poorer in the number of annual species, but numerous species of *Leptaleum, Senecio, Astragalus, Eremopyrum,* and *Bromus* are found there around moist springs.

In favorable years, the clay desert has many saltworts with a summer–fall vegetation period; these include *Salsola, Gamanthus, Climacoptera,* and *Halimocnemis.* These are excellent fall–winter ranges. Data for the consumption and nutritive value of the main forage-plant species are presented in Table III.

2. PLANT PHENOLOGY

Desert plants vary widely with respect to times and duration of vegetative growth (Table IV). The majority of the arboreous plants have a very long vegetative period. Most possess a deep root system that reaches the different soil layers and provides a small but constant moisture source. Additionally, many of these plants also possess a number of morphological adaptations, such as microphylly, aphylly, and heat and cold resistance. More than half the shrubs, including *Haloxylon* and *Salsola,* grow from early spring to winter. Consumption of these plants often is associated with vegetative stages; the juicy shoots of spring and summer growth are rarely consumed, but the dry, fruit-bearing branches of fall and winter are eaten readily. A number of shrubs, such as *Calligonum* and *Astragalus,* vegetate in spring and early summer; shoots can be eaten green from the plant at that time or at a later time, after the shoots fall. Members of the genus *Ephedra* vegetate all through the year but are consumed mainly in winter and early spring.

Semishrubs fall into three main groups with respect to vegetation times and consumption rates: (1) species vegetating from early spring to late fall include mostly sagebrushes (*Artemisia*), and major consumption occurs in fall and winter, although these plants can be grazed slightly in the spring and summer; (2) species that vegetate from spring to winter are primarily the saltworts (*Salsola orientalis, S. gemmascens,* and *Anabasis salsa*), which are consumed only after growth is completed in the fall and winter; at that time, a rearrangement of the chemical composition takes place in these plants; precipitation is conducive to this process; and (3) semishrubs that vegetate and are consumed mostly in summer and spring, including *Astragalus, Smirnovia, Convolvulus,* and *Kochia.*

The majority of the perennial herbs are ephemeral with respect to their developmental rhythm in that their life activity is pronounced only in winter and spring. Such plants are the Liliaceae and Cyperaceae, which have bulbs, tubers, rhizomes, and other underground storage organs. Perennial herbs such as sedges

Table IV

Vegetative Periods of Plant Life Forms in the Karakum Southern Desert Subzone[a]

Life form	Year-round		Winter–spring		Spring		Spring–early summer		Spring–summer		Spring–summer–fall		Total	
	Number	%	Number	%	Number	%	Number	%	Number	%	Number	%	Number of species	%
Trees									1	0.4	4	1.4	5	1.8
Shrubs	2	0.7					3	1.1	10	3.6	13	4.7	28	10.1
Dwarf shrubs	1	0.4									3	1.1	4	1.4
Semishrubs	1	0.4					6	2.2	4	1.4	3	1.1	14	5.1
Dwarf semishrubs	10	3.6							7	2.5	8	2.9	25	9.0
Perennial herbs			14	5.1	25	9.0			12	4.3	9	3.2	60	21.6
Biannual herbs									1	0.4	1	0.4	2	0.8
Annual herbs	—	.	80	28.8	—	.	31	11.2	15	5.4	13	4.7	139	50.2
	14	5.1	94	33.9	25	9.0	40	14.5	50	18.0	54	19.5	277	100

[a] Nechayeva *et al.* (1973).

are either consumed green or preserved dry and are eaten by livestock throughout the year. Perennials, which vegetate in summer and fall, are highly specialized plants with a xeromorphic structure and a robust root system. These plants grow in specific ecological conditions, such as barchan dunes (*Aristida, Tournefortia,* and *Heliotropium*). The times of consumption differ because of the stiff green leaves; *A. karelinii* and *A. pennata* are consumed only infrequently during winter, when the plants are dry, and specifically after precipitation. *Tournefortia* and *Heliotropium,* which vegetate in summer when other green plants are few, are consumed readily at this time and also later in a dry state.

The majority of annuals vegetate in winter and spring and dry up early from April to early May. These ephemerals are species predominantly from the families Gramineae, Cruciferae, Papaveraceae, and Compositae. They are readily consumed in the vegetative as well as in the dry state. A number of annuals vegetate during the summer and fall hot months. The halophytic *Salsola, Gamanthus,* and *Halimocnemis* grow on salted soils; sclerophytic *Agriophyllum* and *Horaninovia* grow on barchan dunes. These plants are consumed only after vegetative growth is completed in late fall and winter.

Seasonal suitability of ranges is a function of the times of plant vegetation and consumption. Desert ranges are classified as year-round or seasonal (spring–summer, fall–winter).

3. STRUCTURE OF PLANT COMMUNITIES

Plant communities are characterized by various patterns of vertical structure, which is determined mainly by dominant life forms. Different biomorphs are distinguished by the mass and size of the aboveground and underground organs. Trees and shrubs contribute to numerous microenvironmental associations; one example is the "under-crown spot." This, as well as other conditions, helps determine a horizontal heterogeneity of plant cover.

In accordance with the domination of different life forms, desert ranges fall into three principal types:

a. Large Shrub. Composed of nearly all the large-shrub life forms (*Haloxylon* and *Calligonum*), this type also includes middle-sized shrubs (*Salsola, Calligonum,* and *Ephedra*), semishrubs (*Astragalus* and *Mausolea*), perennial herbs, and annuals. In these communities, aboveground and underground organs of the plants are located at different levels, fully utilizing air and soil resources. Crown heights are from 3 to 5 m high, and root depth, from 10 to 20 m.

b. Semishrub. Two to three semishrub species (*Artemisia, Salsola,* and *Anabasis*) and a small number of herbs are in this community. The aboveground organs do not exceed 50 cm in height, and the roots lie at 1.5 to 2 m.

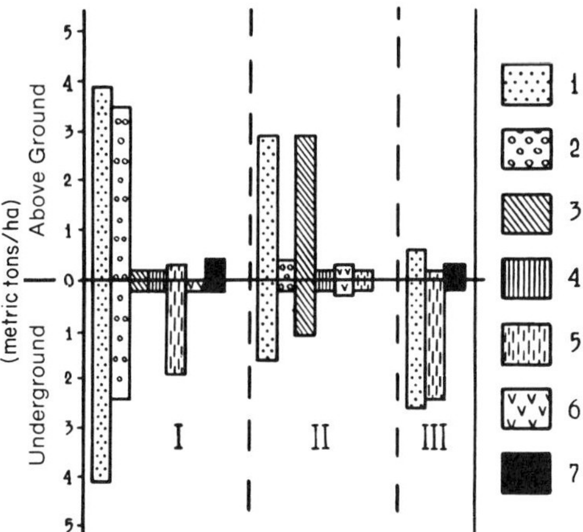

Fig. 2. Dynamic changes of forage yield under the rest regime of grazing (dry weight, metric tons/ha).

c. Herbaceous Ranges. This group includes both perennial (*Carex* and *Poa*) and annual (*Bromus, Astragalus,* and *Malcolmia*) herbs. These communities have the most simple pattern. The aboveground plant parts are normally 25–40 cm high; the bulk of the roots are found in the upper meter of soil.

Large-shrub communities utilize natural resources very efficiently and produce a large biomass; herbaceous phytocoenoses are the least productive (Fig. 2; Table V).

Life-form composition of plants, the pecularities of their vegetation, and livestock consumption are all important for determining the seasonability of ranges; ranges with a considerable proportion of shrubs are classified as year-round, those with a predomination of semishrubs are classified as fall–winter and less frequently as year-round, and herbaceous ranges are largely classified as spring–summer.

Resistance of range forage crops to varying meteorological factors and to grazing is a function of the morphology of aboveground and underground plant organs. All trees, shrubs, and semishrubs have an adequately branched and deep root system. In these plants, the mass of the root system is roughly equal to the mass of the shoots. In perennials, however, the underground organs (roots, rhizomes, and bulbs), in most cases, exceed the aboveground organs in size and mass. For instance, in the forage herbs *Carex physodes* and *C. pachystylis,* the mass of roots and rhizomes is 15–25 times greater than that of the aboveground shoots. In annuals, the roots are shallow, weakly developed, and 5–10 times smaller than the aboveground organs.

Table V

Phytomass Structure (in Metric Tons per Hectare) in Desert Ranges with Different Plant Biomorphs[a]

Species groups	Aboveground			Underground shoots	Total phytomass
	Total yield of shoots (current year)	Perennial parts	Total		
Tall-shrub, sandy deserts					
Haloxyn persicum, Calligonum sp., Carex physodes	0.47	3.28	3.75	4.32	8.07
Calligonum rubens, Mausolea eriocarpa, Carex physodes	0.51	2.09	2.60	4.81	7.41
Low-shrub, sandy deserts					
Salsola arbuscula, Artemisia kemrudica, Carex physodes	0.46	2.34	2.80	3.94	6.74
Dwarf-semishrub, clay deserts					
Salsola gemmscens, Artemisia kemrudica, Gamanthus gamocarpus	0.45	2.80	3.25	1.87	5.12
Dwarf-semishrub, gypsum deserts					
Artemisia turanica, A. diffusa	0.36	0.74	1.10	4.60	5.70
Herbaceous, foothill loess deserts					
Carex pachystillis, Poa bulbosa	0.45	0	0.45	2.47	2.92

[a] Values represent the means of many years (Nechayeva, 1977; Momotov, 1977).

In the aboveground part of trees and shrubs, perennial parts predominate; annual shoots comprise only 6–20% of the entire aboveground plant. In semishrubs, however, 31–40% of the aboveground parts are annual shoots; perennial parts are proportionally less. Thus, the portion of the plant mass that can be grazed on an annual basis is slight in trees and shrubs, much larger in semishrubs, and fully consumed in perennial and annual herbs.

The crop production by shrubs and perennial herbs varies the least and is the least affected by grazing. On the other hand, crop production by annuals varies widely, and these plants are affected by grazing to a correspondingly greater extent. In shrubs with a large perennial mass, such as *Haloxylon, Calligonum,*

Table VI

Yield Fluctuations (in Metric Tons per Hectare) in Years and Seasons in the Southern Desert Subzone

Species groups	Total annual yield: Multiannual means	Total annual yield: Yield fluctuations (years)	Seasonal yields (numerator = gross, denominator = consumed): Spring	Summer	Fall	Winter
Tall-shrub ranges						
Haloxylon persicum, *Calligonum* sp., *Carex physodes*	0.47	0.30–0.75	0.33/0.26	0.39/0.14	0.33/0.16	0.16/0.10
Calligonum rubens, *Mausolea eriocarpa*, *Carex physodes*	0.51	0.31–0.92	0.45/0.35	0.39/0.19	0.21/0.10	0.16/0.08
Low-shrub ranges						
Salsola arbuscula, *Artemisia kemrudica*, *Carex physodes*	0.46	0.20–0.86	0.39/0.28	0.36/0.23	0.24/0.11	0.17/0.09
Semishrub ranges						
Salsola gemmascens, *Artemisia kemrudica*, *Gamanthus gamocarpus*	0.45	0.22–1.05	0.20/0	0.43/0	0.40/0.22	0.28/0.17
Herb ranges						
Carex pachystylis, *Poa bulbosa*	0.45	0.13–1.28	0.45/0.35	0.43/0.26	0.24/0.12	0.18/0.07

and *Salsola*, the weight of the annual shoots or forage crop fluctuates within a range of 50 to 180% in different years with a multiyear mean of 0.19 metric tons/ha. In semishrubs (*Artemisia*, *Salsola*, and *Anabasis*), the values are 50–260% with a mean crop value of 0.1 metric tons/ha. In spring perennial herbs (*Carex*), the range is 70 to 200% with a mean crop of 0.1 metric tons/ha. Annuals vary widely in different years; for example, small winter–spring herb crops range from 12 to 260% with a mean of 0.1 metric tons/ha for grasses and 10 to 400% for herbs. *Salsola* and *Gammanthus* are almost absent from the plant cover in certain years, but occasionally their crops reach a yield of 0.3 metric tons/ha. With a mean of 0.06 metric tons/ha, their crops fluctuate from 17 to 517% in different years. Patterns of crop fluctuation in plants of different biomorphs also manifest themselves in the plant communities or range types where they dominate (Table VI). Years of bad harvest in sand deserts have been recorded in 26% of the cases where shrub ranges predominate, and in 40% of the cases in foothills where herbs are grazed.

The composition of sheep diet in a range is a function of the range type, season, and period of grazing on a given plot. Herb pastures of foothill deserts offer green or dry herbs as a year-round sheep diet.

On semishrub, such as sagebrush–herb ranges, herbs constitute 54% of the sheep diet. Sagebrush constitutes 15% in spring when grazing begins, but after a few days the herbs constitute only 17% of the diet, and sagebrush 53%. In summer the diet contribution of herbs is 33 to 57%, and the share of sagebrush is 22 to 50%. In fall and winter, sagebrush contributes 51–68% by weight.

In tall-shrub ranges of the sandy desert, the bulk of the sheep diet is green or dry herbs, mostly sedge. The share of shrubs in the diet in spring and summer is 13–25%, and in fall, 33–48%. With moderate sheep grazing, only a negligible part of the annual shoots of shrubs and semishrubs is used—in spring and summer 12–19%, and in fall and winter 18–22% of the total mass of shoots. This ensures the preservation of shrubs and semishrubs under grazing conditions.

The daily diet of a karakul sheep in kilograms (dry weight) is as follows: spring, 2.5; summer, 2; fall, 1.8; and winter, 1.7. Therefore, 0.81 metric tons (dry weight) of forage are required per sheep, per year. Because of the fluctuations in pasture yield in different years, a reserve of 15 to 17% of the range area is usually planned. Thus the yearly requirement of range forage is 0.95 metric ton per animal. In years of poor yield, additional feeding of coarse and concentrated fodders is used. The potential capacity of the Soviet Union arid zone is ~34.2 million sheep (Table VII); at present there are ~17 million.

The main livestock species are sheep (mainly karakul) and one-humped camels in the southern subzone, and mutton and wool varieties of sheep and two-humped (bactrian) camels in the northern subzone.

Desert husbandry in the Soviet Union long ago turned from a nomadic to a

Table VII

Capacity of Sheep Ranges in Middle Asia and Southern Kazakhstan[a]

Subzone	Area (thousand km^2) Total	Area (thousand km^2) of ranges	Mean annual forage yield (metric tons/ha)	Total forage reserve (million tons)	Number of sheep (million) that can be maintained
Northern (southern Kazakhstan)	744	546	0.31	17.25	18.2
Southern					
(Turkmen SSR,	491	410	0.13	5.41	5.7
Uzbek SSR)	451	335	0.29	9.74	10.3
	1686	1291		32.40	34.2

[a] Nikolaev *et al.* (1977).

more progressive method of using allotted areas. This has permitted a long-term scheme of range use with up-to-date methods of range irrigation and improvement, as well as improvement in the living conditions of breeders. Ranges are used in accordance with their seasonability, and rotation methods are employed. Livestock pressure on ranges is under control.

C. Medicinal Plants

At present, more than 1000 species are being intensively investigated for potential groups of therapeutic and other useful compounds. Some 600 species, which grow in the mountains and valleys, have been found to contain biologically active compounds (alkaloids, saponins, triterpenes, and such). Members of the family Polygonaceae are rich in tanning substances; Amaryllidaceae, Caryophyllaceae, and Leguminosae contain predominantly saponins; Compositae, Labiatae, and Gentianaceae yield alkaloids; and Iridaceae, Rosaceae, and Euphorbiaceae contain triterpenes. The same plant species, depending on ecological conditions, can contain a varying context of useful substances, which accounts for the contradictory evidence that is sometimes found in the literature (Shalyt, 1951; Kudryashova *et al.*, 1979).

Among all the medicinal and other useful plants, species of *Glycyrrhiza* (Leguminosae; licorices) are the most important. Of the seven species found in the Soviet Union, *G. glabra* L., which grows in all the river valleys of middle Asia, and *G. uralensis* Fisch., distributed in the Ural River valley, have long been used.

Glycyrrhiza glabra L. is a perennial herb that grows up to 100 cm high. It forms a great number of large roots and rhizomes, which provide the raw material for radix *Glycyrrhiza*. The root contains glycyrrhizin, a saponin, and glucose, sucrose, and the glucoside liquiritine, which is a flavone.

The quality of radix *Glycyrrhiza* is determined by the amount of glycyrrhizin present (6–16%). These active substances primarily belong to the group of flavonoid and triterpene compounds and are the sources of a number of medicinal preparations. In addition, radix *Glycyrrhiza* has a variety of other uses—as a foaming agent, for brewing and confectioneries, for preparation of soft drinks, and the processing of tobacco. The aboveground part is used as forage. In the Soviet Union, radix *Glycyrrhiza* is obtained from both wild populations and cultivated plantations. It is widely used in the Soviet Union and is exported.

About 40 promising plants, many of which are widely distributed, have a high alkaloid content. The shrub *Salsola richteri* (Chenopodiaceae) is distributed everywhere in sandy deserts. Without forming a dense brushwood, it occurs in large areas as a codominant of the shrub layer. The green parts (flowers and leaves) contain 1.47–2.84% salsoline, a valuable substance that increases in content from spring to autumn. A single collector is able to handpick from 20 to

60 kg of raw material per day. The raw material can also be used for preparing alkalis with an ash content of 17 to 24%.

Peganum harmala L. (Zygophyllaceae), a perennial herb, is distributed everywhere, except on sands near settlements, forming dense undergrowth. It contains harmine alkaloids: harmine ($C_{13}H_{12}N_2O$), harmaline ($C_{13}H_{14}N_2O$), harmalol ($C_{12}H_{12}N_2O$), and peganine ($C_{11}H_{12}N_2O$). The highest content of alkaloid (3–4%) is found in the seeds.

Ephedra (Ephedracene; *E. ciliata* C. A. Mey., *E. equisetina* Bge., and *E. intermedia* Schr.) grow on mountain slopes and in valleys with heavy soils. Only *E. strobilacea* is distributed in sandy deserts. *Ephedra equisetina* is most important with respect to alkaloid content. This species contains a number of alkaloids: ephedrine ($C_{10}H_{15}NO$), *N*-methylephedrine ($C_{11}H_{17}NO$), and catine ($C_9N_{13}NO$); total content is 0.23–0.36%.

Of practical interest are *Heliotropium arguzioides* and *H. dasycarpum* (Boraginaceae), which grow in sandy deserts. These contain the alkaloids heliotrine and trichodesmine, which are used in experimental medicine.

High concentrations of alkaloids also have been found in such plants as *Delphinium biternatum* Huth., *D. creophilum* Huth., *D. semibarbatum* Bienert and Boiss, *Thalictrum minus* L. (Ranunculaceae), *Gentiana kaufmanniana* Rgl. and Smalh, *G. turkestanorum* Gand., and *Swertia graciflora* Gontsch (Gentianaceae).

Large amounts of saponins have been found in 50 plant species, which are located mainly in mountain areas. Especially rich in saponins are members of the family Caryophyllaceae; the genus *Acanthophyllum* (*A. glandulosum* Bge., *A. microcephallum* Boiss., and *A. paniculatum* Rgl.) consists of perennial herbs and dwarf semishrubs, dwelling on the slopes of dry mountains distributed in Kopetdag. These species yield a valuable raw material, radix saponariae turkestanikae. Members of *Gypsophila,* mainly *G. bicolor,* yield radix saponariae albae (All-Union Committee for Standards, 1948); saponin (22.5%) and sapogenin (4%) are prepared from the roots.

It is expedient to increase the productivity of *Acanthophyllum* by sowing additional seeds to make the brushwood more dense. These plants can be successfully cultivated both on irrigated and unirrigated lands, although the saponin content is decreased in the latter case.

In the mountain regions of Tajik SSR are a number of other plants of the same family (Caryophyllaceae) that contain saponins. These include *Saponaria griffithiana* Boiss., *Silene brahuica* Boiss., *S. scabrifolia* Kom., *S. hunganica* B. Fedtsch., and *S. walliciana* Klodsch.

Saponins also are found in some Liliaceae species, including *Asparagusbucha bucharicus* Tljin., *Allium divarsicum* Rgl., *A. elatum* Rgl., *A. hissaricum* Vved., and *A. rosenbachianum* Rgl.

Several Leguminosae are also saponin-rich (*Astragalus xanthomeloides*

Korov. and M. Pop., *Caragana turkestanika* Kom., *Melilotus albus* Desr.), as are some Rhamnaceae (*Zizyphus jujuba* Mill.), Malvaceae (*Althaea rhyticarpa* Trautv.), and Umbelliferae [*Athamanta macrophylla* (Rgl. and Schmalk.) Korov., *Ferula jaeschkiana* Vatke., *Korshinskaya olgae* (Rgl. and Schmalk.) Lipsky]. In the desert zone, species that yield triterpene saponins predominate among saponin-containing plants. These include some Chenopodiaceae, Caryophyllaceae, Ranunculaceae, Papaveraceae, Berberidaceae, Brassicaceae, and Leguminosae. Steroid saponins are contained in *Tribulus terestris* L., *Solanum depilatum* Kitagawa, and 11 species of *Allium*.

Cardiac glycosides are found in some Cruciferae, including *Erysimum czernjajevii* N. Busch, *E. diffusum* Ehrh., *Syrenia sessiliflora* Ledeb., *S. siliculosa* (M.B.) Andrz., and *Berteroa incana* (L.) DC. These species are readily cultivated in unirrigated lands.

D. Industrial Plants

1. FIBER AND PAPER PLANTS

Of industrial significance are those plants that occur as dense formations in areas close to processing sites and that require the least number of processes in manufacturing paper–cellulose products.

Phragmites communis Trin. (Gramineae) is a tall grass that grows in river valleys at moist sites and in dense formations, with no sites yielding more than 50 metric tons/ha (dry weight). Cellulose yield in the manufacture of wrapping paper is 62–70% of the raw material (dry weight); the yield of bleached cellulose is 39% (dry weight) (Protopopov, 1942). Floor mats are produced from young stalks. Split stalks are used in the production of brushes.

Saccharum spontaneum L. (Gramineae) is a tall grass that is distributed in river valleys in low, annually flooded plots. It grows in formations, sometimes with various species of *Typha* and *Phragmites communis*. The yield is 6–16 metric tons/ha (dry weight). The stalks contain 43%, and the leaves, 31%, cellulose, which is used in the production of top-quality paper. The leaves and stalks are used in the production of mats.

Erianthus purpurascens Anderss. (Gramineae) grows in large turfs, mainly in river valleys. Yields are ~10 metric tons/ha (dry weight) and sometimes are as high as 18 to 20 metric tons/ha. Cellulose content is more than 40% in the stalks and ~31% in the leaves. It frequently grows together with *Phragmites communis, Glycyrrhiza glabra,* and other large plants that can be used for the production of cellulose. The leaves are suitable for the production of ropes and packing fabric, and the stalks are used for mats, wicker articles, and as a building material.

Typha gorssheimii Pobed., *T. elephantina* Roxb., and *T. angustata* Bary. and

Chaub. (Typhaceae) grow in moist areas of river valleys. These large plants contain 27.6% cellulose (Protopopov, 1942). Unprocessed plants are used for production of mats, baskets, and bast matting. Fiber produced from the *Typha* leaves undergoes special processing to make an excellent stuffing material for oil filters.

2. RESIN PLANTS

Plants containing resins, oils, and gums are widely distributed in the desert, foothills, and mountain areas. These are the large herbaceous genera of Umbelliferae, *Ferula* and *Dorema*. Their roots contain 20–30% resin (fresh weight), which is of practical significance. Resin can be extracted from roots and collected from stalks, where it accumulates and congeals at damaged sites or man-made cuts. The resin content of the milk is 40 to 63%, dry weight. Essential oil is extracted from the fruits (2–3%, fresh weight).

In the foothill subzone, there are distributed *Ferula badrakema* K.-Pol., *F. badhysi* Eug. Kor., and *Dorema aitchisonii* Eug. Kor.; in the mountains, *F. oopoda* (Boiss. and Busche) Boiss., *F. kopetdagensis* Eug. Kor., and *D. hyrcanum* K.-Pol.; in the desert, *F. assa-foetida* L. The resin of ferulas is used in medicine (halbane, assafetida, and sumbul) and perfumery.

Pistacia vera L. resin is obtained by tapping. The average resin yield per tree is 42 g. With respect to composition and use, this resin is similar to the product obtained from *Pistacia mutica* Fisch. and C. A. Mey. and to the resin *sakkyz,* which is obtained in Iran from the same plants (Shalyt, 1951).

3 INSECTICIDE AND RATICIDE PLANTS

Anabasis aphylla L. (Chenopodiaceae) is one of the most important wild insecticide plants. This semishrub is distributed on clay soils in deserts and on long, fallow lands in irrigated zones. It yields the raw material for anabasine sulfate, one of the best pest poisons. Five alkaloids have been identified: anabasine, $C_{10}H_{14}N_2$; aphyllidine, $C_{15}H_{22}N_2O$; lupinine, $C_{10}H_{19}NO_2$; anaphylline, $C_{15}H_{22}N_2O$; and aphyilline, $C_{14}H_{24}N_2O$. Anabasine is the most important, being close to nicotine in structure and biological reaction. The highest alkaloid content, 9%, is found in green shoots at the beginning of vegetative growth. In summer, the alkaloid content drops 3–5%, and the amount of raw material increases. Hence, *A. aphylla* is harvested and stored in late summer to early fall. *Anabasis aphylla* contains 7–17% oxalic acid and 16–24% ash, which is used in the soap manufacturing industry.

Various alkaloids also are contained in the following plants: *Sophora pachycarpa* L., *Ammothamnus lehmanni* Bge. (Leguminosae), and *Halostachys caspica* C. A. Mey.

A promising raticide is *Merendera robusta* Bge. (Liliaceae), which is common in the foothill subzone in spring. Its bulbs contain colchicine.

E. Other Plants

1. SAND STABILIZERS

Members of native flora are used for stabilizing drift sands. These plants root readily since they are well adapted to the specific ecology of barchan dunes.

The plants used as sand stabilizers are *Calligonum caput-medusae, C. rubens, C. microcarpum, Haloxylon persicum, Salsola paletzkiana, S. richteri,* and *Ammodendron conolyi* (shrubs); and *Aristida karelinii, A. pennata, Agriophyllum latifolium, A. minus,* and *Salsola paulseni* (herbs).

2. RANGE-IMPROVEMENT PLANTS

To improve various types of range in the southern subzone, native plants are used with special attention given to their ecology. Plants commonly used to improve ranges in sand deserts are as follows: *Haloxylon persicum, H. aphyllum, Calligonum rubens, C. caput-medusae, C. arborescens, C. microcarpum, C. setosum, Ephedra strobilacea,* and *Aellenia subaphylla* var. *arenaria* (trees and shrubs); and *Mausolea oriocarpa, Artemisia kemrudica, Astragalus unifoliolatus,* and *A. paucijugus* (semishrubs).

In clay and gypsum deserts, range improvement is carried out by sowing *Haloxylon aphyllum, Aellenia subaphylla* var. *typica, Kochia prostrata, Eurotia ceratoides,* and *Salsola orientalis.*

Used to improve range in the foothill loess deserts are *Haloxylon persicum, H. aphyllum, Ephedra strobilacea, Aellenia subaphylla* var. *typica, A. subaphylla* var. *arenaria, Kochia prostrata, Eurotia ceratoides, Salsola orientalis, Artemisia kemrudica, A. badhysi, A. turanica, A. arenicola, A. terrae-albae,* and *A. halophylla.*

In the northern desert subzone, man-made ranges and haylands are created by sowing *Kochia prostrata.*

3. DECORATIVE PLANTS

The flora of both valley and mountain regions contain a number of beautiful flowering plants. These include several species of *Eremurus* in the mountains [*E. olgae* Rgl., *E. aitchisonii* Baker., *E. robustus* Rgl., and *E. spektabilis* M.B., with rosy flowers, and *E. stenophyllus* (Boiss. and Bushe) Baker, with golden yellow flowers], and in the deserts [*E. anisopterus* (Kar. et Kir.) Rgl. and *E. inderiensis* (M.B.) Rgl., with white-rosy flowers].

Tulipa has red (*T. lanata* Rgl., *T. tubergeniana* Hoog., *T. kushkensis* B. Fedtsch., *T. hoogiana* B. Fedtsch., and *T. micheliana* Hoog.) or orange-red and yellow flowers (*T. praestanns* Hoog. and *T. lehmanian* Merckl.). *Allium* (A. *karataviense* Rgl., *A. stipitatum* Rgl., *A. elatum* Rgl., and *A. giganteum* Rgl.) also has lovely flowers, as do species of the genus *Iris* (*I. hoogiana* Dyces., *I. korolkovii* Rgl. and Maw., *I. songarica* Schrenk, and *I. linifolia* O. Fedtsch.).

Among other genera with an ornamental value are *Rhinopetalum arianum* Losinsk. and Vved., *Paeonia intermedia* C. A. Mey., *Ostrovskia magnifica* Rgl., *Petilium eduardii* (Rgl.) Vved., *Delphinium semibarbatum* Biem., and others.

All the above plants are readily propagated by seed. Seedlings flower after the third to fifth year. Many of these plants have already been cultivated, and tulips have long been exported to Holland and other countries.

V. SUMMARY

The flora of the Soviet Union arid zone is rich in useful plants that can provide valuable raw materials. Natural dense formations of fruit-bearing plants in mountain regions are used as a genetic pool for breeding novel varieties. In this connection, more attention should be given to the study and conservation of natural formations of fruit-bearing trees and shrubs.

Many plants have a number of useful substances that can be used in applications in technology and medicine. Hence, chemical studies on floras of various regions are necessary; ethnobotany should be considered as well. In estimating reserves and processing raw material, total utilization should be the objective. In addition to the main product, waste material should be used to obtain by-products, which also might be of great value.

In addition to the use of natural populations, natural plant resources should be studied with a view toward domestication to improve the quality of the plant itself, as well as techniques of cultivation. Currently, increasing applications are found by cultivating native shrubs and semishrubs to improve desert ranges.

Thanks to the construction of powerful irrigation canals, conditions have been created for the cultivation of technologically important plants, not only on unirrigated but also on irrigated lands. Of special importance is the expansion of hydrocarbon, medicinal, and subtropical food plants (pomegranate, fig, and olive trees) in the climatically favorable southern regions; however, many plants have not received adequate study, and their resources still remain obscure. A further search for useful plants, elucidation of chemical compositions, and development of effective management and cultivation practices are important problems for the future.

REFERENCES

Blinovsky, K. V., and Mizgireva, O. F. (1959). Wild fruit plants of Kopetdag, their conservation and utilization (in Russia). *Tr. Inst. Biol. Akad. Nauk Turkm. SSR, Izd. Akad. Nauk Turkm. SSR, (Seriya Biologich)* **3,** 27–35. [In Russian.]

Blinovsky, K. V., and Mizgireva, O. F. (1971). "Flora Vegetation of the Gorge of the Varzob River

(On Utilization of Plant Resources of Pamiro–Altai in Russia).'' pp. 239–289. Nauka Publ., Leningrad.

Klyushkin, E. A. (1958). Cultivation of pistach in the Turkmenian SSR. *Tr. Inst. Bot. Akad. Nauk Turkm. SSR, Ashkhabad* **4,** 127–151. [In Russian].

Korovin, E. P. (1961, 1962). Vegetation of Middle Asia and Southern Kaxakhstan. *Isd Akad. Nauk Uzbek. SSR., Tashkent* **1,** 1–452; *ibid* **2,** 1–547. [In Russian].

Kudryashova, O. I., Popov, N. G., Stepanenko, O. G., and Sarnayeva, T. A. (1979). Alkaloid-, saponin-, and terpene-bearing plants of Tajikistan. *In* ''Rastis. Resurs: Tajilistana; Introduktsiga Poleznykh Rasteniy,'' pp. 5–12. Donish, Dushanbe. [In Russian].

Lobova, E. V. (1960). ''Soils of the USSR Desert Zone.'' USSR Acad. Sci. Publ., Moscow. [In Russian].

Medicinal and Technological Raw Material (1948). All-Union Committee for Standards, USSR Council of Ministers, Moscow. [In Russian].

Momotov, I. F. (1977). Kyzylkum desert station. *In* ''Productivity of the Asian Arid Zone,'' pp. 63–89. Nauka Publ., Leningrad. [In Russian].

Nechayeva, N. T. (1975). Productivity patterns in the plant cover of Turkmenistan deserts in relation to life forms composition. *Probl. Osvoeniya Pustyn* **1,** 11–20. [In Russian].

Nechaeyeva, N. T., Vasilevskaya, V. K., and Antonova, K. G. (1973). ''Life Forms of the Karakum Desert Plants.'' Nauka Publ., Moscow. [In Russian].

Nechayeva, N. T. (1977). Lowland Karakum, Karrykul station. *In* ''Productivity of the Asian Arid Zone.'' pp. 90–117, Nauka Publ., Leningrad. [In Russian].

Nikolaev, V. N., Amangeldyev, A. A., and Smetankina, V. A. (1977). ''Desert Ranges, Their Forage Estimation and Classification.'' Nauka Publ., Moscow.

Popov, M. G. (1929). Wild fruit-bearing trees and shrubs of the Middle Asia. *Tr. no Prikladnoi Bot., Genetike i Selektsii* **22,** 3. [In Russian].

Protopopov, G. F. (1942). Paper-cellulose plants. *In* ''Syriyevye Resursy Uzbekistana.'' [In Russian].

Shalyt, M. S. (1951). ''Wild Useful Plants of the Turkmenian SSR.'' Moscow Soc. Naturalists Publ., Moscow. [In Russian].

Solovyev, A. K. (1955). ''Wild Vines of the Western Kopetdag,'' pp. 16–26. Tr. Inst. Biol. ser. Botanich, Turkm. SSR., Acad. Sci. Publ., Ashkhabad. [In Russian].

9

SUMMARY

J. R. Goodin

Department of Biological Sciences
and International Center for Arid and Semiarid Land Studies
Texas Tech University
Lubbock, Texas

David K. Northington

Department of Biological Sciences
Texas Tech University
Lubbock, Texas

I. INTRODUCTION

The purpose of this chapter is to summarize some of the basic information common to each of the preceding chapters and to evaluate the potential for development of plant resources in the arid and semiarid regions of the world. If

Table I

Arid and Semiarid Countries of the World[a]

Country	Total area (km^2)	% Arid and hyperarid	Area	% Semiarid	Area
The Americas					
Argentina	2,680,405	23	616,492	42	1,125,770
Bolivia	1,060,410	10	106,040	15	159,062
Brazil	8,216,0218	—	—	5	410,810
Canada	9,629,522	—	—	2	192,590
Chile	730,645	27	197,272	7	51,145
Colombia	1,099,342	0.5	5,497	1	10,992
Cuba	111,545	—	—	0.5	552
Ecuador	273,595	3	8,212	5	13,685
Mexico	1,904,010	26	495,042	20	380,800
Netherland Antilles	927	97	900	1	10
Paraguay	392,620	—	—	14	54,967
Peru	1,240,560	1	12,405	6	74,432
United States	9,188,862	8	114,862	22	2,024,050
Venezuela	880,360	1	8,802	2	17,607
Europe					
Bulgaria	107,057	—	—	2	2,140
Greece	127,360	—	—	13	16,557
Portugal	88,882	—	—	3	2,667
Spain	487,212	2	9,620	43	209,502
Middle East					
Bahrain	577	100	577	—	—
Iran	1,50,750	54	859,005	32	509,040
Iraq	419,812	74	301,662	35	62,972
Israel	20,047	9	1,802	84	16,842
Jordan	88,995	93	82,765	6	5,340
Kuwait	15,500	100	15,500	—	—
Lebanon	9,875	—	—	23	2,270
Oman	205,070	78	159,957	22	45,117
Qatar	21,250	100	21,250	—	—
Saudi Arabia	2,075,000	98	2,033,500	2	41,500
Syria	178,745	50	89,372	39	69,710
Turkey	753,455	—	—	18	135,607
United Arab Emirates	80,750	100	80,750	—	—
North Yemen	188,250	61	114,832	26	48,945
South Yemen	277,287	95	263,802	5	13,885
Australia	7,419,772	49	3,635,687	20	1,483,955
Asia					
Afghanistan	625,000	36	225,000	38	237,500
China	9,228,750	13	1,199,737	10	922,875
India	3,066,165	4	122,647	17	521,247

Table I (*Continued*)

Country	Total area (km²)	% Arid and hyperarid	Area	% Semiarid	Area
Mongolia	1,510,500	22	332,310	55	830,775
Pakistan	864,387	49	423,547	29	255,670
Soviet Union	21,500,875	9	1,935,080	11	2,365,097
Africa					
Algeria	2,298,987	87	2,000,120	8	183,920
Angola	1,203,382	3	36,102	8	96,270
Benin	108,710	—	—	2	7,610
Botswana	579,512	10	57,952	84	486,790
Cameroon	458,922	—	—	3	13,767
Cape Verde	3,892	9	350	91	3,542
Chad	1,239,500	59	731,305	23	285,085
Djibouti	22,250	100	22,250	—	—
Egypt	967,250	100	967,250	—	—
Ethiopia	1,179,445	35	412,807	28	330,245
Kenya	562,400	49	275,575	37	208,087
Libya	1,698,405	98	1,672,927	1	16,985
Madagascar	566,645	2	11,332	8	45,332
Mali	1,196,637	49	586,352	42	502,587
Mauritania	1,080,457	94	1,015,630	6	64,827
Morocco	602,205	67	403,477	25	150,552
Mozambique	506,927	—	—	15	76,040
Namibia	794,567	56	444,957	39	309,882
Niger	1,223,000	85	1,039,550	15	183,450
Nigeria	891,672	3	26,750	20	178,335
Senegal	198,887	11	20,887	53	100,640
Somalia	615,502	82	504,712	14	86,170
South Africa	1,179,697	31	365,707	22	259,532
Sudan	2,418,750	46	1,112,625	28	677,250
Tanzania	912,250	—	—	21	191,572
Uganda	227,690	—	—	14	31,877
Upper Volta	264,500	5	13,225	52	137,540
Zambia	726,465	—	—	11	79,910
Zimbabwe	377,010	—	—	33	124,412

[a] After Rogers (1981).
[b] Not represented.

for no other reason, the amount of the world's land-surface area that is categorized as arid and semiarid demands an examination of its potential utility for the world's ever-increasing human population. As can be seen from Table I, some countries are essentially all arid, and many others include a substantial percentage of land that is botanically "nonproductive" due to unpredictable moisture. The total for the entire world is ~33%, which is an impressive figure by itself.

In the following geographic summaries we shall briefly synthesize the demographic pressures and desirability for the increased utility of plant resources from each of these areas. In addition, we have attempted to mention categories and even specific plants that might have significant potential for increased productivity in each of the geographical regions represented by the preceding chapters.

II. AREA PROFILES

A. Africa

The population pressures in most of the African countries containing arid and semiarid regions are increasing. These are rural areas, and the most significant impact on the sparse vegetation is the removal of woody species for fuel wood. The denuding of these areas for wood has accelerated the desertification process in many of the African desert areas to the point that nomadic life-styles are increasing.

The potential for increased use of the dry areas of Africa for plant resources is, therefore, greatest for fuel-wood species. Multipurpose plants that might provide both fodder for domesticated animals and fuel wood would be especially desirable. Several species of *Casuarina* have been introduced in the past for fuel-wood production, and localized efforts at increasing the availability of selected native species for fodder and fuel wood are ongoing. Many of these efforts are on a small, regional scale and have not yet changed the overall pattern of desertification. The need for examination of introduced species is significant in Africa.

B. Australia

Because of the relatively small human population, especially when discussed per capita, Australia has essentially no pressing need to increase utilization of her arid and semiarid zones. In addition, the population is stable in its distribution with essentially all of Australia's ±15 million people concentrated in four large, and several smaller cities along the more mesic coastal regions, primarily on the east coast. As a result, the sparse rural population is also fairly stable without a noteworthy net movement either to or from urban areas. Those who reside in the large arid and semiarid parts of Australia, therefore, have sufficient per capita area from which to extract a living so that there is no pressure to intensify plant productivity beyond its natural capacity.

The primary plant industry for the dry areas, historically and currently, is pastoral, with sheep or cattle dominant, depending on the area and the nature of the native forage. Although there is some mining in the dry areas and slightly increasing human impact due to recreational activities and tourism, their greatest potential for escalated utilization is with forage plants. *Acacia, Eucalyptus,* and *Casuarina* species are the most likely candidates for intensified domestication.

Among nonnative plant species, Australia has initiated an exploratory introduction of guayule and jojoba to investigate their potential as commercially viable crops. Australia needs neither natural rubber nor jojoba oil, but both would be attractive exports if they prove to be economically viable as crop plants.

C. Indian Subcontinent

There is an exceptionally high need to utilize the arid and semiarid regions of India for plant cultivation. By Indian demographic standards, the dry areas of India are not heavily populated, but poor past management of native plant resources has been destructive, often irreversibly. The problems of desertification and plant resource management are considerably magnified by the ~300 million cattle, which feed on native desert vegetation and consume the forage grown on arable land that could be producing food for human consumption.

The most suitable use of the dry zones of India are for rangeland and limited fuel-wood production, not for cultivation of food crops. Proper management could allow for increased cattle forage from these areas, partially releasing some of the more suitable areas, currently in fodder cultivation, to crop production.

D. Middle East

This region of the world includes tremendous country-to-country variability in demography, standards of living, and plant resources. For example, it is estimated that Egypt will have a population of 64.9 million by the year 2000, essentially all of whom will be crowded onto only 4% of the land. Several of the oil-producing countries, on the other hand, are underpopulated and have to import labor. With more than 90% of their total surface areas classified as arid or semiarid, the development of plant resources either is, or could some day become, very important to the countries of the Middle East.

Egypt is one of the countries in which the development of such plant resources would be very beneficial. The arable land along the Nile River and especially in the Nile Delta is being placed under great pressure. With a rapidly increasing population and an agricultural system dependent on domestic animals for plowing, turning irrigation wheels, and transportation, the cultivated lands must produce both animal forage and food for the cities. If some of the forage production could be moved to nonproductive land, then more crops could be grown in the arable sections.

Xerophytic plants such as *Atriplex* and *Opuntia* could provide protein and carbohydrate forage sources and could be grown using minimal irrigation, even brackish water. Since the Aswan Dam prevents flooding in the delta region, the silt that is being removed to make bricks for continued urban growth will not be replaced, further stressing the amount of productive land available for cultivation.

E. North America

Population pressure to increase the utilization of arid and semiarid regions exists in Mexico but not in Canada or the United States. Central American countries have very little area that can be considered nonproductive due to moisture unavailability. Human impact in the form of recreational activities and tourism affects the native vegetation in the desert areas of the southwestern United States. Some of the activities, such as dirt-bike racing, result in rapid and long-lasting deterioration. Thus, concern exists for the proper management of these areas.

The greatest potential for increased utilization of plant resources in northern Mexico and in the southwestern and western United States possibly rests in the cultivation of guayule and jojoba for their natural rubber and oil, respectively. For food, *Cucurbita foetidissima* is thought to have high potential. *Aloe vera* and *Asclepias* have potential for medicinal or industrial uses, and *Atriplex* and *Purshia tridentata* (bitterbrush) might provide increased forage for domesticated animals. The potential for biomass production for fuel rests primarily on *Atriplex canescens, Chrysothamnus nauseosus, Sarcobatus vermiculatus,* and *Artemisia tridentata.*

F. People's Republic of China

Although the demographic pressures in China encourage the use of arid and semiarid regions for increased productivity, the total area of China allows for a certain amount of dispersal of this pressure. Current efforts include the use of suitable trees and shrubs for dune stabilization and windbreaks. Combining windbreaks with irrigation canals, cultivation is done primarily in a checkerboard pattern, which best protects the useful native plants and suitable crop plants and maximizes water efficiency. This form of irrigation has resulted in a salinization problem in these areas, requiring the use of halophytic species for dune stabilization and windscreens. *Artemisia* species are among the most suitable plants used in the arid and semiarid areas that employ these production methods.

G. South America

Approximately 90% of the world's population growth since the mid-1970s has occurred in tropical regions. These increases have been most significant in African and especially in South American countries. Because of the wholesale clearing of the tropical forests and thus their probable destruction, these population increases must ultimately affect the arid and semiarid areas of South America.

The increased use of the dry areas for domestic forage and for fuel wood is projected as having the greatest potential for the native plant resources. *Prosopis* and *Acacia* have the greatest potential as native fuel-wood species, while *Chenopodium quinua* and *Amaranthus caudatus* have been mentioned as possible food crops from arid and semiarid regions in South America.

H. Union of Soviet Socialist Republics

While there is no demographic pressure to develop plant resources in the arid and semiarid regions, there are estimated to be ~1600 native plant species available in the deserts. This number of species provides an excellent germ plasm source in general, with fruit- and nut-bearing trees and shrubs being unusually plentiful due to niches in the montane areas. *Juglans, Pistachia vera,* and *Pivera* are among the taxa having the greatest potential in dry regions, but pomegranates, figs, and olives could also prove to be productive if irrigated.

Whether because of population pressures or economic desirability, it is clear that plant resources in the arid and semiarid regions of the world need to be carefully examined as potential crops. This effort is best done in countries like Australia, the United States, and the Soviet Union, where care can be given to proper resource development in a delicate ecosystem. Countries like India, Mexico, and Egypt are already feeling demographic pressures to utilize their desert areas to help alleviate food, forage, and fuel shortages. It is under these circumstances that permanent degradation of the environment could result from overintensive use of plants not best suited to the ecological parameters of the deserts.

III. PHILOSOPHY AND RATIONALE

Any attempt to utilize and exploit the world's arid-land plant resources carries with it an exceptional ecological responsibility for doing so within the framework of a methodical and sound judgment. Unlike other highly buffered ecosystems in which recovery occurs quickly, disturbance in the arid lands may take a lifetime or even hundreds of years for recovery. In some cases, recovery is never possible and the landscape may be permanently changed. On the other hand, the arid zone contains a wealth of biological treasures, and only recently has the scientific community come to realize the importance of the germ plasm found there. Peoples of the arid lands have long lived within their environments, existing on the productivity provided locally. As long as population pressures are small, human existence, although never spectacular in terms of living standards, has managed rather well. Modern-day hunters and gatherers have managed to survive within the framework of a spartan life-style, often escaping the vagaries of nature only by careful planning. Unpredictable precipitation makes a sedentary life-style questionable at best, and nomadism has often served these people to advantage because it has been possible to move on if the rains fail to come.

With increasing population pressures in many parts of the world, the challenges of increasing nomadism become greater. Government programs often declare migratory life-styles to be inherently bad, and attempts are made to develop and fund permanent settlements within the arid lands. In many cases, such programs have met with disastrous results. We have learned too late that nomadic choices were an evolutionary process that served certain peoples better than permanent settlements. Changes in social structure and in the economy have not been particularly satisfactory, and the environmental and biological consequences are sometimes severe and irreversible.

A. Exporting Technology

The developed and industrialized nations have often suggested that all good things are possible for the world if the lesser-developed countries would only follow the lesson that has worked so well elsewhere. Nations with vast natural and human resources have been able to achieve impressive gains in food production, industrial output, and "progress" by following a program of sound management and planning, basic and applied research, and a program of technology transfer to all corners of the land. Some agriculturists in the United States, for example, have suggested that all nations with the right climate, soils, and germ plasm could follow this example of productivity and produce at the same level that has been characteristic of the American success story. Recent experiences in the developing countries have led to a more studied approach to exporting high technology. The socioeconomic and cultural patterns of many countries simply do not lend themselves to modern agricultural technology, as envisioned by developed countries. Instead, there appears to be a need for small-scale farming systems as varied as the countries themselves. The concept of a small portion of the population feeding the remainder is not likely to come about in countries with insurmountable population pressures, minimal natural resources, and little capital. Each of these societies needs a tailor-made system for food production, often at the individual or village level. Even though urban migration continues to be a major problem, much of the population in developing countries continues to be rural. A large portion of the national commitment must be devoted to food production. Size of farms, marketing systems, production systems, and local customs dictate a methodology far removed from the success story in the developed countries.

If one subscribes to the hypothesis above, then there would appear to be a place for the specialized and localized plant resources that have traditionally served a society well. While it is unlikely that most of these plant resources would ever become a world crop, they serve an important purpose under specialized environmental and cultural conditions, and their study and development would appear to be justified.

B. Domestication of Crops

The history of domestication is an interesting one. Those living in industrial societies where food, fiber, industrial, and medicinal choices are varied would assume that the world offers a plethora of variety. An American supermarket stocks literally hundreds of items from which to choose, and it is possible to have nutritious and varied meals on any given day. Given these choices, it is difficult to believe that some 80% of the entire world's food supply is devoted to wheat, rice, and corn. Likewise, it seems impossible that no new major food crops have been domesticated in the past 2000 years, except for coffee and tomatoes. The complications of domestication are far more subtle than the average consumer would imagine. Some of the reasons for lack of domestication include

1. *A complicated infrastructure for the existing agricultural economy that resists change:* A particular crop is grown in an area because the climatic and edaphic factors allow it to be grown there. Over a period of time, the crop must prove itself and become competitive with other alternative endeavors. There may be no market for the new product, labor costs may be too high, the product may not be as palatable as other foods, perishability may be too great, and the list goes on.
2. *A lack of stable production:* Even though a new crop might look particularly promising at any given time, unpredictability of the climate, lack of a reliable seed or propagation source, and other factors may doom the developing species before its potentials are revealed.
3. *A failure to stand the test of time in the genotype–environment interaction:* Both the developed and developing countries know all too well about the promise of a new plant that will solve all problems. Photographs of limitless harvests, which have been produced under all growing conditions, often precede the introduction of some new miracle crop plant. Its promoter, often a public-relations expert rather than an agronomist or horticulturist, sells the product so skillfully that everyone becomes caught up in the hype. Sometimes perfectly good plants for specific uses under localized conditions are oversold and therefore lose favor throughout the world.

These factors, among others, have been the sad tale of attempted domestication throughout the world and throughout history. Consequently, few new species ever pass the test of time and make a contribution to agricultural or industrial productivity.

It would appear that if any "new" species has a chance to become successful as a crop, then it must undergo a rigorous screening with a careful explanation of objectives. An attempt to introduce formally possible new crops has been undertaken by the Board on Science and Technology for International Development of the Advisory Committee on Technology Innovation of the National Academy of Sciences (NAS) of the United States. In 1975, this committee set out to investigate plant resources that showed promise for improving the quality of life in tropical areas. Its report was originally addressed to government administrators, technical-assistance personnel, and researchers in agriculture, nutrition, and related disciplines who were concerned with helping developing countries achieve "a more efficient and balanced exploitation of their biological resources." The product of the committee's work, financed by the Office of Science and Technology for the U.S. Agency for International Development, was published as *Underexploited Tropical Plants with Promising Economic Value* (National Academy of Sciences, 1975). The book includes sections on newly recognized species of cereals, roots and tubers, vegetables, fruits, oilseeds, forage, and others. Interestingly enough, some of the "new" crops recommended in this book are very ancient crops, such as the grain amaranths now cultivated as minor grain crops in Latin America. Once important major food crops in the tropical highlands, they were relegated to an insignificant status following the Spanish conquest, primarily because the bountiful harvest had religious significance for the local Indians. The grain amaranths were displaced by the larger-seeded grains, and they have never regained their former position in agricultural productivity. Although better adapted to the montane conditions, their status as important crops has never again been widely accepted. It would thus appear that social, religious, and political influence was the deciding factor in the continued domestication of one genus.

The NAS book has been reprinted several times since 1975, and it is now widely distributed all over the world. Perhaps no single publication has done so much to publicize the possibilities for domestication of new crops. The committee has continued to publish other books of more detailed information on specific species (*Guayule: An Alternative Source of Natural Rubber,* 1977; *Leucaena: Promising Forage and Tree Crop for the Tropics,* 1977; *Tropical Legumes: Resources for the Future,* 1979; *Firewood Crops: Shrub and Tree Species for Energy Production,* 1980; and *The Winged Bean: A High-Protein Crop for the Tropics,* 1981). While not all of these books have direct application to the arid lands, many of them do; they serve as a model of what can happen if new species are introduced to the agricultural world with a proper perspective (Goodin and Northington, 1979; Theisen *et al.,* 1978).

C. The Case for Jojoba and Guayule

In recent years, partially as a result of the widespread publicity given in NAS publications, two species native to the southwestern United States and the adjacent deserts of northern Mexico have received worldwide attention. Jojoba (*Simmondsia chinensis*) and guayule (*Parthenium argentatum*) have emerged as two species with definite prospects for domestication. To say that they were "discovered" in the last few years would certainly be an abstraction of the truth, because both species have been known for many years as important arid-land plant resources. Jojoba received widespread publicity when chemists discovered that the quality of the oil approximated that of sperm whale oil, an event that coincided with the placement of the sperm whale on the endangered species list. If those industrial processes using sperm whale oil could switch to jojoba oil, it would seem to be the answer to the environmentalist's prayers. Although vast regions of the Sonoran Desert are inhabited by jojoba, genetic diversity and plant density made harvesting from wild stands rather uneconomical. Thus, attempts were made to begin the domestication process, and since the early 1960s a monumental effort has gone into the breeding, cultivation, harvesting, and processing of jojoba. Enticements of extraordinary prices for the oil and waxes, tied primarily not to industrial lubricants, but to the lucrative cosmetic industry, brought many unscrupulous entrepreneurs into the new crops business. Most people who invested in jojoba plantations did so with an honest and sincere objective, but leadership was often more concerned with land speculation and tax write-offs than with solving a problem in international agriculture. Thousands of hectares of desert land was cleared, leveled, planted, irrigated, fertilized, and cultivated in most of the world's arid countries. Word spread quickly from California and Arizona to the Middle East, Africa, South America, and Australia. Planting jojoba became the "in" thing to do in the 1970s.

We contend that the mistake made by most investors in jojoba plantations is that they expected to see a new drought-resistant crop with exceptional production grow from desert soils with little water and fertilizer. Even though jojoba obviously has the genetic potential for survival under extreme environmental conditions, high levels of production require many of the inputs characteristic of mesic crops. Such conditions often lead to new insects, diseases, and weeds that were never a problem in the wild.

Currently, a few jojoba plantations are coming into production around the world. Many others will fail to produce because of mismanagement or a myriad of other environmental, edaphic, and biological factors. The market appears quite uncertain for a crop with a very large worldwide production. Whenever the price goes down, other cropping alternatives will become more appealing; only time and the marketplace will determine the domestication success of jojoba. In addition to the high energy inputs of production, harvesting and processing costs are currently quite expensive. Also, there would appear to be few secondary credits associated with jojoba farming. Although the biomass and seed meal left after oil extraction might provide feed for livestock, there is little else to recommend this plant that might help its survival in the world economy.

In many respects, the case for guayule parallels that of jojoba. Guayule, a desert shrub of the arid southwestern United States and northern Mexico, produces a high-quality latex in its leaves, stems, and roots. This latex has been known by local Indians for perhaps hundreds of years. Guayule's ''discovery'' first came about during World War II when supplies of natural rubber from Southeast Asia became unavailable. At the time, synthetic rubber was just being developed, and the wear characteristics of synthetic rubber proved to be inferior to that of the natural latex produced by *Hevea brasiliensis*. In fact, the virtues of guayule had been known much earlier, and guayule provided much of the natural rubber utilized in the United States at the turn of the century. A tremendous scientific effort was devoted to guayule production as a part of the war effort, but the entire project was abandoned as soon as the war ended. Little consideration was given to its further domestication until the 1973 oil embargo by the oil producing and exporting countries (OPEC). At that time, the United States began to realize that strategic national supplies of certain materials would help in the development of a sound defense posture. As a result, considerable interest was rekindled in guayule, and the interest has spread to all the arid parts of the world.

Although the production potential of guayule appears to be great, and considerable information is available concerning its cultivation and harvest, the economics of guayule production in arid regions is highly questionable. Like jojoba, guayule is adapted to desertic conditions, but survival is one matter and agricultural production is another. In addition, an increase in dry matter in guayule is often accompanied by a decrease in the percentage of latex; that is, drought-stressed guayule plants have a higher latex content per unit dry weight than nonstressed plants. One must determine the best biomass yield for the rubber content. As long as imported rubber continues to be available, it appears to compete more effectively than the domestic source. If guayule is going to become a valuable domestic crop, its production will probably have to be subsidized by the government, not for its agricultural potential, but because it is a strategic national resource. As part of the adaptability of arid-land plants, many woody species synthesize high-quality latex. Among these, *Chrysothamnus nauseosus* (rubber rabbitbrush) would appear to be a likely candidate.

IV. THE FUTURE OF ARID LAND PLANT RESOURCES

The comments of these international experts representing various parts of the world would suggest that the potential for utilization of plant resources from arid and semiarid lands is very great indeed. The discussion for each of the regions suggests that the number of species available is even greater than we had hoped. We are not so naive as to believe that all of these species are candidates for domestication. In many cases, their historical use on a localized basis is an indication of their ultimate potential. In a utopian world with all-encompassing schemes, there is still a place for localized application of food, fiber, forage, industrial, and medicinal plants. Some plants have served a local society well for hundreds or even thousands of years, yet those same plants have little application to other parts of the world. In other cases, however, there is reason to believe that localized plant resources are among the world's best kept secrets, and some of these, such as the grain amaranths and *Chenopodium quinoa,* may have excellent application for many arid regions.

It has been encouraging to see how much interest has been generated in arid-land plant resources throughout the world. Literally hundreds of workers from most countries of the world have expressed interest in learning more about the potential. For too long people who live in arid lands, and scientists who work there, have been the "stepchildren" of society and its educational institutions. Suddenly, the population pressures on the arid lands of most continents has caused a reevaluation of natural resources, migration, and conservation.

It is impossible to generalize the utilization of biological resources in various parts of the world. The day-to-day pressures of survival in India or China are far removed from the carefully planned conservation of resources that is possible in Australia. In North America, the arid-land pressures are not so much for food, as they are for recreation and aesthetics (Stark, 1966). Ecologists agree that arid-land ecosystems are fragile, yet increasing numbers of city dwellers look to the arid lands for their leisure time. Most of them are poor stewards of the land, and educating them to a rational utilization of the arid lands is a slow and agonizing process.

How do we convince society of the importance of conservation of germ plasm? How do we convince society that the arid lands are not some never-ending reservoir of freedom to roam as one pleases? How do we convince society that biology is inherently different in the desert than it is in the mesic zones?

These and similar questions will continue to challenge the planners and scientists who choose to work in the arid and semiarid lands. We fully realize that the hyperarid regions are less likely to become productive, except with massive inputs of energy. Instead, opportunities exist on the semiarid fringes to integrate economic botany with the environment to improve the land for the benefit of all. Poor planning and overutilization, accompanied by an unpredictable climate, leads to desertification, a worldwide problem of enormous magnitude. Yet desertification *can* be reversed. We must find the methodologies for preservation and conservation of arid regions, for manipulation and management for an increase in productivity, and for upgrading the landscape for generations to come. It can be done, but making it happen will take a superhuman effort. Fortunately, arid-land scientists are tough.

REFERENCES

Goodin, J. R., and Northington, D. K. (eds.) (1979). "Arid Land Plant Resources." International Center for Arid and Semi-Arid Land Studies, Texas Tech Univ., Lubbock.

National Academy of Sciences (1975). "Underexploited Tropical Plants with Promising Economic Value." Advisory Committee on Technology Innovation, Natl. Acad. Sci., Washington, D.C.

National Academy of Sciences (1977). "Guayule: An Alternative Source of Natural Rubber." Advisory Committee on Technology Innovation, Natl. Acad. Sci., Washington, D.C.

National Academy of Sciences (1977). "Leucaena: Promising Forage and Tree Crop for the Tropics." Advisory Committee on Technology Innovation, Natl. Acad. Sci., Washington, D.C.

National Academy of Sciences (1979). "Tropical Legumes: Resources for the Future." Advisory Committee on Technology Innovation, Natl. Acad. Sci., Washington, D.C.

National Academy of Sciences (1980). "Firewood Crops: Shrub and Tree Species for Energy Production." Advisory Committee on Technology Innovation, Natl. Acad. Sci., Washington, D.C.

National Academy of Sciences (1981). "The Winged Bean: A High-Protein Crop for the Tropics," 2nd ed. Advisory Committee on Technology Innovation, Natl. Acad. Sci., Washington, D.C.

Rogers, J. A. (1981). Selected dryland areas of the world. *Arid Lands Newsletter* No. 14 (July) pp. 24–25. [P. Paylore, ed.; published by Office of Arid Lands Studies, Univ. of Arizona, Tucson].

Stark, N. (1966). Review of highway planting information appropriate to Nevada. Desert Res. Inst., Univ. of Nevada, Coll. Agric. Bull. No. B-7, Reno.

Theisen, A. A., Knox, E. G., and Mann, F. L. (1978). Feasibility of introducing food crops better adapted to environmental stress, Vol. I and II. Natl. Sci. Found., Directorate for Applied Science and Research Applications, Division of Applied Research, Washington, D.C.

INDEX